T0224059

A Course on Basic Model Theory

Haimanti Sarbadhikari · Shashi Mohan Srivastava

A Course on
Basic Model Theory

Springer

Haimanti Sarbadhikari
Stat-Math Unit
Indian Statistical Institute
Kolkata, West Bengal
India

Shashi Mohan Srivastava
Stat-Math Unit
Indian Statistical Institute
Kolkata, West Bengal
India

ISBN 978-981-13-5318-5   ISBN 978-981-10-5098-5   (eBook)
DOI 10.1007/978-981-10-5098-5

© Springer Nature Singapore Pte Ltd. 2017
Softcover reprint of the hardcover 1st edition 2017
This work is subject to copyright. All rights are reserved by the Publisher, whether the whole or part
of the material is concerned, specifically the rights of translation, reprinting, reuse of illustrations,
recitation, broadcasting, reproduction on microfilms or in any other physical way, and transmission
or information storage and retrieval, electronic adaptation, computer software, or by similar or dissimilar
methodology now known or hereafter developed.
The use of general descriptive names, registered names, trademarks, service marks, etc. in this
publication does not imply, even in the absence of a specific statement, that such names are exempt from
the relevant protective laws and regulations and therefore free for general use.
The publisher, the authors and the editors are safe to assume that the advice and information in this
book are believed to be true and accurate at the date of publication. Neither the publisher nor the
authors or the editors give a warranty, express or implied, with respect to the material contained herein or
for any errors or omissions that may have been made. The publisher remains neutral with regard to
jurisdictional claims in published maps and institutional affiliations.

Printed on acid-free paper

This Springer imprint is published by Springer Nature
The registered company is Springer Nature Singapore Pte Ltd.
The registered company address is: 152 Beach Road, #21-01/04 Gateway East, Singapore 189721, Singapore

To our grandchildren
Pikku, Chikku, Totu and Duggu

Preface

Model theory is a branch of mathematical logic that studies mathematical structures through sentences of a suitable formal language concerning the elements of structures. Since mathematicians rarely exploit the precise syntactical structure of sentences, model theory gives new techniques to mathematics. These have been used very successfully to settle some outstanding conjectures in hardcore mathematics.

Therefore, model theory is a subject in its own right. It has its own deep concepts and is rich in techniques. Currently, it is a very active and challenging area of research. The main purpose of this book is to usher young researchers into this beautiful, challenging and useful subject.

About the Book

This book is an exposition of the most basic ideas of model theory. It is primarily aimed at lower undergraduate students who would like to work in model theory. Knowledge of logic will be helpful, though not essential, for this book. This book is a book on model theory and not mathematical logic. So, we completely avoid proof theoretic approach. Throughout the book, the style is semantic except for introducing languages and interpretations (i.e. structures) formally. This enables us to get into the subject rather quickly.

Chapters 1–6 constitute the core of model theory which all researchers should start with. In Chap. 7, we have presented model theory of valued fields. It also contains the full proof of Ax-Kochen theorem on Artin's conjecture on the field of p-adic reals.

Because we have put equal stress on applications, a good background in algebra is required. In Appendices A–C, we give necessary background material from set theory and algebra that is required for this book. Some background material on algebra has also been presented in the main text as we go along. It is hoped that this will make the book self-contained.

Ideally, the book should be covered in two semesters. The first two chapters contain most of the basic concepts and basic results to get started. These two chapters contain standard materials that are traditionally covered in the first intro-duction to the subject. It can also be taught to senior undergraduate students.

Chapter 1 mainly concentrates on basic concepts. It includes first-order lan-guage, its terms and formulas and its structures, homomorphisms, embeddings and elementary embeddings, Skolemization of a theory, definability, etc. In order to handle definable equivalence classes, a brief introduction of many-sorted logics, imaginary elements and elimination of imaginaries is presented. In the first reading, if time does not permit, sections on many-sorted logics and elimination of imagi-naries can be skipped.

Chapter 2 contains most of the introductory techniques and results. Ultraproduct of structures and Łoś fundamental lemma, compactness theorem and its conse-quences, upward Löwenheim Skolem theorem, quantifier elimination and model completeness are some of the most basic results presented in the chapter. These are used to present the model theory of dense, linearly ordered sets without end points, torsion-free divisible abelian groups and ordered divisible abelian groups, alge-braically closed fields and real closed fields. The chapter concludes with some applications in algebra and geometry such as Hilbert Nullstellensatz, Ax's theorem on polynomials, Chevalley's projection lemma on algebraically closed fields and its real counterpart Artin–Seidenberg theorems on real closed fields, solution of Hilbert's seventeenth problem, etc.

Both Chaps. 1 and 2 end up with a large number of exercises. They are an integral part of the subject. Several concepts are introduced in the exercises. Readers should work out all the exercises. Much of the material presented in the exercises are used later.

Chapters 3–6 can be termed as the beginning of modern model theory and require a bit of sophistication. In Chap. 3, we make a systematic study of types. Types are used to define most of the modern concepts and are essential for the development of modern model theory. Chapters 4 and 5 form the bedrock of modern model theory. In Chap. 4, we introduce important subclasses of structures and theories. Of particular interest are topics on saturated structures and stable theory. We introduce Morley rank and Morley degree as well as forking inde-pendence in Chap. 4. In Chap. 5, we introduce indiscernibles and prove Morley categoricity theorem. In Chap. 6, we initiate the study of strong types. Strong types are equivalence classes of the so-called bounded, invariant, equivalence relations. We introduce mainly Lascar strong types and Kim–Pillay strong types. Strong types are important for stable theories, simple theories, independence, etc. These are also interesting mathematical objects. To illustrate this, we show their connection with descriptive set theory.

There are many important topics such as stability theory, simple theory and independence, NIP theories, etc. that are not covered in this book. This is primarily because we want this to remain an introductory graduate level text book to get started in model theory.

Acknowledgements

We are greatly indebted to Krzysztof Krupiński for a series of lectures he gave us in Wroclaw on strong types and their connection with descriptive set theory. In the process, he introduced us to some other concepts needed for strong types. All these are included in this book. We are also grateful to Anand Pillay for his encouragement and help while we were learning model theory. We also thank him for his comments on the original manuscript of this book. At the end, we would like to confess that we are rather new to this subject. Over a period of time, we have developed a great admiration for model theory in its own right as well as for its deep interaction with many other branches of mathematics. We have used several books including Chang and Keisler [9], Hodges [20], Marker [41], Pillay [46], Delzell and Prestel [10], Tent and Ziegler [64], Wagner [67], etc. to learn model theory. These books have greatly shaped our view and understanding of the subject. Thus, there is a natural influence of these books on our book. We may not have given credits where they are due. This is more a reflection of our ignorance than anything else. Needless to say that no result and no proof presented in this book are due to us. We thank Ms. Paramita Adhya for providing various computer related services. Finally, it is a pleasure to express our appreciation for the patience and cooperation of our daughter Rosy, our son Ravi, our son-in-law Suraj and our daughter-in-law Deepali during the period we were busy writing this book.

Kolkata, India   Haimanti Sarbadhikari
April 2017   Shashi Mohan Srivastava

Contents

About the Authors

Haimanti Sarbadhikari graduated from the Indian Statistical Institute, Kolkata, in 1979 under the supervision of Prof. A. Maitra. She was a Ford Foundation Fellow at the University of California, Los Angeles, in 1979, and Visiting Assistant Professor at the University of Illinois, Chicago, in 1980. She joined the Indian Statistical Institute, Kolkata, in 1980, as the permanent faculty and retired from the position of professor in 2010. Her primary research interest is descriptive set theory.

Shashi Mohan Srivastava graduated from the Indian Statistical Institute, Kolkata, in 1980, under the supervision of Prof. A. Maitra. He was a member of the Institute for Advanced Studies, Princeton, during the academic year 1979–80. He joined the Indian Statistical Institute, Kolkata, in 1980, as the permanent faculty and is currently a higher academic grade professor. His primary research interest lies in descriptive set theory. He is the recipient of the Indian National Science Academy Medal for young scientists. He is the author of the books *A Course on Borel Sets* and *A Course on Mathematical Logic* (both published by Springer). He is an enthusiastic teacher and has written several expository articles aimed at popularising mathematical logic and set theory in India.

Notations

$\omega, \mathbb{N}$	The set of natural numbers $0, 1, 2, \ldots$
$\mathbb{Z}$	The set of all integers
$\mathbb{Q}$	The set of all rational numbers
$\mathbb{R}$	The set of real numbers
$\mathbb{C}$	The set of complex numbers
$\mathbb{R}_{\mathrm{alg}}$	The field of real algebraic numbers
$\subset$	Subset
$\supset$	Superset
$\mathcal{P}(X)$	Power set of the set X
$\lvert s \rvert$	Length of the expression s
$\lvert L \rvert$	Cardinality of the language L
t^M	The value of variable free term t in the structure M
$M \vDash \varphi$	φ true in M
$\mathrm{Th}(M)$	The set of all sentences true in M, Theory of M
$\mathrm{Th}(\mathcal{M})$	The set of all sentences true in all $M \in \mathcal{M}$
$M \vDash T$	M a model of T
$T \vDash \varphi$	φ a theorem of T
$\equiv_T$	Formulas equivalent in theory T
DLO	The theory of dense linearly ordered sets without end points
DAG	The theory of torsion-free divisible abelian groups
ODAG	The theory of ordered divisible abelian groups
ACF	The theory of algebraically closed fields
$\mathrm{ACF}(p)$	The theory of algebraically closed fields of characteristic $p, p = 0$ or a prime
OF	The theory of ordered fields
RCF	The theory of real closed fields
RCOF	The theory of algebraically closed ordered fields
PA	Peano arithmetic
$\sqsubseteq$	Substructure
$\sqsupseteq$	Extension of a structure

$\mathrm{End}(M)$	Set of endomorphisms of M
$\equiv$	Isomorphic
$\preceq$	Elementary substructure
$\succeq$	Elementary extension
$\simeq$	Elementarily equivalent
$\mathrm{Diag}(M)$	The atomic diagram of the structure M
$\mathrm{Aut}(M)$	Automorphism group of M
$\mathrm{Aut}(M)_A$	Subgroup of pointwise stabiliser of A of automorphisms of M, $A \subset M$
$\mathrm{Aut}(M)_{(A)}$	Subgroup of setwise stabiliser of A of automorphisms of M, $A \subset M$
$\mathrm{Diag}_{\mathrm{el}}(M)$	The elementary diagram of the structure M
$\mathcal{H}(X)$	Skolem hull of X, Ehrenfeucht–Mostowski model
$\mathrm{dcl}(A)$	Definable closure of A
$\mathrm{DCA}(A)$	Definable closure of A in algebraic sense
$\mathrm{acl}(A)$	Algebraic closure of A
$\mathrm{ACL}(A)$	Algebraic closure of A in algebraic sense
$\mathrm{tp}^M(\bar{a})$	Type of the tuple $\bar{a}$ in the structure M
$\mathrm{tp}^M(\bar{a}/A)$	Type of the tuple $\bar{a}$ in the structure M over $A \subset M$
$\bar{a} \models p$	$\bar{a}$ realises the type p
$p(M)$	The set of all realisations of the type p in M
$S_n(T)$	The set of all complete n-types of the theory T
$S_n(M)$	The set of all complete n-types of the structure M
$S_n(M/A)$	The set of all complete n-types of the structure M over $A \subset M$
$\mathrm{EM}(X)$	Ehrenfeucht–Mostowski type of X
$\mathrm{EM}(X/A)$	Ehrenfeucht–Mostowski type of X over A
$\dim(A)$	Dimension of A
$\mathrm{MR}^M(\varphi)$	Morley rank of φ in M
$\mathrm{MD}(X)$	Morley degree of X
$\mathrm{MR}(\bar{a}/A)$	Morley rank of $\bar{a}$ over A
$\mathrm{MD}(\bar{a}/A)$	Morley degree of $\bar{a}$ over A
$\underset{A}{\downarrow}$	Forking independence over A
E_{Sh}	Shelah equivalence relation
E_{KP}	Kim–Pillay equivalence relation
E_L	Shelah equivalence relation
$\mathrm{Aut}\,f_L(\mathbb{M})$	The subgroup of $\mathrm{Aut}(\mathbb{M})$ generated by $\mathrm{Aut}_M(\mathbb{M})$, $M \preceq \mathbb{M}$ small
$\mathrm{Gal}_L(\mathbb{M})$	The group $\mathrm{Aut}(\mathbb{M})/\mathrm{Aut}\,f_L(\mathbb{M})$
$\mathrm{Gal}_L(T)$	Galois group of T
$|E| =_B |F|$	E and F have same Borel cardinality
$\mathrm{char}(\mathbb{F})$	Characteristic of field $\mathbb{F}$
$\overline{\mathbb{F}}$	Algebraic closure of field $\mathbb{F}$
WOP	Well-ordering principle
ON	Class of all ordinals
AC	Axiom of choice
ZL	Zorn's lemma
CH	Continuum hypothesis

GCH	Generalised continuum hypothesis
$\mathbb{F}^{\sim}$	Residue field of the valued field $\mathbb{F}$
$\mathbb{Q}_p$	Field of p-adic reals
$\mathbb{Z}_p$	ring of p-adic integers
$|E| =_B |F|$	E and F have same Borel cardinality

Chapter 1
Introductory Concepts

Abstract In this chapter, we present most of the introductory concepts. Numerous examples and exercises are given as we go along. In a sense, it sets up our vocabulary. Readers new to logic should read this chapter carefully and work out all the exercises.

1.1 Languages, Terms and Formulas

In this section, we present the syntax of first-order logic.

The *signature S of a first-order language L* consists of

(i) a set of *constant symbols* $\{c_i : i \in I\}$,
(ii) for each positive integer n, a set of *n-ary function symbols* $\{f_j : j \in J_n\}$, and
(iii) for each $n \geq 1$, a set of *n-ary relation symbols* $\{p_k : k \in K_n\}$.

Above sets may not necessarily be non-empty. Besides these symbols, to make a first-order statement, each language L also has

(iv) a sequence of *variables* $x_0, x_1, x_2, \ldots$,
(v) connectives $\neg$ (negation) and $\vee$ (disjunction),
(vi) $\exists$ (existential quantifier) and
(vii) the *equality symbol* $=$, a binary relation symbol.

All these constitute a *first-order language*. We shall be a bit informal and also use letters x, y, z, u, v, w with or without suffixes for variables. The ordering $x_0, x_1, \ldots$ of variables will be called the alphabetical order of the variables. A finite sequence of elements in L will be called an *expression* in L. For an expression s, $|s|$ will denote the *length* of s. Sometimes we shall write $|y| = n$ to indicate that $y = (y_0, \ldots, y_{n-1})$ is a n-tuple of variables. If S and S' are signatures of L and L' respectively and $S \subset S'$, i.e. each constant symbol in S is a constant symbol in S', each n-ary function symbol in S is a n-ary function symbol in S' and each n-ary relation symbol in S is a n-ary relation symbol in S', we call L' an *extension* of L or L a *restriction* of L'.

Let L be a first-order language with signature S. We set

$$|L| = \max\{|S|, \aleph_0\}.$$

© Springer Nature Singapore Pte Ltd. 2017
H. Sarbadhikari and S.M. Srivastava, *A Course on Basic Model Theory*,
DOI 10.1007/978-981-10-5098-5_1

If κ is an infinite cardinal and $|L| \le \kappa$, we call L a κ-language. $\aleph_0$-languages will simply be called *countable languages*.

Let L be a first-order language. The set of all *terms* of L (also called L-terms) is the smallest set $\mathcal{T}$ of expressions in L that contains all variables and constant symbols and is closed under the following operation: whenever $t_1, \ldots, t_n \in \mathcal{T}$, $f t_1 \ldots t_n \in \mathcal{T}$, where f is any n-ary function symbol of L. We shall write $t[x_0, \ldots, x_{n-1}]$ to indicate that t is a term in which no variable other than $x_0, \ldots, x_{n-1}$ occurs.

Exercise 1.1.1 Let L be a first-order language.

1. If L is countable, show that the set of all L-terms is of cardinality $\aleph_0$.
2. If κ is an infinite cardinal and L a κ-language, show that the set of all L-terms is of cardinality $\le \kappa$.

To avoid confusion, we shall follow usual mathematical convention of using parentheses, commas, etc., to express a term. For instance, if f and g are binary function symbols, instead of writing $g f x y z$, we shall write $g(f(x, y), z)$. On the other hand, we shall also drop parentheses when there is no possibility of confusion. Further, we shall adopt the convention of association to the right for omitting parentheses. For instance, instead of writing $t_1 + (t_2 + (t_3 + t_4))$, we shall write $t_1 + t_2 + t_3 + t_4$.

If $t[x_0, \ldots, x_{n-1}], t_0, \ldots, t_{n-1}$ are terms, then $t[t_0, \ldots, t_{n-1}]$ will denote the term obtained from t by simultaneously replacing each occurrence of x_i in t by t_i, $i < n$.

Exercise 1.1.2 Show that if $t[x_0, \ldots, x_{n-1}]$, $t_0, \ldots, t_{n-1}$ are terms, then $t[t_0, \ldots, t_{n-1}]$ is a term.

(Hint: Use induction on the length of t.)

We define the set of all *subterms* of a term t by induction on the length of t as follows: t is a subterm of t. If $f t_1 \ldots t_n$ is a subterm of t, so is each t_i, $1 \le i \le n$. An expression is a subterm of t if and only if it is obtained as above.

Expressions of the form $p t_1 \ldots t_n$, where p is an n-ary relation symbol (including the equality symbol $=$) and $t_1, \ldots, t_n$ are terms, are called *atomic formulas*. In this case p is the equality symbol, we write $t_1 = t_2$ instead of $= t_1 t_2$.

A *formula* of L, also called an L-formula, is inductively defined as follows: every atomic formula is a formula—these are all the formulas of *rank* 0; if φ and ψ are formulas of rank $\le k$ and v is a variable, then $\neg \varphi$ (the negation of φ); $\exists v \varphi$ (an instantiation of φ) and $\varphi \lor \psi$ (the disjunction of φ and ψ) are formulas of rank $\le k + 1$. The set of expressions so obtained are all the formulas of L. The rank of a formula φ is the least k such that φ is of rank $\le k$. We shall follow standard convention and use parentheses to avoid ambiguities.

Exercise 1.1.3 Let L be a first-order language.

1. If L is countable, show that the set of all L-formulas is of cardinality $\aleph_0$.
2. If κ is an infinite cardinal and L a κ-language, show that the set of all L-formulas is of cardinality $\le \kappa$.

There are more logical connectives and quantifiers that are commonly used. But they are all defined in terms of $\neg$, $\vee$ and $\exists$ as follows:

$\forall v \varphi$ is an abbreviation of $\neg \exists v \neg \varphi$, $\varphi \wedge \psi$ abbreviates $\neg(\neg\varphi \vee \neg\psi)$, $\varphi \rightarrow \psi$ is an abbreviation of $(\neg\varphi) \vee \psi$ and $\varphi \leftrightarrow \psi$ abbreviates $(\varphi \rightarrow \psi) \wedge (\psi \rightarrow \varphi)$. The connective $\wedge$ is called *conjunction* and the quantifier $\forall$ the *universal quantifier*.

As in the case of terms, we adopt the convention of association to the right for omitting parentheses. This means that $\varphi \vee \psi \vee \xi$ is to be read as $\varphi \vee (\psi \vee \xi)$; $\varphi \vee \psi \vee \xi \vee \eta$ is to be read as $\varphi \vee (\psi \vee (\xi \vee \eta))$ and so on. Further, $\varphi \rightarrow \psi \rightarrow \xi$ is to be read as $\varphi \rightarrow (\psi \rightarrow \xi)$; $\varphi \rightarrow \psi \rightarrow \xi \rightarrow \eta$ is to be read as $\varphi \rightarrow (\psi \rightarrow (\xi \rightarrow \eta))$ and so on.

If $\varphi_1, \ldots, \varphi_n$ are formulas, we shall often write $\vee_{i=1}^n \varphi_i$ for $\varphi_1 \vee \cdots \vee \varphi_n$ and $\wedge_{i=1}^n \varphi_i$ for $\varphi_1 \wedge \cdots \wedge \varphi_n$. Also, we shall often write $t \neq s$ instead of $\neg(t = s)$, where t and s are terms. Further, we shall often write $\exists \overline{v} \varphi$ instead of $\exists v_0 \ldots \exists v_{n-1} \varphi$ and $\forall \overline{v} \varphi$ instead of $\forall v_0 \ldots \forall v_{n-1} \varphi$.

The set of all *subformulas* of a formula φ is the smallest set $\mathcal{S}(\varphi)$ of formulas of L that contains φ and satisfies the following conditions: whenever $\neg\psi$ or $\exists v \psi$ is in $\mathcal{S}(\varphi)$, so is ψ, and whenever $\psi \vee \xi$ is in $\mathcal{S}(\varphi)$, so are ψ and ξ.

An occurrence of a variable v in a formula φ is *bound* if it occurs in a subformula of the form $\exists v \psi$; otherwise, the occurrence is called *free*. A variable is said to be free in φ if it has a free occurrence in φ. We shall write $\varphi[v_0, \ldots, v_n]$ if φ is a formula all of whose free variables belong to the set $\{v_0, \ldots, v_n\}$. Note that this does not mean that each of $v_0, \ldots, v_n$ has a free occurrence in φ. Let $\varphi[x_0, \ldots, x_n]$ be an L-formula in which x_n has a free occurrence. Then the formula $\forall x_0 \ldots \forall x_n \varphi$ is called the *closure* of φ.

A formula with no free variable is called a *closed formula* or a *sentence*. A formula that contains no quantifiers is called an *open formula* or *quantifier free*. A formula of the form $\exists \overline{v} \varphi$, φ open, will be called an *existential formula* or a $\exists$-formula and those of the form $\forall \overline{v} \varphi$, φ open, *universal formulas* or $\forall$-formulas. Likewise formulas of the form $\forall \overline{x} \exists \overline{y} \varphi$ ($\exists \overline{x} \forall \overline{y} \varphi$), φ open, will be called $\forall \exists$-formulas (respectively $\exists \forall$ formulas).

Let t be a term, v a variable and φ a formula of a language L. We say that *the term t is substitutable for v in φ* if for each variable w occurring in t, no subformula of φ of the form $\exists w \psi$ contains an occurrence of v that is free in φ. If terms $t_1, \ldots, t_n$ are substitutable for $v_1, \ldots, v_n$ respectively in φ, then $\varphi_{v_1,\ldots,v_n}[t_1, \ldots, t_n]$, or $\varphi[t_1, \ldots, t_n]$ when there is no possibility of confusion, called an *instance* of φ, will denote the expression obtained from φ by simultaneously replacing all free occurrences of $v_1, \ldots, v_n$ in φ by $t_1, \ldots, t_n$ respectively. Note that whenever we shall talk of $\varphi_{v_1,\ldots,v_n}[t_1, \ldots, t_n]$, it will be assumed that $t_1, \ldots, t_n$ are substitutable in φ for $v_1, \ldots, v_n$, respectively.

Exercise 1.1.4 Assume that terms $t_1, \ldots, t_n$ are substitutable for $v_1, \ldots, v_n$ respectively in a formula φ. Show that $\varphi[t_1, \ldots, t_n]$ is a formula.

If $\varphi[x]$ is a formula, the sentence

$$\exists x_0 \ldots \exists x_{n-1} (\forall x (\varphi[x] \rightarrow \vee_{i=0}^{n-1} (x = x_i)))$$

will be abbreviated by

$$\exists_{\leq n} x \varphi \text{ or by } \exists_{<n+1} x \varphi.$$

Also,

$$\exists x \varphi \wedge \neg \exists_{\leq n} x \varphi$$

will be abbreviated by

$$\exists_{>n} x \varphi \text{ or by } \exists_{\geq n+1} x \varphi.$$

Finally, the formula

$$\exists_{=n} x \varphi$$

will stand for

$$\exists_{\leq n} x \varphi \wedge \exists_{\geq n} x \varphi.$$

1.2   Structures and Truth in a Structure

Let L be a first-order language. A *structure* for L or an *L-structure* consists of

* a non-empty set M,
* for each constant symbol c, an element c^M of M,
* for each n-ary function symbol f, a n-ary map $f^M : M^n \to M$ and
* for each n-ary relation symbol p, a n-ary relation $p^M \subset M^n$ on M.

It is customary to denote a structure like this as $\mathcal{M}$ and call M the *universe* of $\mathcal{M}$. However, we shall call M itself the structure. c^M, f^M and p^M are called *interpretations* of c, f and p respectively in M. Further, when the underlying structure is understood, we shall often use the same symbol for constant, function and relation symbols and their respective interpretations in M. Thus, often we shall write c, f, p for c^M, f^M and p^M, respectively.

Let L' be an extension of L and M' be an L'-structure. By ignoring the interpretations of symbols in L' which are not symbols in L, we get an L-structure, say M. In this case, we call M a *reduct* of M' to L or M' an *expansion* of M to L'.

Let M be an L-structure and $N \subset M$. Suppose for every constant symbol c, $c^M \in N$ and N is closed under each f^M, f a function symbol. Then N can be canonically made into an L-structure by defining $c^N = c^M$, c a constant symbol, and f^N and p^N to be the restrictions of f^M and p^M to N, where f and p are function and relation symbols, respectively. Such an $N \subset M$ is called a *substructure* of M. In this case, we also call M an *extension* of N. We shall write $N \sqsubseteq M$ or $M \sqsupseteq N$ if N is a substructure of M. It is emphasised that we have reserved the notation $\subset$ for the set theoretic subset relation and are using $\sqsubseteq$ for substructure relation among structures of a language.

If M is an L-structure and $A \subset M$, L_A will denote the extension of L obtained by adding each $a \in A$ as a new constant symbol. Thus, elements of A have dual role—

as elements of A as well as constant symbols of L_A. This will cause no confusion because from the context the role of $a \in A$ will be clear. L_M-formulas will be called *formulas with parameters* and L_A-formulas formulas with parameters in A. Every formula with parameters will be thought of in the form $\varphi[\bar{x}, \bar{a}]$ where $\varphi[\bar{x}, \bar{y}]$ is an L-formula and $\bar{a} \in M$. Often we shall suppress parameters $\bar{a}$ and simply say that $\varphi[\bar{x}]$ is a formula with parameters. We shall regard M as an L_M-structure by interpreting each constant symbol $a \in M$ by a itself.

For a variable free L_M-term t, we define the *value* of t in M, denoted by t^M, by induction on the length of t as follows.

1. $t^M = c^M$, if $t = c$, c a constant symbol.
2. $t^M = f^M(t_1^M, \ldots, t_n^M)$, where f is a n-ary function symbol, $t_1, \ldots, t_n$ variable free terms and $t = f(t_1, \ldots, t_n)$.

Let M be an L-structure. By induction on the rank of L_M-sentences φ, we now define when is φ *true* in M, written $M \models \varphi$. We shall write $M \not\models \varphi$ if φ is not true in M. In this case, we also say that φ is false in M.

1. If t and s are variable free L_M-terms,

$$M \models t = s \Leftrightarrow t^M = s^M.$$

2. If $pt_1 \ldots t_n$ is a variable free atomic formula, then

$$M \models pt_1 \ldots t_n \Leftrightarrow p^M(t_1^M, \ldots, t_n^M).$$

3. For L_M-sentences φ and ψ,

$$M \models \neg\varphi \Leftrightarrow M \not\models \varphi \ \& \ M \models \varphi \vee \psi \Leftrightarrow M \models \varphi \text{ or } M \models \psi.$$

4. $M \models \exists x \varphi \Leftrightarrow M \models \varphi_x[a]$ for some $a \in M$.

A formula with free variables is said to be true in M if its closure is true in M.

Exercise 1.2.1 Let φ and ψ be closed formulas. Show the following.

1. $M \models \varphi \wedge \psi$ if and only if $M \models \varphi$ and $M \models \psi$.
2. $M \models \varphi \rightarrow \psi$ if and only if either $M \not\models \varphi$ or $M \models \psi$.
3. $M \models \varphi \leftrightarrow \psi$ if and only if either both φ and ψ are true in M or both are false in M.
4. $M \models \forall v \varphi$ if and only if $M \models \varphi[a]$ for all $a \in M$.

If M is an L-structure, $Th(M)$ denotes the set of all L-sentences true in M and is called *the theory of M*. More generally, if $\mathcal{M}$ is a class of L-structures, then $Th(\mathcal{M})$ will denote the set of all L-sentences true in all $M \in \mathcal{M}$.

If the signature of L is finite, we call M *decidable* if there is an algorithm which decides whether a closed L-formula φ is true in M or not, i.e. whether $\varphi \in Th(M)$ or not. See [59, Chap. 6] for details.

If an L-formula φ is true in all L-structures, we call it a *tautology*. We shall write $\models \varphi$ if φ is a tautology. We call φ and ψ *tautologically equivalent* if $\varphi \leftrightarrow \psi$ is a tautology.

Exercise 1.2.2 Show the following.

1. $\neg\neg\varphi$ and φ are tautologically equivalent.
2. $\neg(\varphi \vee \psi)$ and $\neg\varphi \wedge \neg\psi$ are tautologically equivalent.
3. $\neg(\varphi \wedge \psi)$ and $\neg\varphi \vee \neg\psi$ are tautologically equivalent.
4. $\varphi \wedge (\psi \vee \xi)$ and $(\varphi \wedge \psi) \vee (\varphi \wedge \xi)$ are tautologically equivalent.
5. $\varphi \vee (\psi \wedge \xi)$ and $(\varphi \vee \psi) \wedge (\varphi \vee \xi)$ are tautologically equivalent.
6. $\varphi \rightarrow \psi \rightarrow \xi$ and $\psi \rightarrow \varphi \rightarrow \xi$ are tautologically equivalent.
7. $\neg\exists v\varphi$ and $\forall v\neg\varphi$ are tautologically equivalent.
8. $\neg\forall v\varphi$ and $\exists v\neg\varphi$ are tautologically equivalent.
9. $\varphi \vee \exists v\psi$ and $\exists v(\varphi \vee \psi)$ are tautologically equivalent if v is not free in φ.
10. $\varphi \vee \forall v\psi$ and $\forall v(\varphi \vee \psi)$ are tautologically equivalent if v is not free in φ.
11. If w does not occur in φ, then $\forall v\varphi$ and $\forall w\varphi_v[w]$ are equivalent.

Exercise 1.2.3 Let φ, ψ and ξ be closed formulas. Show that $\varphi \rightarrow \psi \rightarrow \xi$ and $(\varphi \rightarrow \psi) \rightarrow \xi$ are not tautologically equivalent.

Exercise 1.2.4 A formula is called a *literal* if it is either atomic or the negation of an atomic formula. A formula φ is said to be in *disjunctive normal form* (DNF in short) if it is in the form $\vee_{i=1}^{k} \wedge_{j=1}^{n_i} \varphi_{ij}$ with each φ_{ij} a literal. A formula φ is said to be in *conjunctive normal form* (CNF in short) if it is in the form $\wedge_{i=1}^{k} \vee_{j=1}^{n_i} \varphi_{ij}$ with each φ_{ij} a literal. A formula φ is said to be in *prenex normal form* if it is in the form $Q_0 v_0 \ldots Q_{n-1} v_{n-1} \psi$, where each Q_i is either an existential quantifier or a universal quantifier and ψ is open. Show the following.

1. Every open formula φ is tautologically equivalent to a formula ψ in DNF as well as to a formula ξ in CNF.
2. Every formula φ is tautologically equivalent to a formula ψ in prenex normal form.

A *first-order theory* or simply a *theory* T consists of a first-order language $L(T)$ or simply L and a set of L-formulas, called the *axioms* of T. For a cardinal $\kappa \geq \aleph_0$, T is called a *κ-theory* if $|L(T)| \leq \kappa$. $\aleph_0$-theories are simply called *countable theories*. A *model* of T is an L-structure M in which each axiom of T is true. If M is a model of T, we shall write $M \models T$. A formula φ is called a *theorem* of T, written $T \models \varphi$, if it is true in all models of T.

A set of L-sentences T' is said to *axiomatise* T if every formula in T is a theorem of T' and also every $\varphi \in T'$ is a theorem of T.

Exercise 1.2.5 Show that T' axiomatises T if and only if T and T' have the same class of models.

A theory T is called *finitely axiomatisable* if there is a finite set of sentences that axiomatises T. Theories T and T' in the same language will be considered to be the same if each axiomatises the other one.

Exercise 1.2.6 Show that T is finitely axiomatisable if and only if there is an $L(T)$-sentence φ such that for every structure M of $L(T)$, $M \models \varphi \Leftrightarrow M \models T$.

For a theory T, $T_\forall$ ($T_\exists$, $T_{\forall\exists}$, $T_{\exists\forall}$) will denote the set of all universal (respectively existential, $\forall\exists$, $\exists\forall$) sentences which are theorems of T. A theory T is called *universal* (*existential*, $\forall\exists$, $\exists\forall$) if $T_\forall$ (respectively $T_\exists$, $T_{\forall\exists}$, $T_{\exists\forall}$) axiomatises T.

Assume that the signature of L is finite. An L-theory T is called *decidable* if there is an algorithm to decide whether an L-sentence φ is a theorem of T or not. If T is not decidable, it is called *undecidable*. An L-structure M is called *decidable* if there is an algorithm to decide if an L-sentence is true in M or not. See [59, Chap. 6] for relevant definitions.

A theory T' is called an *extension* of T if $L(T')$ is an extension of $L(T)$ and every axiom of T is a theorem of T'. If T' is an extension of T with $L(T') = L(T)$, then we call T' a *simple extension* of T. If T is a theory and Γ a set of $L(T)$-formulas, then $T[\Gamma]$ will denote the simple extension of T obtained by adding Γ to the set of axioms. If T' is an extension of T such that every $L(T)$-formula that is a theorem of T' is also a theorem of T, then we call T' a *conservative extension* of T.

If T' is an extension of T and $M \models T'$, we can regard M as a model of T by ignoring the interpretations of symbols in $L(T')$ that are not in $L(T)$. This model of T is called the *restriction of M to T*.

Exercise 1.2.7 If T' is an extension of T, show that every theorem of T is a theorem of T'.

Exercise 1.2.8 Let T' be an extension of an L-theory T obtained by adding new constant symbols $c_0, \ldots, c_{n-1}$ and no new axiom. For any L-formula $\varphi[v_0, \ldots, v_{n-1}]$ show that

$$T \models \forall \bar{v}\varphi[\bar{v}] \Leftrightarrow T' \models \varphi[c_0, \ldots, c_{n-1}].$$

Conclude that T' is a conservative extension of T.

Exercise 1.2.9 For closed $L(T)$ formulas $\varphi_1, \ldots, \varphi_n$ show that

$$T[\varphi_1, \ldots, \varphi_n] \models \varphi \Leftrightarrow T \models \varphi_1 \to \ldots \varphi_n \to \varphi,$$

where φ is any $L(T)$-formula.

A closed formula φ is said to be *decidable in T* if either φ or $\neg\varphi$ is a theorem of T. If φ is not decidable in T, we say that φ is *undecidable in T* or φ is *independent of the axioms of T*.

A theory T is called *consistent* if it has a model. The theory T is called *complete* if it is consistent and if every closed formula is decidable in T. Otherwise, T is called *incomplete*.

A class $\mathcal{E}$ of L-structures is called *elementary* if there is an L-theory T such that $M \in \mathcal{E}$ if and only if $M \models T$. Moreover, if T is finite, we call $\mathcal{E}$ *finitely axiomatisable*.

Formulas $\varphi[\overline{x}]$ and $\psi[\overline{x}]$ are called *equivalent in* T, written $\varphi \equiv_T \psi$ or simply $\varphi \equiv \psi$, if

$$T \models \forall \overline{x}(\varphi \leftrightarrow \psi).$$

Exercise 1.2.10 Show that for any theory T, $\equiv_T$ is an equivalence relation on the set of all formulas of T.

Exercise 1.2.11 Let T be a first-order L-theory. Show that the following statements are equivalent.

1. Every existential L-formula is equivalent in T to a universal L-formula.
2. Every universal L-formula is equivalent in T to an existential L-formula.
3. Every L-formula is equivalent in T to a universal L-formula.
4. Every L-formula is equivalent in T to an existential L-formula.

Exercise 1.2.12 If T' is a conservative extension of T and T is consistent, show that T' is consistent.

Exercise 1.2.13 Show that a class $\mathcal{M}$ of L-structures is elementary if and only if $\mathcal{M} = \{M : M \models Th(\mathcal{M})\}$.

Remark 1.2.14 So far we have followed the tradition in presenting syntax and semantics of first-order logic where there are only $\aleph_0$-many variables. This is sufficient because terms, formulas, proofs, etc., are of finite length. But for model theory, it is at times convenient to have uncountably many variables, say $\{x_i : i \in I\}$ where I is uncountable and x_is distinct. Most of the definitions clearly make sense even when the number of variables is uncountable.

Exercise 1.2.15 Let L be a first-order language with $\aleph_0$-many variables and L' be obtained from L by adding uncountably many variables. Show that every L'-sentence is tautologically equivalent to an L-sentence.

1.3   Examples of Theories

Example 1.3.1 The language for the *theory of linearly ordered sets* LO has one binary relation symbol $<$. The axioms of LO are the following:

(1.1) $\forall x \neg (x < x)$.
(1.2) $\forall x \forall y \forall z((x < y \wedge y < z) \rightarrow x < z)$.
(1.3) $\forall x \forall y(x < y \vee x = y \vee y < x)$.

Sometimes we shall write $x > y$ in place of $y < x$; $x \leq y$ as well as $y \geq x$ will abbreviate the formula $x < y \vee x = y$.

The *theory of dense linearly ordered sets without end points*, denoted by DLO, has the same language as that of LO. In addition to the axioms of LO it has following axioms

(1.4) $\forall x \forall y ((x < y) \rightarrow \exists z (x < z \wedge z < y))$.
(1.5) $\forall x \exists y (y < x)$.
(1.6) $\forall x \exists y (x < y)$.

A linearly ordered set D is called *discrete* if every element of D which is not the first element has an immediate predecessor and every element which is not the last element has an immediate successor.

Exercise 1.3.2 Show that the class of all discrete linearly ordered sets with no end points is elementary.

Example 1.3.3 The set of all rational numbers $\mathbb{Q}$ with usual order is a model of DLO and the set $\mathbb{Z}$ of all integers with usual order is model of the theory of discrete linearly ordered sets with no end points.

Example 1.3.4 The language for *the theory of groups* has a constant symbol e, two binary function symbols $+$ and $-$ and the following axioms:

(2.1) $\forall x \forall y \forall z (x + (y + z) = (x + y) + z)$.
(2.2) $\forall x (x + e = x \wedge e + x = x)$.
(2.3) $\forall x \exists y (x + y = e \wedge y + x = e)$.
(2.4) $\forall x \forall y \forall z (x - y = z \leftrightarrow x = z + y)$.

The *theory of abelian groups* has in addition the following axiom:

(2.5) $\forall x \forall y (x + y = y + x)$.

In case of abelian groups, the identity symbol is taken to be 0 instead of e. In this language, for any $n \geq 1$ and any variable x, nx will stand for the term

$$\underbrace{x + \cdots + x}_{n \text{ times}}.$$

The theory of *torsion-free abelian groups* has, besides (2.1)–(2.5), for each $n \geq 1$, the following axiom:

(2.6) $\forall x (x \neq 0 \rightarrow nx \neq 0)$.

The theory of *torsion-free divisible abelian groups*, denoted by DAG, has, besides (2.1)–(2.6), for each $n \geq 1$, the following axiom:

(2.7) $\forall x \exists y (ny = x)$.

The language for the theory of *ordered abelian groups* is an extension of the language of groups by a binary relation symbol $<$. Its axioms are axioms of LO (1.1)–(1.3), axioms of abelian groups (2.1)–(2.5) and the following axiom:

(2.8) $\forall x \forall y \forall z (x < y \rightarrow x + z < y + z)$.

The language for the *theory of divisible ordered abelian groups*, denoted by $ODAG$, is the same as that of ordered abelian groups. Its axioms are axioms of ordered abelian groups and (2.7).

Example 1.3.5 The group of integers $\mathbb{Z}$ is a model of the theory of ordered abelian groups and $\mathbb{Q}$ and $\mathbb{R}$ are models of $ODAG$. The group $\mathbb{Z}/n\mathbb{Z}$ of integers modulo n is an abelian group which is not torsion free. The group S_n of permutations of n-elements, $n \geq 3$, is a group which is not abelian.

Exercise 1.3.6 Show that a substructure of a group is a group.

Exercise 1.3.7 Show that every nonzero ordered abelian group is infinite.

Exercise 1.3.8 Show that every ordered abelian group is torsion free.

Exercise 1.3.9 Show that every ordered divisible abelian group is a dense linearly ordered set without end points.

Example 1.3.10 The language for *the theory of commutative rings with identity* is an extension of the language of groups by having one more constant symbol 1, and one more binary function symbol $\cdot$. The axioms of this theory are the axioms (2.1)–(2.5) of abelian groups together with the following axioms:

(3.1) $\forall x \forall y \forall z (x \cdot (y \cdot z) = (x \cdot y) \cdot z)$.
(3.2) $\forall x (x \cdot 1 = x \wedge 1 \cdot x = x)$.
(3.3) $\forall x \forall y \forall z (x \cdot (y + z) = x \cdot y + x \cdot z)$.
(3.4) $\forall x \forall y (x \cdot y = y \cdot x)$.

For any variable x and $n > 1$, x^n will stand for the term

$$\underbrace{x \cdot \cdots \cdot x}_{n \text{ times}}.$$

Note that a substructure of a commutative ring R with identity is a subring of R. *The theory of integral domains* has one more axiom:

(3.5) $\forall x \forall y (x \cdot y = 0 \rightarrow (x = 0 \vee y = 0))$

A ring R is called *ordered* if it is equipped with a linear order $<$ on R such that for every $x, y, z \in R$ the following conditions are satisfied.

(3.6) $0 < x$ and $0 < y$ imply $0 < x \cdot y$.
(3.7) $x < y$ implies $x + z < y + z$.

The theory of fields has the same language as that of the theory of commutative rings with identity. Its axioms are the axioms (2.1)–(2.5) of abelian groups, axioms (3.1)–(3.4) of the theory of commutative rings with identity and

(3.8) $0 \neq 1$.
(3.9) $\forall x (x \neq 0 \rightarrow \exists y (x \cdot y = 1))$.

The *theory of algebraically closed fields*, denoted by ACF, has in addition to the axioms of fields, for each $n \geq 1$ the following axiom:

(3.10) $\forall x_0 \ldots \forall x_n (x_n \neq 0 \rightarrow \exists x_{n+1} (x_0 + x_1 \cdot x_{n+1} + \cdots + x_n \cdot x_{n+1}^n = 0))$.

The *theory of fields of characteristic p*, p a prime, in addition to field axioms, has the following axiom:

(3.11) $\underline{p} = 0$,

where $\underline{n}$ denotes the term

$$\underbrace{1 + \cdots + 1}_{n \text{ times}},$$

$n > 1$.

The *theory of fields of characteristic 0* in addition to the field axioms has the axiom:

(3.12) $\underline{n} \neq 0$,

for each $n > 1$. The theory of algebraically closed fields of characteristic p, p a prime, will be denoted by $ACF(p)$ and that of characteristic 0 by $ACF(0)$.

A field $\mathbb{F}$ of characteristic 0 is called a *differential field* if there is a unary function δ satisfying the following axioms:

(3.13) $\delta(x + y) = \delta(x) + \delta(y)$.
(3.14) $\delta(x \cdot y) = \delta(x) \cdot y + x \cdot \delta(y)$.

The function δ is called a *derivation*. Since this theory has no relation symbol (except, of course, equality), its atomic formulas are polynomial expressions in powers of δ. We call a differential field $\mathbb{F}$ *differentially closed* if whenever a conjunction of finitely many literals with parameters in $\mathbb{F}$ has a solution in an extension of $\mathbb{F}$, it has a solution in $\mathbb{F}$. We shall not prove here that the class of all differentially closed fields is elementary.

A field $\mathbb{F}$ is called *ordered* if in addition it is equipped with a linear order $<$ making it into an ordered ring.

The theory of ordered fields is denoted by OF. A field $\mathbb{F}$ is called *orderable* if there is a linear order $<$ on $\mathbb{F}$ making it into an ordered field.

The *theory of real closed fields* has two equivalent definitions.

(i) It is a field satisfying the following additional axioms:

(3.15) $\forall x \exists y (x = y^2 \lor x + y^2 = 0)$.
(3.16) For every $n \geq 1$,

$$\forall x_1 \ldots \forall x_n (1 + x_1^2 + \cdots + x_n^2 \neq 0).$$

(3.17) For every odd $n \geq 1$,

$$\forall x_0 \ldots \forall x_n (x_n \neq 0 \rightarrow \exists x_{n+1} (x_0 + x_1 \cdot x_{n+1} + \cdots + x_n \cdot x_{n+1}^n = 0).$$

(ii) Equivalently, a real closed field is an ordered field that satisfies axioms (3.15) and (3.17).

The theory without order relation and having (3.15)–(3.17) as additional axioms will be denoted by RCF and that with order relation will be denoted by $RCOF$. The field of real numbers and that of real algebraic numbers are models of RCF and of $RCOF$.

Exercise 1.3.11 Show that every algebraically closed field is infinite.

Exercise 1.3.12 Let $\mathbb{F}$ be an orderable field. Show the following.

1. There does not exist $x_1, \ldots, x_n$ such that $-1 = x_1^2 + \cdots + x_n^2$.
2. $\sum_{i=1}^{n} x_i^2 = 0 \rightarrow \wedge_{i=1}^{n} (x_i = 0)$.

Exercise 1.3.13 Show that an algebraically closed field is not orderable. Also show that no finite field is orderable

Example 1.3.14 Let $(\mathbb{F}, 0, 1, +', \cdot)$ be a field. *The theory of vector spaces over $\mathbb{F}$* is an extension of the theory of abelian groups with an additional unary function symbol $r\cdot$ for each $r \in \mathbb{F}$, and the following additional axioms:

(4.1) $\forall x (1 \cdot x = x)$,
(4.2) $\forall x \forall y (r \cdot (x + y) = r \cdot x + r \cdot y)$,
(4.3) $\forall x ((r +' s) \cdot x = r \cdot x + s \cdot x)$,
(4.4) $\forall x (r \cdot (s \cdot x) = (r \cdot s) \cdot x)$,

where $r, s \in \mathbb{F}$.

It is easily seen that the class of left R-modules, R a commutative ring with identity, is elementary.

Exercise 1.3.15 Show that the class of torsion-free divisible abelian groups is precisely the class of all vector spaces over the field of rational numbers $\mathbb{Q}$.

Example 1.3.16 Let G be a group. A *G-space* is a set X with a map $\cdot : G \times X \rightarrow X$ (we shall write $g \cdot x$ for $\cdot(g, x)$) satisfying the following axioms:

(5.1) $\forall x (e \cdot x = x)$, where $e \in G$ is the identity element of G.
(5.2) For every $g, h \in G$, $g \cdot (h \cdot x) = (gh) \cdot x$.

In this case, we say that G *acts* on X and the map $\cdot : G \times X \rightarrow X$ the *action* of G on X. Clearly, the class of all G-spaces is elementary. The action is called *free* if for all $g \neq e$, $g \cdot x \neq x$ for all x. For each $x \in X$, $\{g \cdot x : g \in G\}$ is called the *orbit* of x. Note that orbits partition X. We denote the space of all orbits by X/G.

Example 1.3.17 The signature for *Peano Arithmetic PA* contains a constant symbol
0, a unary function symbol S (which designates the successor function), two binary
function symbols $+$ and $\cdot$, and a binary relation symbol $<$. Its axioms are:

(6.1) $\forall x(\neg(Sx = 0))$.
(6.2) $\forall x \forall y(Sx = Sy \rightarrow x = y)$.
(6.3) $\forall x(x + 0 = x)$.
(6.4) $\forall x \forall y(x + Sy = S(x + y))$.
(6.5) $\forall x(x \cdot 0 = 0)$.
(6.6) $\forall x \forall y(x \cdot Sy = (x \cdot y) + x)$.
(6.7) $\forall x(\neg(x < 0))$.
(6.8) $\forall x \forall y(x < Sy \leftrightarrow (x < y \lor x = y))$.
(6.9) For every formula $\varphi[x]$, the formula

$$\varphi[0] \rightarrow \forall x(\varphi \rightarrow \varphi[Sx]) \rightarrow \forall x \varphi.$$

Example 1.3.18 The language for *the theory of (undirected) graphs* has a binary
relation symbol E and following axioms:

(7.1) $\forall x \neg E(x, x)$.
(7.2) $\forall x \forall y(E(x, y) \rightarrow E(y, x))$.

A *random graph* is a graph with following axioms:

(7.3) $\exists x \exists y(x \neq y)$
(7.4) For every $n \geq 1$,

$$\forall \overline{x} \forall \overline{y}(\wedge_{i<n} \wedge_{j<n} (x_i \neq y_j) \rightarrow \exists z(\wedge_{i<n} E(x_i, z) \wedge_{j<n} \neg(E(y_j, z) \wedge z \neq y_j))).$$

Note that random graphs are all infinite.

An L-structure where L has only a binary relation symbol E is called a *directed
graph*. A directed graph with no loop and no cycle is called a *tree*. So its axioms are

(7.5) $\forall x \neg E(x, x)$.
(7.6) For every $n > 1$,

$$\neg \exists \overline{x}(x_0 = x_{n-1} \wedge \wedge_{i<n-1} E(x_i, x_{i+1})).$$

Exercise 1.3.19 Show that the class of all bipartite graphs is elementary.

Exercise 1.3.20 Show that the class of all infinite sets is elementary.

1.4 Homomorphism

Let M and N be L-structures. A *homomorphism* from M to N is a map $h : M \rightarrow N$
satisfying the following conditions:

1. For every constant c, $h(c^M) = c^N$.
2. For every n-ary function symbol f and every $\bar{a} \in M^n$, $h(f^M(\bar{a})) = f^N(h(\bar{a}))$.
3. For every n-ary relation symbol p and every $\bar{a} \in M^n$,

$$M \models p[\bar{a}] \Rightarrow N \models p[h(\bar{a})],$$

i.e.

$$p^M(\bar{a}) \Rightarrow p^N(h(\bar{a})).$$

For any M, id_M will denote the identity morphism on M. If M_1, M_2 and M_3 are L-structures and $h_1 : M_1 \to M_2$ and $h_2 : M_2 \to M_3$ are homomorphisms, then their composition $h_2 \circ h_1 : M_1 \to M_3$ is a homomorphism. Thus, it is easy to see that the class of all L-structures and homomorphisms form a category under the composition. If a homomorphism is one-to-one, it is called a *monomorphism* and it is called an *epimorphism* if it is surjective. A homomorphism $f : M \to M$ is called an *endomorphism* of M. $End(M)$ will denote the set of all endomorphisms of M.

Proposition 1.4.1 *Let M and N be L-structures and $h : M \to N$ a homomorphism. Suppose f and p are k-ary function and relation symbols respectively and $t[x_0, \ldots, x_{n-1}], t_0[x_0, \ldots, x_{n-1}], \ldots, t_{k-1}[x_0, \ldots, x_{n-1}]$ are L-terms. Then for every $\bar{a} \in M^n$,*

(a) $h(t^M[\bar{a}]) = t^N[h(\bar{a})]$.
(b) $M \models p[t_0^M[\bar{a}], \ldots, t_{k-1}^M[\bar{a}]] \Rightarrow N \models p[t_0^N[h(\bar{a})], \ldots, t_{k-1}^N[h(\bar{a})]]$.

Proof (a) is proved by induction on the length $|t|$ of t. (b) is straightforward from (a).   □

Exercise 1.4.2   1. Let M, N be L-structures and $f, g : M \to N$ homomorphisms. Show that $\{a \in M : f(a) = g(a)\}$ is a substructure of M.
   2. Let M be an L-structure and $X \subset M$.

   (a) Show that there is a smallest substructure of M containing X.
   (b) We say that X is a *generator* of M if M is the only substructure of M containing X. Let N be another L-structure and $f, g : M \to N$ homomorphisms such that $f|X = g|X$. Show that $f = g$.

1.5   Embedding

A homomorphism $h : M \to N$ is called an *embedding* if instead of condition (3) of the definition of homomorphism in Sect. 1.4, the following condition is satisfied:

(4) For every n-ary relation symbol p and every $\bar{a} \in M^n$,

$$M \models p[\bar{a}] \Leftrightarrow N \models p[h(\bar{a})].$$

Since $=$ is a relation symbol, an embedding is one-to-one. However, a one-to-one homomorphism need not be an embedding.

Example 1.5.1 Let the signature of L have only a binary relation symbol $\leq$. Take $M = N = \{0, 1\}$. Define

$$x \leq^M y \Leftrightarrow x = y$$

and

$$x \leq^N y \Leftrightarrow (x = y) \vee (x = 0 \wedge y = 1).$$

Then the identity map from M to N is a bijective homomorphism but not an embedding.

Example 1.5.2 If M is a substructure of N, then the inclusion map $i : M \hookrightarrow N$ is an embedding.

Example 1.5.3 From standard algebra argument, we know that substructures of fields are precisely integral domains. Indeed, every integral domain D is embedded into its quotient field $\mathbb{Q}(D)$.

Example 1.5.4 Every countable linearly ordered set $(M, <)$ is embeddable in the set of all rational numbers $\mathbb{Q}$ with usual order. Enumerate $M = \{a_0, a_1, a_2, \ldots\}$. Set $h(a_0) = 0$. Suppose for $n \geq 1$, we have defined an order-preserving map $h : \{a_0, \ldots, a_{n-1}\} \to \mathbb{Q}$. If $a_n < a_i$ for all $i < n$, define $h(a_n)$ to be any rational number less than every $h(a_0), \ldots, h(a_{n-1})$. On the other hand, if $a_n > a_i$ for all $i < n$, define $h(a_n)$ to be any rational number greater than every $h(a_0), \ldots, h(a_{n-1})$. Otherwise, there exist $a_i < a_j, i, j < n$, such that $a_i < a_n < a_j$ and for no $k < n, a_i < a_k < a_j$. Then we define $h(a_n)$ to be any rational number r such that $h(a_i) < r < h(a_j)$. Inductively, we have thus defined an embedding $h : M \to \mathbb{Q}$.

Example 1.5.5 Let H be a torsion-free abelian group. Then there is a torsion-free, divisible abelian group G and an embedding $\alpha : H \to G$ such that for every torsion-free, divisible abelian group G' and every embedding $\beta : H \to G'$, there is a unique embedding $\gamma : G \to G'$ such that $\beta = \gamma \circ \alpha$.

To see this, set

$$E = \{(h, n) : h \in H, n > 0\}.$$

Define an equivalence relation $\sim$ on E by

$$(h, n) \sim (h', n') \Leftrightarrow n'h = nh'.$$

Let $\frac{h}{n}$ denote the equivalence class containing $(h, n) \in E$ and set

$$G = E/ \sim \ = \{\frac{h}{n} : (h, n) \in E\},$$

the quotient space,

$$0 = \frac{0}{1},$$

$$\frac{h}{n} + \frac{h'}{n'} = \frac{n'h + nh'}{nn'},$$

$$\frac{h}{n} - \frac{h'}{n'} = \frac{n'h - nh'}{nn'}$$

and

$$\alpha(h) = \frac{h}{1}, h \in H.$$

These are well defined. Note that $m\frac{x}{mn} = \frac{x}{n}$. It is fairly easy to see that these make G into a torsion-free divisible abelian group with $\alpha : H \to G$ an embedding.

Now given a torsion-free, divisible abelian group G' and an embedding $\beta : H \to G'$, define $\gamma : G \to G'$ by

$$\gamma(\frac{h}{n}) = \frac{\beta(h)}{n}, \quad \frac{h}{n} \in G,$$

where $\frac{\beta(h)}{n}$ is the unique element g of G' such that $ng' = \beta(h)$.

The group G obtained above is unique upto isomorphism and is called the *divisible hull* of H.

Example 1.5.6 Let H be an ordered abelian group. Then there is a divisible, ordered, abelian group G and an embedding $\alpha : H \to G$ such that for every divisible, ordered, abelian group G' and every embedding $\beta : H \to G'$, there is a unique embedding $\gamma : G \to G'$ such that $\beta = \gamma \circ \alpha$.

Let $<$ denote the ordering on H. Every ordered abelian group is torsion-free. We proceed as in the last example, and define

$$\frac{h}{n} < \frac{h'}{n'} \Leftrightarrow n'h < nh'.$$

The ordered abelian group G thus defined is unique upto isomorphism and is called the *ordered divisible hull* of H.

Example 1.5.7 Let D be an ordered integral domain and $\mathbb{K}$ its quotient field. Every element of $\mathbb{K}$ can be expressed in the form $\frac{c}{d} \in \mathbb{K}$ with $d > 0$ For $\frac{a}{b}, \frac{c}{d} \in \mathbb{K}$ with $b, d > 0$, define

$$\frac{a}{b} < \frac{c}{d} \Leftrightarrow a \cdot d < b \cdot c,$$

and

$$\alpha(a) = \frac{a}{1}, \quad a \in D.$$

This makes the quotient field $\mathbb{K}$ into an ordered field with $\alpha : D \to \mathbb{K}$ an (order-preserving) embedding. Further, for every ordered field $\mathbb{F}$ and every order-preserving embedding $\beta : D \to \mathbb{F}$, there is a unique order-preserving embedding $\gamma : \mathbb{K} \to \mathbb{F}$ such that $\gamma \circ \alpha = \beta$.

Proposition 1.5.8 *Let M, N be L-structures. Then a map $h : M \to N$ is an embedding if and only if for every open formula $\varphi[x_0, \dots, x_{n-1}]$ and every $\overline{a} \in M$,*

$$M \models \varphi[\overline{a}] \Leftrightarrow N \models \varphi[h(\overline{a})]. \tag{$*$}$$

Proof We first prove the if part. For any constant symbol c, $M \models c^M = c$. Hence, by $(*)$, $N \models h(c^M) = c$, i.e. $h(c^M) = c^N$.

Now let f be an n-ary function symbol, $\overline{a} \in M$ and $b = f^M(\overline{a})$. Since $y = f x_0 \dots x_{n-1}$ is an open formula and $M \models b = f(\overline{a})$, by $(*)$, $N \models h(b) = f(h(\overline{a}))$, i.e. $h(b) = f^N(h(\overline{a}))$.

For a n-ary relation symbol p, $p[x_0 \dots x_{n-1}]$ is an open formula. Clearly for every $\overline{a} \in M$,

$$M \models p[\overline{a}] \Leftrightarrow N \models p[h(\overline{a})]$$

is a special case of $(*)$. Thus, if part is proved.

Only if part follows because

$$\{\varphi[x_0, \dots, x_{n-1}] : \forall \overline{a} \in M(M \models \varphi[\overline{a}] \Leftrightarrow N \models \varphi[h(\overline{a})])\}$$

contains all atomic formulas and is closed under $\neg$ and $\vee$. $\square$

Exercise 1.5.9 Let M, N be L-structures and $f : M \to N$ an embedding. Show the following:

1. For every existential formula $\varphi[\overline{x}]$ and every $\overline{a} \in M$,

$$M \models \varphi[\overline{a}] \Rightarrow N \models \varphi[f(\overline{a})].$$

2. For every universal formula $\varphi[\overline{x}]$ and every $\overline{a} \in M$,

$$N \models \varphi[f(\overline{a})] \Rightarrow M \models \varphi[\overline{a}].$$

Exercise 1.5.10 Let T be a theory and $\varphi[\overline{x}]$ a formula. Assume that there is a universal formula $\psi[\overline{x}]$ such that $\varphi \equiv_T \psi$. Show that whenever $M, N \models T, N \subseteq M$ and $\overline{a} \in N$, $M \models \varphi[\overline{a}] \Rightarrow N \models \varphi[\overline{a}]$.

Exercise 1.5.11 A substructure N of an L-structure M is called *existentially closed* in M if whenever an existential L_N-sentence $\exists x \varphi[x]$ is true in M, it is true in N. Let $M \models T$. Show the following:

1. Every substructure of M is a model of $T_\forall$.
2. Every extension of M is a model of $T_\exists$.
3. Every existentially closed substructure of a model of T is a model of $T_{\forall\exists}$.

Let $(I, \leq)$ be a linearly ordered set and $\{M_i : i \in I\}$ a family of L-structures such that $i < j \Rightarrow M_i \sqsubseteq M_j$. Such a family $\{M_i : i \in I\}$ is called a *chain* of L-structures. Set $M = \cup_{i \in I} M_i$.

* We put $c^M = c^{M_i}$ for some $i \in I$ where c is a constant symbol.
* If $\overline{a} \in M^n$, then $\overline{a} \in M_i^n$ for some $i \in I$. Let f be a n-ary function symbol and p a n-ary relation symbol. We define

$$f^M(\overline{a}) = f^{M_i}(\overline{a}) \ \& \ p^M(\overline{a}) \Leftrightarrow p^{M_i}(\overline{a}).$$

These are well defined and make M into an L-structure such that $M_i \sqsubseteq M$ for each $i \in I$. The structure M so defined is called the union of $\{M_i : i \in I\}$.

Proposition 1.5.12 *Let T be a $\forall\exists$ theory. Then the union of a chain of models of T is a model of T.*

Proof Let $\{M_i : i \in I\}$ be a chain of models of T and $M = \cup_i M_i$. Take an axiom $\forall\overline{x}\exists\overline{y}\varphi[\overline{x}, \overline{y}]$, φ open, of T and $\overline{a} \in M$. Then $\overline{a} \in M_i$ for some $i \in I$. Since $M_i \models T$, $M_i \models \varphi[\overline{a}, \overline{b}]$ for some $\overline{b} \in M_i \subset M$. Since $M_i \sqsubseteq M$, $M \models \varphi[\overline{a}, \overline{b}]$. $\square$

The converse of this result is true and will be proved in Corollary 2.4.6.

A theory T is called *inductive* if the union of a chain of models of T is a model of T.

If M is an L-structure, then the *atomic diagram* of M, denoted by $Diag(M)$, is

$$\{p[\overline{a}] : M \models p[\overline{a}], p \text{ a relation symbol}$$

$$\cup\{\neg p[\overline{a}] : M \not\models p[\overline{a}], p \text{ a relation symbol}\}.$$

Theorem 1.5.13 (Atomic Diagram Theorem.) *Let M be an L-structure. Then $N \models Diag(M)$ if and only if there is an embedding $h : M \to N$.*

Proof Let $N \models Diag(M)$. Define $h : M \to N$ by

$$h(a) = a^N, \ a \in M.$$

Then for every atomic formula $p[x_0 \ldots x_{n-1}]$ and every $\overline{a} \in M$,

$$M \models p[\overline{a}] \Leftrightarrow N \models p[h(\overline{a})].$$

By the arguments contained in the proof of Proposition 1.5.8, this implies that h is an embedding.

Conversely, let $h : M \to N$ be an embedding. To avoid ambiguity, let i_a stand for the constant symbol a in L_M, $a \in M$. Set $\varphi[x]$ to be the atomic formula $i_a = x$. Then $M \models \varphi[a]$. Since h is an embedding, $N \models \varphi[h(a)]$. Hence, $a^N = h(a)$. Using once again the fact that h is an embedding, we have $N \models Diag(M)$. $\square$

1.6 Isomorphism and Categoricity of Theories

Let M, N be L-structures. An *isomorphism* $h : M \to N$ is an embedding which is also a bijection. Two L-structures M and N are called *isomorphic* if there is an isomorphism $h : M \to N$. It is easily seen that the inverse of an isomorphism and the composition of two isomorphisms are isomorphisms. If $h : M \to N$ is an embedding, then $h(M)$ is a substructure of N isomorphic to M.

We write $M \equiv N$ if there is an isomorphism from M to N. It is easily seen that $\equiv$ is an equivalence relation on the class of all L-structures. Sometimes we shall write $h : M \equiv N$ to say that h is an isomorphism from M onto N.

An isomorphism $h : M \to M$ is called an *automorphism* of M. The set of all automorphisms of M is denoted by $Aut(M)$. It forms a group under composition. The group $Aut(M)$ acts on M canonically by

$$\sigma \cdot x = \sigma(x), \quad x \in M, \sigma \in Aut(M).$$

Also, note that for each $n \geq 1$, $Aut(M)$ acts on M^n by

$$g \cdot \overline{a} = g(\overline{a}), \quad g \in Aut(M), \overline{a} \in M^n.$$

For $A \subset M$, we define

$$Aut_A(M) = \{\sigma \in Aut(M) : \forall x \in A(\sigma(x) = x)\}.$$

So, $Aut_A(M)$ is the subgroup of pointwise stabilisers of A. Elements of $Aut_A(M)$ are called *automorphisms of M over A*. Further, we define

$$Aut_{(A)}(M) = \{\sigma \in Aut(M) : \sigma(A) = A\},$$

the subgroup of setwise stabilisers or simply stabilisers of A.

If $\mathbb{K}$ is a field and $\mathbb{L}$ a subfield, then $Aut(\mathbb{K})_{\mathbb{L}}$ is generally denoted by $G(\mathbb{K}, \mathbb{L})$. It is called the *Galois group of $\mathbb{K}$ over $\mathbb{L}$*.

Proposition 1.6.1 *Let M, N be L-structures and $h : M \to N$ an isomorphism. Then for every formula $\varphi[x_0, \ldots, x_{n-1}]$ and every $\overline{a} \in M$,*

$$M \models \varphi[\overline{a}] \Leftrightarrow N \models \varphi[h(\overline{a})]. \tag{$*$}$$

Proof Set

$$\Phi = \{\varphi[x_0, \ldots, x_{n-1}] : \forall \overline{a} \in M(M \models \varphi[\overline{a}] \Leftrightarrow N \models \varphi[h(\overline{a})])\}.$$

Clearly Φ contains all atomic formula $\varphi[\overline{x}]$ and is closed under negation and disjunction. Therefore, it is sufficient to show that whenever $\varphi[x, x_0, \ldots, x_{n-1}] \in \Phi$, so is $\exists x \varphi$. Take any $\overline{a} \in M^n$. Then

$$\begin{aligned}
M \models \exists x \varphi[x, \overline{a}] &\Leftrightarrow M \models \varphi[a, \overline{a}] &&\text{for some } a \in M \\
&\Leftrightarrow N \models \varphi[h(a), h(\overline{a})] \\
&\Leftrightarrow N \models \varphi[b, h(\overline{a})] &&\text{for some } b \in N \\
&\Leftrightarrow N \models \exists x \varphi[x, h(\overline{a})]
\end{aligned}$$

Second equivalence holds by our assumption and third equivalence holds because h is a surjection. The proof is complete now. $\qquad\square$

Let $\kappa \geq \aleph_0$ be a cardinal number. A theory T is called κ-*categorical* if any two models of T of cardinality κ are isomorphic.

Example 1.6.2 The theory of infinite sets is κ-categorical for every infinite cardinal κ.

Example 1.6.3 For every $\kappa > \aleph_0$, DAG is κ-categorical.

Proof Let $G_1, G_2 \models DAG, |G_1| = |G_2| > \aleph_0$. Then G_1, G_2 are vector spaces over $\mathbb{Q}$ of the same dimension. Hence, they are isomorphic as vector spaces. In particular, they are isomorphic as models of DAG. $\qquad\square$

Exercise 1.6.4 Let G be a group of cardinality $\leq \kappa$. Show that the theory of free G-spaces is λ-categorical for all $\lambda > \max\{\aleph_0, \kappa\}$.

Exercise 1.6.5 Show that DAG has exactly $\aleph_0$-many pairwise non-isomorphic countable models such that any other countable model of DAG is isomorphic to one of them.

Example 1.6.6 For every $\kappa > \aleph_0$, $ACF(p)$, $p = 0$ or prime, is κ-categorical.

Proof Note that if $\mathbb{F}$ is an algebraically closed field and $|\mathbb{F}| = \kappa > \aleph_0$, then $\mathbb{F}$ is of transcendence degree κ. Our claim follows from the fact that any two algebraically closed fields of the same characteristic and same transcendence degree are isomorphic. See [31, Chap. VIII, Sect. 1] for relevant definitions and result. $\qquad\square$

Example 1.6.7 DLO is $\aleph_0$-categorical.

Proof Let $\mathbb{Q}_1, \mathbb{Q}_2 \models DLO$ be countable. Enumerate $\mathbb{Q}_1 = \{r_n\}$ and $\mathbb{Q}_2 = \{s_m\}$. Set $n_0 = 0$ and $m_0 = 0$. Suppose for some i, $n_0, \ldots, n_{2i}$ and $m_0, \ldots, m_{2i}$ have been defined so that the map f defined by

$$f(r_{n_j}) = s_{m_j}, \ 0 \le j \le 2i,$$

is injective and order-preserving.

Now let m_{2i+1} be the first natural number k such that s_k is different from each of $s_{m_j}, \ j \le 2i$. Since $\mathbb{Q}_1 \models DLO$, there is a natural number l such that r_l is different from each of $r_{n_j}, \ j \le 2i$ and the extension of f sending r_l to $s_{m_{2i+1}}$ is order-preserving. Set n_{2i+1} to be the first such l. Then the map $f(r_{n_j}) = s_{m_j}, \ j \le 2i + 1$, is injective and order-preserving.

Now define n_{2i+2} to be the first natural number l such that r_l is different from each of $r_{n_j}, \ j \le 2i + 1$. Again observe that there is a natural number k such that s_k is different from each of $s_{m_j}, \ j \le 2i + 1$, and the extension of the above map by defining $f(r_{n_{2i+2}}) = s_k$ is order-preserving. Set s_{2i+2} to be the least such k. It is easily checked that $f : \mathbb{Q}_1 \to \mathbb{Q}_2$ thus defined is an isomorphism $\square$

Exercise 1.6.8 Show that DLO is not κ-categorical for any $\kappa > \aleph_0$.

Remark 1.6.9 Using compactness theorem (which will be proved in the next section), it will be easy to show that for every infinite cardinal κ, there is an abelian group G_1 and a non-abelian group G_2 such that $|G_1| = |G_2| = \kappa$. It follows that the theory of groups is not κ-categorical for any infinite cardinal κ. Thus, so far we have seen examples of the following possibilities for a theory T:

1. T is κ-categorical for all $\kappa \ge \aleph_0$.
2. T is κ-categorical for no $\kappa \ge \aleph_0$.
3. T is $\aleph_0$-categorical but not κ-categorical for any $\kappa > \aleph_0$.
4. T is not $\aleph_0$-categorical but is κ-categorical for all $\kappa > \aleph_0$.

Łos conjectured that these are all the possibilities, i.e. if T is κ-categorical for some $\kappa > \aleph_0$, it is λ-categorical for all $\lambda > \aleph_0$. In a remarkable contribution to model theory, Morley [43] proved the conjecture of Łos. This paper of Morley contains some of the most significant concepts of model theory and heralded a new era in the subject. We shall prove Morley's theorem later in the book.

Remark 1.6.10 The argument contained in the last proof, known as back and forth argument, is very useful in model theory. We shall prove many results using this technique.

Exercise 1.6.11 Let $M, N \models DLO$, $|M| = |N| = \aleph_0$, $A \subset M$ finite and $f : A \to N$ an order-preserving injection. Then there is an isomorphism $g : M \to N$ extending f.

Exercise 1.6.12 Consider the ordered space $\mathbb{Q}$ of rational numbers. Show that for each $n \ge 1$, the number of orbits in $\mathbb{Q}^n$ under the action of $Aut(\mathbb{Q})$ is finite.

Example 1.6.13 The theory of random graphs is $\aleph_0$-categorical.

Proof The proof uses the back and forth argument as in the case of the last Example 1.6.7. Let $G_1 = (V_1, E_1)$ and $G_2 = (V_2, E_2)$ be two countable random graphs. Enumerate $V_1 = \{a_i : i \in \omega\}$ and $V_2 = \{b_i : i \in \omega\}$. We define a sequence of partial, one-to-one finite functions $\{f_n : n \in \omega\}$ from V_1 into V_2 satisfying the following conditions:

1. $f_0(a_0) = b_0$.
2. $a_i \in domain(f_{2i}), i \in \omega$.
3. $b_i \in range(f_{2i+1}), i \in \omega$.
4. $n < m \Rightarrow f_n \subset f_m$.
5. For every $n \in \omega$ and every $a, a' \in domain(f_n)$, $(a, a') \in E_1 \Leftrightarrow (f_n(a), f_n(a')) \in E_2$.

This will complete the proof because then $\cup_n f_n : V_1 \to V_2$ will be an isomorphism.

Suppose $i \in \omega$ and f_{2i} have been defined. If $b_i \in range(f_{2i})$, take $f_{2i+1} = f_{2i}$. Otherwise, set $g = f_{2i}^{-1}$,

$$X = \{g(b_j) : b_j \in domain(g) \wedge (b_i, b_j) \in E_2\}$$

and

$$Y = \{g(b_j) : b_j \in domain(g) \wedge (b_i, b_j) \notin E_2\}.$$

Since V_1 is a random graph, there is a $a_k \in V_1 \setminus (X \cup Y)$ such that $(a_j, a_k) \in E_1$ whenever $a_j \in X$ and $(a_j, a_k) \notin E_1$ for all $a_j \in Y$. Set $h = g \cup \{(b_i, a_k)\}$ and take $f_{2i+1} = h^{-1}$.

Suppose $i \in \omega$ and f_{2i+1} has been defined. If $a_{i+1} \in domain(f_{2i+1})$, take $f_{2i+2} = f_{2i+1}$. Otherwise, set

$$X = \{f_{2i+1}(a_j) : a_j \in domain(f_{2i+1}) \wedge (a_{i+1}, a_j) \in E_1\}$$

and

$$Y = \{f_{2i+1}(a_j) : a_j \in domain(f_{2i+1}) \wedge (a_i, a_j) \notin E_1\}.$$

Clearly, X and Y are finite disjoint subsets of V_2. Since V_2 is a random graph, there is a $b_k \in V_2 \setminus (X \cup Y)$ such that $(b_j, b_k) \in E_2$ whenever $b_j \in X$ and $(b_j, b_k) \notin E_2$ for all $b_j \in Y$. Take $f_{2i+2} = f_{2i+1} \cup \{(a_{i+1}, b_k)\}$. □

For the following exercise, see [58] for relevant definitions and results.

Exercise 1.6.14 Let M be an L-structure. Equip M^M with the product of discrete topologies on M and $Aut(M) \subset M^M$ with the subspace topology. So, basic open sets of $Aut(M)$ are of the form

$$\Sigma(\bar{a}, \bar{b}) = \{f \in Aut(M) : f(\bar{a}) = \bar{b}\},$$

where $\bar{a}, \bar{b} \in M^n, n \geq 1$. Note that for any $\bar{a} \in M$, $Aut_{\bar{a}}(M) = \Sigma(\bar{a}, \bar{a})$. Show the following:

1. Show that $H \subset Aut(M)$ is closed if and only if every $g \in Aut(M)$ such that for every finite tuple $\overline{a}$ in M there is a $h \in H$ with $h(\overline{a}) = g(\overline{a})$ belongs to H.
2. For $A \subset M$, $Aut_A(M)$ is a closed set in $Aut(M)$. In particular, Galois groups of a field $\mathbb{K}$ over its subfields are closed in $Aut(\mathbb{K})$.
3. $Aut(M)$ is a topological group.
4. A subgroup G of $Aut(M)$ is open if and only if it contains $Aut_{\overline{a}}(M)$ for some finite tuple $\overline{a} \in M$. (Hint: Note that $\Sigma(\overline{a}, \overline{b}) \subset G \Rightarrow \Sigma(\overline{b}, \overline{a}) \subset G$.)
5. A subgroup G of $Aut(M)$ is open if and only if it is a union of sets of the form $Aut_{\overline{a}}(M)$, $\overline{a}$ a finite tuple of elements in M of fixed length.
6. A subgroup G is dense if and only if for each $n \geq 1$ and each $\overline{a} \in M^n$, the orbits

$$\{g \cdot \overline{a} : g \in G\} = \{g \cdot \overline{a} : g \in Aut(M)\}.$$

7. If M and the signature of L are countable, then $Aut(M)$ is a G_δ set, i.e. a countable intersection of open sets, in M^M. Conclude that in this case $Aut(M)$ is a Polish space, i.e. a completely metrisable, second countable space.

1.7 Elementary Embedding

So far nothing significantly different from usual approach to mathematics has been done. We have formally presented the syntax and semantics of first-order logic followed by mathematicians. Notions of homomorphisms, embeddings, isomorphisms are quite standard. Because logic deals with formulas, we get new and very useful notion of embedding and equivalence of models which we are going to introduce now. This can be viewed as the first gift of logic to mathematics. Most of the concepts introduced in this section are due to Tarski and Vaught [63].

The converse of Proposition 1.6.1 is not true in general. More specifically, let M, N be L-structures and $h : M \rightarrow N$ be such that for every L-formula $\varphi[\overline{x}]$ and every $\overline{a} \in M$,

$$M \models \varphi[\overline{a}] \Leftrightarrow N \models \varphi[h(\overline{a})]. \tag{$*$}$$

Then h is an embedding but may not be surjective. (See Example 1.7.8.)

A map $h : M \rightarrow N$ satisfying condition $(*)$ of Proposition 1.6.1 is called an *elementary embedding*. A substructure M of N is called an *elementary substructure* of N if the inclusion map $i : M \hookrightarrow N$ is elementary. In this case, we also call N an *elementary extension* of M and write $M \preceq N$ or $N \succeq M$. We shall also write $h : M \preceq N$ if $h : M \rightarrow N$ is elementary.

Two L-structures M and N are called *elementarily equivalent*, written $M \simeq N$, if they satisfy the same closed L-formulas. Clearly $\simeq$ is an equivalence relation on the class of all L-structures. With these definitions we have the following result.

Theorem 1.7.1 *A theory T is complete if and only if it is consistent and models of T are pairwise elementarily equivalent.*

Let M, N be L-structures and there be an elementary embedding from M to N. Then $M \simeq N$. Further, $M \simeq N$ implies that for every L-theory T, $M \models T \Leftrightarrow N \models T$.

Proposition 1.7.2 *Let T be a complete theory. If T has an infinite model, all its models are infinite. In fact, if there is a finite model M of T, then for all $N \models T$, $|N| = |M|$.*

Proof Let $M \models T$ and $|M| = n$. Then $M \models \exists_{=n} x(x = x)$. Since $N \simeq M$, $N \models \exists_{=n} x(x = x)$, i.e. $|N| = n$.   □

If M is an L-structure, we define the *elementary diagram* of M by

$$Diag_{el}(M) = \{\varphi : \varphi \text{ an } L_M \text{ sentence } \& \ M \models \varphi\}.$$

By slightly modifying the proof of atomic diagram theorem 1.5.13, we get

Proposition 1.7.3 $N \models Diag_{el}(M)$ *if and only if there is an elementary embedding $h : M \to N$.*

The proof is left to the reader as a simple exercise.

Proposition 1.7.4 (Tarski–Vaught Test) *Let M be an L-structure and N a subset of M. Then N is the universe of an elementary substructure of M if and only if for every L-formula $\varphi[x, \overline{x}]$, $|x| = 1$, and every $\overline{a} \in N$*

$$M \models \exists x \varphi[x, \overline{a}] \Rightarrow \exists b \in N(M \models \varphi[b, \overline{a}]) \qquad (*)$$

Proof If part: We first show that if N satisfies $(*)$, N is a substructure of M.

Let c be a constant symbol. Then $M \models \exists x(x = c)$. Hence, $M \models b = c$ for some $b \in N$. So, $c^M = b \in N$.

Next let f be a n-ary function symbol and $a_1, \ldots, a_n \in N$. Since $M \models \exists x(x = f(a_1, \ldots, a_n))$, $M \models b = f(a_1, \ldots, a_n)$ for some $b \in N$. This implies that $f^M(a_1, \ldots, a_n) \in N$. Thus, we can regard N as a substructure of M by taking $p^N = p^M | N$, where p is a relation symbol.

By the given condition,

$$\Phi = \{\varphi[\overline{x}] : \forall \overline{a} \in N(M \models \varphi[\overline{a}] \Leftrightarrow N \models \varphi[\overline{a}])\}$$

contains all atomic formula and is closed under $\neg$, $\vee$ and $\exists$. Hence, N is an elementary substructure of M.

Only if part: Since $N \preceq M$, $M \models \exists x \varphi[x, \overline{a}]$ implies $N \models \exists x \varphi[x, \overline{a}]$. Hence $N \models \varphi[b, \overline{a}]$ for some $b \in N$. As $N \preceq M$, $M \models \varphi[b, \overline{a}]$.   □

Let $(I, \leq)$ be a linearly ordered set and $\{M_i : i \in I\}$ a family of L-structures such that $i < j \Rightarrow M_i \preceq M_j$. Such a family $\{M_i : i \in I\}$ is called an *elementary chain* of L-structures.

Proposition 1.7.5 *If $\{M_i : i \in I\}$ is an elementary chain of L-structures, then $M = \cup_{i \in I} M_i$ is an elementary extension of each of M_i.*

Proof Let Φ be

$$\{\varphi[\overline{x}] : \varphi \text{ is an } L - \text{formula } \forall i \in I \forall \overline{a} \in M_i (M_i \models \varphi[\overline{a}] \Leftrightarrow M \models \varphi[\overline{a}])\}.$$

Clearly, Φ contains all atomic formulas and whenever $\varphi, \psi \in \Phi$, $\neg\varphi, \varphi \vee \psi \in \Phi$.

Now let $\varphi[x, \overline{x}] \in \Phi$ and $\overline{a} \in M_i$. Suppose $M_i \models \exists x \varphi[x, \overline{a}]$. Get $a \in M_i$ such that $M_i \models \varphi[a, \overline{a}]$. Then, by induction hypothesis, $M \models \varphi[a, \overline{a}]$. Hence, $M \models \exists x \varphi[x, \overline{a}]$.

Next suppose $M \models \exists x \varphi[x, \overline{a}]$. Then there is a $j > i$ and $a \in M_j$ such that $M \models \varphi[a, \overline{a}]$. Hence, by induction hypothesis, $M_j \models \varphi[a, \overline{a}]$. Since $M_j \models \exists x \varphi[x, \overline{a}]$ and $M_i \preceq M_j$, $M_i \models \exists x \varphi[x, \overline{a}]$. $\square$

A weaker form of the following result was proved by Löwenheim in [38] and for countable L and countable X it was Skolem in [56]. In the most general form as stated below it was proved by Tarski and Vaught in [63].

Theorem 1.7.6 (Downward Löwenheim–Skolem Theorem.) *Let M be an L-structure and $X \subset M$. Then there is an elementary substructure N of M containing X such that $|N| \leq \max\{|L|, |X|\}$.*

Proof Inductively we define a sequence of $\{N_k\}$ of subsets of M as follows: $N_0 = X$. Suppose N_k has been defined and $|N_k| \leq \max\{|L|, |X|\}$. Then the set of L_{N_k}-formulas $\varphi[x]$ is of cardinality $\leq \max\{|L|, |X|\}$. Whenever $\varphi[x]$, $|x| = 1$, is an L_{N_k}-formula and $M \models \exists x \varphi[x]$, we choose a $b_\varphi \in M$ such that $M \models \varphi[b_\varphi]$. Let N_{k+1} consist of elements of N_k and all these b_φ. Then $|N_{k+1}| \leq \max\{|L|, |X|\}$.

Now take $N = \cup_k N_k$. Then $|N| \leq \max\{|L|, |X|\}$. It is routine to check that N is an elementary substructure of M. $\square$

Corollary 1.7.7 (Skolem) *Every countable, consistent theory has a countable model.*

Example 1.7.8 Let T be a countable theory with an uncountable model, say M. For instance, we may take $\mathbb{R} \models DLO, \mathbb{R} \models ODAG$ or $\mathbb{C} \models ACF(0)$. Since T is countable, M has a countable, elementary substructure, say N. Thus, we get an elementary embedding, namely the inclusion map $i : N \hookrightarrow M$ which is not surjective.

We see that downward Löwenheim–Skolem theorem is a technique to build small models. In the next chapter, we shall prove so-called upward Löwenheim–Skolem theorem which will give us a technique to build large models.

A linearly ordered set $(A, <)$ is called complete if every set $B \subset A$ which is bounded above has a least upper bound.

Proposition 1.7.9 (Cantor) *Every complete, order dense, linearly ordered set $(A, <)$ with more than one point is uncountable.*

Proof If possible, suppose A is countable. Since $|A| > 1$, $|A| = \aleph_0$. Enumerate $A = \{a_n\}$ with $a_0 < a_1$. Set $x_0 = a_0$ and $y_0 = a_1$. Note that $x_0 < y_0$. Let $x_0 < \cdots < x_n < y_n < \cdots < y_0$ have been defined. Let l_n be the first integer such that $x_n < a_{l_n} < y_n$. Set $x_{n+1} = a_{l_n}$. Now let r_n be the first integer such that $x_{n+1} < a_{r_n} < y_n$. Set $y_{n+1} = a_{r_n}$.

Thus, we have defined $x_0 < x_1 < x_2 < \cdots < y_2 < y_1 < y_0$. Since A is complete, $\sup\{x_n\}$ exists in A. But by our construction for no p, $x_n < a_p < y_n$ for all n. This contradiction completes the proof.   □

Corollary 1.7.10 *The class of all complete, order dense, linearly ordered sets is not elementary.*

This shows a limitation of expressibility power of first-order languages—"that a linearly ordered set is complete" is not expressed by any set of first-order sentences of a first-order language whose signature is countable.

Exercise 1.7.11 Show that the class of all complete ordered fields is not elementary.

Exercise 1.7.12 Let N be an elementary substructure of an L-structure M. Show that for every $g \in Aut(M)$, $g(N)$ is an elementary substructure of M.

Exercise 1.7.13 Show that if there is an extension M' of M in which N is elementarily embedded, then N is existentially closed in M.

Exercise 1.7.14 Let T be a first-order theory such that every formula is equivalent in T to an existential formula. Show that every submodel N of a model M of T is an elementary submodel of M.

Exercise 1.7.15 Let M, N be L-structures and $|N| < \aleph_0$. Show that every elementary embedding $h : N \to M$ is an isomorphism.

Exercise 1.7.16 Let $(I, \leq)$ be a directed set. Suppose $\{M_i : i \in I\}$ is a family of L-structures and whenever $i \leq j$, there is a homomorphism $f_{ji} : M_i \to M_j$ satisfying the following properties:

(a) f_{ii} is the identity morphism on M_i for each $i \in I$, and
(b) $i \leq j \leq k \Rightarrow f_{kj} \circ f_{ji} = f_{ki}$.

We call $(\{M_i : i \in I\}, \{f_{ji} : i \leq j\})$ a *direct system* of L-structures. The *direct limit* of this system consists of an L-structure M and for each $i \in I$, a homomorphism $f_i : M_i \to M$ satisfying the following conditions:

(c) $i \leq j \Rightarrow f_j \circ f_{ji} = f_i$, and
(d) Whenever N is an L-structure and $g_i : M_i \to N$, $i \in I$, a homomorphism such that $i \leq j \Rightarrow g_j \circ f_{ji} = g_i$, there is a unique homomorphism $g : M \to N$ with $g \circ f_i = g_i$ for all $i \in I$.

(i) Show that if $(M, \{f_i : M_i \to M\})$ and $(N, \{g_i : M_i \to N\})$ are direct limits of $(\{M_i : i \in I\}, \{f_{ji} : i \leq j\})$, then there is a unique isomorphism $h : M \to N$ such that $g_i = h \circ f_i$ for each $i \in I$.

To show the existence of the direct limit of $(\{M_i : i \in I\}, \{f_{ji} : i < j\})$, set M' to be the disjoint sum of $\{M_i : i \in I\}$. (Note that, without any loss of generality, we can assume that M_i's are pairwise disjoint.) For $a_i \in M_i$ and $a_j \in M_j$, define

$$a_i \sim a_j \Leftrightarrow \exists k \geq i, j \, (f_{ki}(a_i) = f_{kj}(a_j)).$$

(ii) Show that $\sim$ is an equivalence relation on M'.

Let $M = M'/\sim$ denote the set of all $\sim$-equivalence classes, $q : M' \to M$ the quotient map and $f_i = q|M_i, i \in I$.

(iii) If c is a constant symbol, show that $c^{M_i} \sim c^{M_j}$ for all $i, j \in I$.

(iv) For any constant symbol c, define $c^M = q(c^{M_i})$ for any $i \in I$.

(v) Given $(a_1, \ldots, a_n) \in M'$, show that there is an $i \in I$ and $(b_1, \ldots, b_n) \in M_i$ such that $a_l \sim b_l$ for all $1 \leq l \leq n$.

If f is a n-ary function symbol and p a n-ary relation symbol, define

$$f^M(q(a_1), \ldots, q(a_n)) = q(f^{M_i}(b_1, \ldots, b_n))$$

and

$$p^M(q(a_1), \ldots, q(a_n)) \Leftrightarrow p^{M_i}(b_1, \ldots, b_n),$$

where $(b_1, \ldots, b_n)$ are as in (v).

(vi) Show that these are well defined and makes M into an L-structure so that $(M, \{f_i : M_i \to M\})$ is the direct limit of $(\{M_i : i \in I\}, \{f_{ji} : i \leq j\})$.

(vii) Show that if each f_{ji} is an embedding, so is each f_i, and if each f_{ji} is an epimorphism, so is each f_i.

(ix) If each f_{ji} is elementary, each f_i is elementary. Further, in (d) above if each g_i is elementary, the unique $g : M \to N$ satisfying $g \circ f_i = g_i$ for all $i \in I$ is elementary.

(x) Assume that T is a $\forall \exists$ theory, each $M_i \models T$ and each f_{ij} is an embedding. Show that $M \models T$.

1.8 Skolemization of a Theory

The main idea contained in this section was initiated by Skolem in [55].

A theory T is called a *Skolem theory* or that it has *built in Skolem functions* if for every formula $\varphi[x, y_1, \ldots, y_n]$, T has a n-ary function symbol f such that

$$T \models \forall \overline{y}(\exists x \varphi[x, \overline{y}] \to \varphi[f(\overline{y}), \overline{y}]).$$

Theorem 1.8.1 *Every theory T can be extended to a theory T^+ such that*

1. *T^+ has built in Skolem functions.*
2. *$|L(T^+)| = |L(T)|$.*
3. *Any model M of T can be expanded to be a model of T^+.*

Proof Set $T_0 = T$ and $M_0 = M$. Having defined T_i and M_i, let T_{i+1} be the extension of T_i obtained as follows: For each formula $\varphi[x, y_1, \ldots, y_n]$ of T_i, introduce a new n-ary function symbol f_φ and an axiom

$$\forall \overline{y}(\exists x \varphi[x, \overline{y}] \to \varphi[f_\varphi(\overline{y}), \overline{y}]).$$

We call functions f_φ *Skolem functions*.

For any formula $\varphi[x, \overline{y}]$ and any $\overline{a} \in M_i$, if $M_i \models \exists x \varphi[x, \overline{a}]$, define $f_\varphi^{M_{i+1}}(\overline{a}) = b$ such that $M_i \models \varphi[b, \overline{a}]$. Otherwise, define $f_\varphi^{M_{i+1}}(\overline{a}) \in M_i$ arbitrarily. Note that the universe of M_{i+1} remains $M_0 = M$. We have expanded M_i to a model M_{i+1} of T_{i+1} by giving interpretations of new function symbols.

Now set $T^+ = \cup_i T_i$ and $M^+ = \cup_i M_i$. It is easily checked that (1)–(3) are satisfied. $\square$

We say that T has *definable Skolem functions* if for every formula $\varphi[x, \overline{y}]$ there is a formula $\psi[x, \overline{y}]$ such that

$$T \models \forall \overline{y}(\exists_{=1} x \psi[x, \overline{y}] \wedge (\exists x \varphi[x, \overline{y}] \to \forall x(\psi[x, \overline{y}] \to \varphi[x, \overline{y}]))).$$

In this case, we can introduce a new function symbol f_φ that picks up a x such that $\psi[x, \overline{y}]$, i.e. we introduce an axiom

$$x = f_\varphi[\overline{y}] \leftrightarrow \psi[x, \overline{y}].$$

Take an L-theory T with built in Skolem functions. Let $M \models T$ and $X \subset M$. We define
$$\mathcal{H}(X) = \{t^M[\overline{a}] : \overline{a} \in X, t[\overline{v}] \text{ an } L\text{-term}\}.$$

We call $\mathcal{H}(X)$ the *Skolem hull* of X. We now get a differently worded proof of downward Löwenheim–Skolem Theorem.

Theorem 1.8.2 $\mathcal{H}(X)$ *is an elementary substructure of M containing X and $|\mathcal{H}(X)| \leq \max\{|X|, |L|\}$.*

Proof Take an L-formula $\varphi[x, \overline{y}]$. Since T has built in Skolem functions, there is a n-ary function symbol f such that

$$T \models \forall \overline{y}(\exists x \varphi[x, \overline{y}] \to \varphi[f(\overline{y}), \overline{y}]).$$

Let $\overline{b} \in \mathcal{H}(X)$. Then $a = f^M(\overline{b}) \in \mathcal{H}(X)$. Further,

$$M \models \exists x \varphi[x, \overline{b}] \Rightarrow M \models \varphi[a, \overline{b}].$$

The result now follows from Proposition 1.7.4.   □

An L-structure A is said to have definable Skolem functions if $Th(A)$ has definable Skolem functions. This is the same as saying that for every L-formula $\varphi[x, \overline{y}]$ there is an L-formula $\psi[x, \overline{y}]$ such that

$$A \models \forall \overline{y}(\exists_{=1} x \psi[x, \overline{y}] \wedge (\exists x \varphi[x, \overline{y}] \rightarrow \forall x(\psi[x, \overline{y}] \rightarrow \varphi[x, \overline{y}]))).$$

Example 1.8.3 Let T be the Peano Arithmetic PA. Given any formula $\varphi[x, \overline{y}]$, let $\psi[x, \overline{y}]$ be the formula

$$(\forall z \neg \varphi[z, \overline{y}] \wedge x = 0) \vee (\varphi[x, \overline{y}] \wedge \forall z(z < x \rightarrow \neg \varphi[z, \overline{y}])).$$

Then

$$PA \models \forall \overline{y}(\exists_{=1} x \psi[x, \overline{y}] \wedge (\exists x \varphi[x, \overline{y}] \rightarrow \forall x(\psi[x, \overline{y}] \rightarrow \varphi[x, \overline{y}]))).$$

Thus, PA has definable Skolem functions.

Exercise 1.8.4 If T is a Skolem theory, show that for every formula $\varphi[\overline{x}]$ there is an open formula $\psi[\overline{x}]$ such that

$$T \models \forall \overline{x}(\varphi[\overline{x}] \leftrightarrow \psi[\overline{x}]).$$

Exercise 1.8.5 Let T be a Skolem theory and $M, N \models T$. Show that $M \subseteq N \Rightarrow M \preceq N$.

Exercise 1.8.6 Show that the ring of integers $\mathbb{Z}$ has definable Skolem functions. (Hint Use Lagrange's theorem: Every nonnegative integer is a sum of squares of four integers.)

1.9 Definability

Let M be an L-structure. A subset $X \subset M^n$ is called *definable* if there is an L_M-formula $\varphi[\overline{x}]$ such that
$$X = \{\overline{a} \in M : M \models \varphi[\overline{a}]\}.$$

The set X defined by $\varphi[\overline{x}]$ will be denoted by $\varphi(M)$. Moreover, if φ is an L_A-formula, $A \subset M$, we call X A-*definable*. Note that if X is A-definable, there is a finite $B \subset A$

such that X is B-definable. We call $f : M^n \to M^m$ A-definable, $A \subset M$, if its graph is A-definable.

Example 1.9.1 The set of all nonnegative real numbers is an $\emptyset$-definable subset of the field $\mathbb{R}$. It is defined by the formula $\exists y(x = y^2)$.

Example 1.9.2 For every $\bar{a} \in M^n$, the singleton $\{\bar{a}\}$ is definable. It is defined by the L_M-formula $\wedge_{i<n}(x_i = a_i)$.

Remark 1.9.3 It is important to note that the notion of definability is very much dependent on the language. For instance, there are at most continuum many definable subsets of the field $\mathbb{R}$. Let $X \subset \mathbb{R}$ be not definable in the language of rings. Introduce a new unary relation symbol p to the language of rings and interpret $p^{\mathbb{R}} = X$. Then X is defined by the formula $p[x]$ in the new language.

A family $\mathcal{C}$ of subsets of M^n, $n \geq 1$, is called a *pointclass*. Using the fact that the set of all L-formulas is the smallest set of expressions containing all atomic formulas and closed under $\neg$, $\vee$ and $\exists$, the reader can easily prove the following result.

Theorem 1.9.4 *Let M be an L-structure. The pointclass of all definable sets in M is the smallest pointclass $\mathcal{D}$ satisfying the following conditions:*

1. *If $\varphi[\bar{x}]$ is an atomic formula, then the set*

$$\{\bar{a} \in M : M \models \varphi[\bar{a}]\} \in \mathcal{D}.$$

2. *If $A, B \subset M^n$ are in $\mathcal{D}$, so are $A \cup B$ and $M^n \setminus A$.*
3. *If $A \subset M^{n+1}$ is in $\mathcal{D}$, so is its projection*

$$\pi(A) = \{\bar{a} \in M^n : \exists a \in M((\bar{a}, a) \in A)\}.$$

4. *If $A \subset M^{n+m}$ is in $\mathcal{D}$ and $\bar{b} \in M^m$, the section*

$$A_{\bar{b}} = \{\bar{a} \in M^n : (\bar{a}, \bar{b}) \in A\} \in \mathcal{D}.$$

Exercise 1.9.5 Show that the pointclass $\mathcal{D}$ of definable sets in M is closed under finite unions, finite intersections and under substitutions by definable functions, i.e. if $A \subset M^n$ is in $\mathcal{D}$ and $f_1, \ldots, f_n : M^m \to M$ are definable, so is the set $B \subset M^m$ defined by

$$\bar{a} \in B \Leftrightarrow (f_1(\bar{a}), \ldots, f_n(\bar{a})) \in A.$$

In particular, if A is definable, so is $M \times A$.

Exercise 1.9.6 Show that if $A \subset M^{n+1}$ is definable, so is its *co-projection* $B \subset M^n$ defined by

$$\bar{a} \in B \Leftrightarrow \forall a \in M((\bar{a}, a) \in A).$$

Exercise 1.9.7 Show that $f = (f_1, \ldots, f_l) : M^k \to M^l$ is definable if and only if each of $f_1, \ldots, f_l$ is definable.

Exercise 1.9.8 Show that if $f : M^k \to M^l$ and $g : M^l \to M^m$ are definable, so is their composition $g \circ f : M^k \to M^m$.

Exercise 1.9.9 Let $D \subset \mathbb{R}^n$ be definable. Show that its closure $\overline{D}$ with respect to the usual topology is definable in the language of ordered fields.

Exercise 1.9.10 Let $\mathbb{K}$ be a field and $M_{m \times n}(\mathbb{K})$ denote the set of all $m \times n$ matrices over $\mathbb{K}$. We identify $M_{m \times n}(\mathbb{K})$ with $\mathbb{K}^{mn}$ in a canonical way. We shall follow usual convention and write $M_n(\mathbb{K})$ in place of $M_{n \times n}(\mathbb{K})$. Show the following:

1. The determinant function $A \to |A|$, $A \in M_n(\mathbb{K})$ is $\emptyset$-definable.
2. The set of all $n \times n$ non-singular matrices $GL_n(\mathbb{K})$ is $\emptyset$-definable.
3. The matrix multiplication $M_{m \times n}(\mathbb{K}) \times M_{n \times k}(\mathbb{K}) \to M_{m \times k}(\mathbb{K})$ is $\emptyset$-definable.

Exercise 1.9.11 If $\mathcal{D}$ is the pointclass of all definable sets in M, then show that $|\mathcal{D}| \leq \max\{|L|, |M|\}$.

Example 1.9.12 Let $\mathbb{F}$ be a field and $R = \mathbb{F}[X_1, \ldots, X_n]$ the ring of polynomials over $\mathbb{F}$. We regard $\mathbb{F}$ as the set of all polynomials of degree 0. Then $\mathbb{F}$ is an $\emptyset$-definable subset of the ring R. It is defined by

$$x \in \mathbb{F} \Leftrightarrow x = 0 \vee \exists y (x \cdot y = 1).$$

Example 1.9.13 If $\mathbb{F}$ is a real closed field, then $<$ on $\mathbb{F}$ is definable in the language of rings. It is defined by

$$x < y \Leftrightarrow \exists z (z \neq 0 \wedge y = x + z \cdot z).$$

Example 1.9.14 It was proved by Lagrange that every positive integer is a sum of squares of four integers. (See [21, Chap. 17, Sect. 7]). From this it follows that $\mathbb{N}$ is an $\emptyset$-definable subset of the ring $\mathbb{Z}$:

$$x \in \mathbb{N} \Leftrightarrow \exists z_1 \exists z_2 \exists z_3 \exists z_4 (x = z_1^2 + \cdots + z_4^2).$$

Example 1.9.15 Let $\mathbb{K}$ be a field. A subset X of $\mathbb{K}^n$ is defined by an atomic formula if and only if it is the set of all zeros of a polynomial $f \in \mathbb{K}[X_1, \ldots, X_n]$.

Example 1.9.16 It is known that if $\mathbb{K}$ is an algebraically closed field of characteristic 0, the ring $R = \mathbb{K}[X_1, \ldots, X_n]$ of polynomials over $\mathbb{K}$ satisfies Fermat's last theorem, i.e. for every $k > 2$ whenever $f, g, h \in \mathbb{K}[X_1, \ldots, X_n]$, f, g, h all nonzero, $f^k + g^k = h^k$ implies that $f, g, h \in \mathbb{K}$ [31, Chap. IV, Sect. 7].

If $\mathbb{K}$ is an algebraically closed field of characteristic zero, $\mathbb{K}$ is a $\emptyset$-definable subset of the field of rational functions $\mathbb{K}(X_1, \ldots, X_n)$. For instance, it is defined by the formula

$$\exists y (x^3 = 1 + y^3).$$

Proposition 1.9.17 *Let M be an L-structure, $A \subset M$ and $D \subset M^n$ A-definable. Then $f(D) = D$ for every $f \in Aut_A(M)$.*

Proof Let $\varphi[\overline{x}, \overline{a}], \overline{a} \in A$, define D. For every $\overline{b} \in M$, we have

$$\begin{aligned}
\overline{b} \in D &\Leftrightarrow M \models \varphi[\overline{b}, \overline{a}] \\
&\Leftrightarrow M \models \varphi[f(\overline{b}), f(\overline{a})] \\
&\Leftrightarrow M \models \varphi[f(\overline{b}), \overline{a}] \\
&\Leftrightarrow f(\overline{b}) \in D
\end{aligned}$$

The second equivalence holds because f is an automorphism of M. The first and the last equivalence holds because $\varphi[\overline{x}, \overline{a}]$ defines D whereas the third equivalence holds because $f \in Aut_A(M)$. Our proof is complete. $\square$

For every finite sequence $\overline{a}$ of complex numbers, there is a real number r and a complex number s not in $\mathbb{R}$ such that there is a field isomorphism $f : \mathbb{C} \to \mathbb{C}$ fixing $\overline{a}$ pointwise and mapping r to s: To see this consider the field extension $\mathbb{F} = \mathbb{Q}(\overline{a})$. Note that $\mathbb{F}$ is countable. Take a real r transcendental over $\mathbb{F}$ and a complex number s which is not real and transcendental over $\mathbb{F}$. Then there is an isomorphism $f : \mathbb{F}(r) \to \mathbb{F}(s)$ which fixes $\mathbb{F}$ pointwise and for which $f(r) = s$. Again note that $\mathbb{F}(r)$ and $\mathbb{F}(s)$ are countable. Now take transcendence bases $\{b_\alpha : \alpha < \mathfrak{c}\}$ and $\{c_\alpha : \alpha < \mathfrak{c}\}$ of $\mathbb{C}$ over $\mathbb{F}(r)$ and $\mathbb{F}(s)$ respectively. Using standard arguments we can show that there is an automorphism g of $\mathbb{C}$ such that $g|\mathbb{F}(r) = f$ and $g(b_\alpha) = c_\alpha$ for all $\alpha < \mathfrak{c}$. This g has the desired properties.

This observation and Proposition 1.9.17 imply the following:

Example 1.9.18 The set of all real numbers $\mathbb{R}$ is not a definable subset of the field of complex numbers $\mathbb{C}$.

Let M be an L-structure, $N \sqsubseteq M$ definable, defined by an L_M-formula, say $\varphi[x]$. For L_N-formulas ψ we define its *relativisation to N*, denoted ψ^N, by induction on $|\psi|$ as follows: if ψ is atomic, ψ^N is ψ,

$$(\neg\psi)^N = \neg\psi^N, (\psi \vee \eta)^N = \psi^N \vee \eta^N$$

and

$$(\exists y\psi)^N = \exists y(\varphi[y] \wedge \psi^N).$$

Proposition 1.9.19 *For every L_N-formula $\psi[\overline{x}]$ and every $\overline{b} \in N$,*

$$N \models \psi[\overline{b}] \Leftrightarrow M \models \psi^N[\overline{b}].$$

Proof We prove the result by induction on the rank of ψ. The result is clearly true for atomic ψ and it is true for $\neg\psi$ ($\psi \vee \eta$) if it is true for ψ (respectively ψ and η).

Now suppose the result is true for $\psi[x, \overline{x}]$. Set $\eta[\overline{x}] = \exists x\psi$. Take any $\overline{b} \in N$.

Suppose $M \models \eta^N[\overline{b}]$. Then there is a $b \in M$ such that $M \models \varphi[b]$ as well as $M \models \psi^N[b, \overline{b}]$. Since φ defines N, $b \in N$. By induction hypothesis, $N \models \psi[b, \overline{b}]$. Thus, $N \models \eta[\overline{b}]$.

Now assume that $\overline{b} \in N$ and $N \models \eta[\overline{b}]$. So, there is a $b \in N$ such that $N \models \psi[b, \overline{b}]$. By induction hypothesis, $M \models \psi^N[b, \overline{b}]$. Since φ defines N, $M \models \varphi[b]$. This proves that $M \models \eta^N[\overline{b}]$. $\square$

Remark 1.9.20 In a remarkable result Julia Robinson proved that $\mathbb{N}$ is a $\emptyset$-definable subset of the ordered field $\mathbb{Q}$. This very interesting result is remarkable for many reasons. First Julia Robinson had to prove a deep result in diophantine number theory to produce a formula $\xi[x]$ *in the language of rings with identity such that for a rational number q, $\mathbb{Q} \models \xi[q]$ if and only if q is a natural number.* This result also implies that **the field $\mathbb{Q}$ is undecidable** [13, 52]. Note that if the signature of L is finite and $N \sqsubseteq M$ definable, then there is an algorithm to compute the function $\psi \to \psi^N$ defined above. This implies that if M is decidable, then N is decidable. Equivalently, if N is undecidable, then so is M. The famous first incompleteness theorem of Gödel states that $\mathbb{N}$ is undecidable [59, Theorem 7.2.1]. Undecidability of $\mathbb{Q}$ follows now. We remark that Tarski showed that $\mathbb{R}$ as an ordered field and $\mathbb{C}$ as a field are decidable [62].

1.10 Definable and Algebraic Closures

Let M be an L-structure and $A \subset M$. Elements of the orbit of $b \in M$ under the action of $Aut_A(M)$, the pointwise stabiliser of A, are called *conjugates of b over A.*

An element $a \in M$ is *algebraically definable* over A if its orbit under the action of $Aut_A(M)$ is a singleton, namely $\{a\}$. This is the same as saying that any automorphism of M that fixes A pointwise, fixes a. We set

$$DCL(A) = \{a \in M : a \text{ algebraically definable over } A\}$$

and call it *the definable closure of A in algebraic sense.* A set $A \subset M$ is called *definably closed in algebraic sense* if $DCL(A) = A$. Elements of $DCL(\emptyset)$ will be called *definable elements of M in algebraic sense.*

The following statements are easy to prove.

1. $A \subset DCL(A)$.
2. $A \subset B \Rightarrow DCL(A) \subset DCL(B)$.
3. $DCL(DCL(A)) = DCL(A)$
4. $B \subset DCL(A) \Rightarrow DCL(B) \subset DCL(A)$.
5. $DCL(A)$ is a substructure of M containing A.

(3) follows from the fact that $Aut_A(M) \subset Aut_B(M)$, where $B = DCL(A)$.

For every constant symbol c, $c^M \in DCL(\emptyset)(\subset DCL(A))$ because $\sigma(c^M) = c^M$ for every endomorphism σ of M. Let f be an n-ary function symbol and $\overline{a} \in$

$DCL(A)$. Then for every $\sigma \in Aut_A(M)$,

$$\sigma(f^M(\overline{a})) = f^M(\sigma(\overline{a})) = f^M(\overline{a}).$$

Hence, $DCL(A)$ is closed under f^M for every function symbol f. These observations imply that $DCL(A)$ is canonically a substructure of M containing A.)

An element $a \in M$ is *algebraic over A in algebraic sense* if it has only finitely many conjugates over A, i.e. the orbit of a under the action of $Aut_A(M)$ is finite. We set

$$ACL(A) = \{a \in M : a \text{ algebraic in algebraic sense over } A\}$$

and call it *the algebraic closure of A in algebraic sense*. A set $A \subset M$ is called *algebraically closed in algebraic sense* if $ACL(A) = A$. Also elements in $ACL(\emptyset)$ are called *algebraic elements of M in algebraic sense*. We have

1. $A \subset DCL(A) \subset ACL(A)$.
2. $A \subset B \Rightarrow ACL(A) \subset ACL(B)$.

An L_A-formula $\varphi[x]$ will be called a *definitional formula* if $M \models \exists_{=1} x \varphi[x]$. An element $a \in M$ is *definable* over A if $\{a\}$ is A-definable, i.e. there is a definitional L_A-formula $\varphi[x]$ such that $M \models \varphi[a]$. We set

$$dcl(A) = \{a \in M : a \text{ definable over } A\}$$

and call it *the definitional closure of A*. A set $A \subset M$ is called *definably closed* if $dcl(A) = A$. Elements of $dcl(\emptyset)$ are called *definable elements of M*.

Proposition 1.10.1 *For every $A, B \subset M$,*

(i) $A \subset B \Rightarrow dcl(A) \subset dcl(B)$.
(ii) $A \subset dcl(A) \subset DCL(A)$
(iii) $dcl(dcl(A)) = dcl(A)$.
(iv) $dcl(A)$ *is a substructure of M containing A.*

Proof (i) is entirely trivial.

For any $a \in A$, the L_A-formula $x = a$ witnesses that $a \in dcl(A)$. Take any $a \in dcl(A)$. So, there exists a definitional L_A-formula $\varphi[x]$ such that $M \models \varphi[a]$. Now take any $\sigma \in Aut_A(M)$. Then $M \models \varphi[\sigma(a)]$. These two facts imply that $\sigma(a) = a$. Thus, (ii) is proved.

Since $A \subset dcl(A), dcl(A) \subset dcl(dcl(A))$. Now take any $a \in dcl(dcl(A))$. Then there exists an L-formula $\varphi[x, \overline{y}]$ and a $\overline{b} \in dcl(A)$ such that

$$M \models \varphi[a, \overline{b}] \wedge \exists_{=1} x \varphi[x, \overline{b}].$$

There exists an $\overline{a} \in A$ and for each i, an L-formula $\varphi_i[y_i, \overline{z}]$ such that

$$M \models \varphi_i[b_i, \overline{a}] \wedge \exists_{=1} y_i \varphi_i[y_i, \overline{a}].$$

Take

$$\psi[x, \overline{z}] = \exists \overline{y}(\varphi[x, \overline{y}] \wedge \wedge_i \varphi_i[y_i, \overline{z}]).$$

Then

$$M \models \psi[a, \overline{a}] \wedge \exists_{=1} x \psi[x, \overline{a}].$$

This shows that $dcl(dcl(A)) \subset dcl(A)$.

The formula $x = c$ witnesses that $c^M \in dcl(A)$, c a constant symbol. Now let $a_1, \ldots, a_n \in dcl(A)$ and f a n-ary function symbol. Suppose $\varphi_1[y_1], \ldots, \varphi_n[y_n]$ are L_A-formulas witnessing $a_1, \ldots, a_n \in dcl(A)$, respectively. Then the formula

$$\cdot \ \exists \overline{y}(\wedge_i \varphi_i[y_i] \wedge x = f y_1 \ldots y_n)$$

witnesses that $f^M(a_1, \ldots, a_n) \in dcl(A)$. Now it is easy to see $dcl(A)$ is a substructure of M containing A. $\square$

An L_A-formula $\varphi[x]$ is called *algebraic* if there exists a positive integer n such that

$$M \models \exists_{=n} x \varphi[x].$$

The integer n is called the *degree* of φ and is denoted by $deg(\varphi)$. If no such n exist, φ is called *non-algebraic*. An element $a \in M$ is *algebraic* over A if there is an algebraic L_A-formula $\varphi[x]$ such that $M \models \varphi[a]$.

We set

$$acl(A) = \{a \in M : a \text{ algebraic over } A\}$$

and call it the *algebraic closure of A*. Note that

$$|acl(A)| \leq \max\{|L|, |A|\}.$$

A set $A \subset M$ is called *algebraically closed* if $acl(A) = A$. Elements of $acl(\emptyset)$ are called *algebraic elements* of M.

The following result is easily seen.

Proposition 1.10.2 *For $A, B \subset M$,*

  (i) $A \subset B \Rightarrow acl(A) \subset acl(B)$.
 (ii) *If $a \in acl(A)$, there is a finite $A_0 \subset A$ such that $a \in acl(A_0)$.*
(iii) $A \subset dcl(A) \subset acl(A) \subset ACL(A)$. *In particular, every $b \in acl(A)$ has only finitely many conjugates over A.*
 (iv) $acl(acl(A)) = acl(A)$.
  (v) $B \subset acl(A) \Rightarrow acl(B) \subset acl(A)$.
 (vi) $acl(A)$ *is a substructure of M.*
(vii) *Let $A \subset M$ and N an elementary extension of M. Then the algebraic closure of A in N equals the algebraic closure of A in M. In particular, M is algebraically closed in N.*

Proof (i) and (ii) are entirely trivial.

Let $a \in acl(A)$. Get an L-formula $\varphi[x, \overline{y}]$, an $n \geq 1$ and $\overline{b} \in A$ such that

$$M \models \varphi[a, \overline{b}] \wedge \exists_{=n} x \varphi[x, \overline{b}].$$

Let $a_1, \ldots, a_n$ be all the elements $a' \in M$ such that $M \models \varphi[a', \overline{b}]$. For every $\sigma \in Aut_A(M)$,

$$M \models \varphi[\sigma(a), \sigma(\overline{b})],$$

i.e.

$$M \models \varphi[\sigma(a), \overline{b}].$$

But then $\sigma(a) = a_i$ for some $1 \leq i \leq n$. Thus, $a \in ACL(A)$. (iii) is easily seen now.

Next let $a \in acl(acl(A))$. Get an L-formula $\varphi[x, \overline{y}]$ and $\overline{a} \in acl(A)$ such that for some positive integer k,

$$M \models \varphi[a, \overline{a}] \wedge \exists_{=k} x \varphi[x, \overline{a}].$$

Let $\overline{b} \in A$ be such that for each $a_i \in \overline{a}$, there is an L-formula $\varphi_i[y_i, \overline{z}]$ and a positive integer k_i such that

$$M \models \varphi_i[a_i, \overline{b}] \wedge \exists_{=k_i} y_i \varphi_i[y_i, \overline{b}].$$

Then the L_A-formula

$$\psi[x, \overline{b}] = \exists \overline{y}(\varphi[x, \overline{y}] \wedge \exists_{=k} x \varphi[x, \overline{y}] \wedge \wedge_i \varphi_i[y_i, \overline{b}] \wedge \wedge_i \exists_{=k_i} y_i \varphi_i[y_i, \overline{b}])$$

witnesses that $a \in acl(A)$. (iv) follows. (v) is easily seen.

If f is an n-ary function symbol and $a_1, \ldots, a_n \in acl(A)$, then the formula $x = f(a_1, \ldots, a_n)$ witnesses that

$$f^M(a_1, \ldots, a_n) \in dcl(a_1, \ldots, a_n) \subset acl(acl(A)) = acl(A).$$

(vi) is easily seen now.

(vii) Let $\varphi[x]$ be an L_A-formula. Then for every $n \geq 1$,

$$M \models \exists_{=n} x \varphi[x] \Leftrightarrow N \models \exists_{=n} x \varphi[x].$$

Hence, any $a \in N$ such that $N \models \varphi[a]$ must belong to M. $\qquad\square$

Exercise 1.10.3 Let L and L' be first-order languages, M an L-structure and N an L'-structure. We say that N is *interpretable* in M if there is an L'-structure $N' \subset M^k$ for some $k \geq 1$ such that

  (i) N' is definable,
  (ii) N' is isomorphic to N and

(iii) the interpretations of L'-symbols in N' are definable in M.

1. Show the group $GL_n(\mathbb{K})$ of non-singular $n \times n$ matrices over $\mathbb{K}$ is interpretable in the field $\mathbb{K}$.
2. Show that the groups $O(n)$ and $SO(n)$ are interpretable in the field $\mathbb{R}$ of real numbers.
3. Show that the groups $U(n)$ and $SU(n)$ are interpretable in the field $\mathbb{C}$ of complex numbers.

1.11 Many-Sorted Logic and Imaginary Elements

In mathematics, one builds various structures from a given structure or one simultaneously considers several structures. For instance, one simultaneously considers the quotients of a structure, orbit spaces of group actions or varieties over a field, etc. Thus, one has structures of many sorts and functions and relations on their products. Traditional model theory that we have described so far is a special case of many-sorted model theory having only one sort.

We now briefly describe the formal system of many-sorted logic. Many of the argument of one-sorted logic can easily be seen to hold in the many-sorted case also.

A *many-sorted language* consists of

(∗) a non-empty set $\mathcal{S}$ of *sorts*,
(∗) for each sort s, a set of constant symbols, generically denoted by c_s, of sort s,
(∗) for each finite sequence $\bar{s} = (s_1, \ldots, s_n)$ of sorts, a set $\mathcal{R}_{\bar{s}}$ of relation symbols of sort $\bar{s}$, and
(∗) for each finite sequence $\bar{s} = (s_1, \ldots, s_n)$ of sorts and a sort s, a set $\mathcal{F}_{\bar{s}s}$ of function symbols of sort $\bar{s}s$.

Besides these, the language also has

(∗) for each sort s, a sequence of variables $x_s^0, x_s^1, \ldots,$,
(∗) for each sort s, a binary equality symbol $=_s$, and
(∗) $\neg, \vee, \exists_s, s \in \mathcal{S}$.

We shall often drop sort suffixes on variables, quantifiers and equality symbols. This will cause no confusion.

The set of terms of sort s, $s \in \mathcal{S}$, is the smallest collection of expressions that contains each variable and each constant symbol of sort s and expressions of the form $f(t_0, \ldots, t_{n-1})$, where f is a function symbol of sort $(s_0, \ldots, s_{n-1}, s)$ and $t_0, \ldots, t_{n-1}$ are terms of sorts $s_0, \ldots, s_{n-1}$ respectively.

Atomic formulas are expressions $t_1 =_s t_2$, where t_1, t_2 are terms of sort s and $R(t_1, \ldots, t_n)$, where R is a relation symbol of sort $(s_1, \ldots, s_n)$ and $t_1, \ldots, t_n$ terms of sorts $s_1, \ldots, s_n$, respectively. We build up L-formulas $\varphi[x_0, \ldots, x_{n-1}]$ with variables $x_0, \ldots, x_{n-1}$ of sorts $s_0, \ldots, s_{n-1}$ respectively from atomic formulas using connectives and quantifiers in the usual way. In particular, if $\varphi[x_0, \ldots, x_{n-1}]$ is a formula and x_i is a variable of sort s_i, then $\exists_{s_i} x_i \varphi[x_0, \ldots, x_{n-1}]$ is a formula.

If L is a many-sorted language, an L-structure $\mathcal{M}$ consists of

(∗) for each sort s, a non-empty set $s(M)$,
(∗) for each constant symbol c_s of sort s, an element $c_s^M \in s(M)$,
(∗) for each relation symbol R of sort $(s_1, \ldots, s_n)$, a set $R^M \subset s_1(M) \times \cdots \times s_n(M)$, and
(∗) for each function symbol f of sort $(s_1, \ldots, s_n, s)$, a function $f^M : s_1(M) \times \cdots \times s_n(M) \to s(M)$.

If x is a variable of sort s, sometimes one writes M_x for $s(M)$ and calls it the universe of sort x or simply the universe of x. We define the truth of a formula in $\mathcal{M}$ in the usual way.

Let $\mathcal{M}$ and $\mathcal{N}$ be L-structures. A homomorphism $h : \mathcal{M} \to \mathcal{N}$ consists of maps $s(h) : s(\mathcal{M}) \to s(\mathcal{N})$, $s \in \mathcal{S}$, satisfying

(∗) for each constant c of sort s, $h(c^M) = c^N$,
(∗) for each relation symbol R of sort $(s_1, \ldots, s_n)$ and $(a_1, \ldots, a_n) \in s_1(M) \times \cdots \times s_n(M)$,

$$\mathcal{M} \models R^M(a_1, \ldots, a_n) \Rightarrow \mathcal{N} \models R^M(h(a_1), \ldots, h(a_n)),$$

and
(∗) for each function symbol f of sort $(s_1, \ldots, s_n, s)$ and $(a_1, \ldots, a_n) \in s_1(M) \times \cdots \times s_n(M)$,

$$h(f^M(a_1, \ldots, a_n)) = f^N(h(a_1), \ldots, h(a_n)).$$

Definitions of embedding, elementary embedding, isomorphism, substructure, elementary substructure, etc., should be clear to the reader now. Most of the results proved so far (and many to be proved later) for one-sorted logic will easily generalise to many-sorted cases.

Many-sorted language is a useful device to treat equivalence classes as elements of a structure of a many-sorted language. It is also used to code definable sets by an element of a multisorted structure.

To demonstrate these, let us fix an L-theory T. Let $\varphi[\overline{x}, \overline{y}]$ be an L-formula, $|\overline{x}| = |\overline{y}| = n$, such that

$$T \models \text{``}\varphi[\overline{x}, \overline{y}]\text{ is an equivalence relation.''}$$

Such a φ will be called an *equivalence formula in T*. For each equivalence formula φ in T, we introduce a sort S_φ. These sorts include a sort $S_=$ corresponding to the formula $x = y$, x, y of length 1.

The many-sorted language L^{eq} (which depends on T also) has S_φ, φ an equivalence formula in T, as sorts. For each equivalence formula φ in T as above, L^{eq} has a function symbol f_φ of sort $(\underbrace{S_=, \ldots, S_=}_{n \text{ times}}, S_\varphi)$.

The theory T^{eq} has language L^{eq} and following axioms:.

(∗) T.

(∗) For each equivalence formula φ in T,

$$\varphi[x_1, \cdots, x_n, y_1, \ldots, y_n] \Leftrightarrow f_\varphi(x_1, \ldots, x_n) = f_\varphi(y_1, \ldots, y_n).$$

(∗) For each equivalence formula φ in T,

$$\forall y \text{ of sort } S_\varphi \exists x_1, \ldots, x_n \text{ of sort } S_=(y =_{S_\varphi} f_\varphi(x_1, \ldots, x_n)).$$

Let $M \models T$. Then we get a model M^{eq} of T^{eq} canonically:

(∗) $S_=(M^{eq}) = M$.

(∗) For each equivalence formula $\varphi[\bar{x}, \bar{y}]$ in T with $|\bar{x}| = |\bar{y}| = n$, $S_\varphi^{M^{eq}} = M^n/E_\varphi$, where E_φ is the equivalence relation on M^n defined by φ. In this case we interpret f_φ by the quotient map $f_\varphi^{M^{eq}} : M^n \to M^n/E_\varphi$.

Thus, an equivalence class of a $\emptyset$-definable equivalence relation on M^n now has become an element of M^{eq}. For this reason elements of M^{eq} are called *imaginary elements*. The concept of imaginary elements was introduced by Shelah in [54, Sect. 3.6] as a means to consider imaginary elements such as equivalence classes as genuine elements.

Each definable set in M^k has a natural code in M^{eq}: Let $\varphi[\bar{x}, \bar{y}]$ be an L-formula, $\bar{a} \in M$, $|\bar{x}| = k$ and $|\bar{a}| = n$ and $X = \varphi[M, \bar{a}] \subset M^k$. Now consider the formula

$$\theta[\bar{y}_1, \bar{y}_2] = \forall \bar{x}(\varphi[\bar{x}, \bar{y}_1] \leftrightarrow \varphi[\bar{x}, \bar{y}_2]).$$

Clearly θ is an equivalence formula in T. Let $\bar{a}/E_\theta \in S_\theta(M^{eq})$ be the equivalence class containing $\bar{a}$. Then $\bar{a}/E_\theta$ is the unique point such that $X = \varphi'[M^{eq}, \bar{a}/E_\theta]$, where

$$\varphi'[\bar{x}, y] = \exists \bar{y}(f_\theta(\bar{y}) = y \wedge \varphi[\bar{x}, \bar{y}]).$$

Here variables in $\bar{x}$ and $\bar{y}$ are of sort $S_=$ and variable y is of sort S_θ.

This code is clearly not unique because it depends on the formula that defines X. But each code is definable in terms of any other code: Let $\psi[\bar{x}, \bar{b}]$ be another formula defining X,

$$\theta'[\bar{z}_1, \bar{z}_2] = \forall \bar{x}(\psi[\bar{x}, \bar{z}_1] \leftrightarrow \psi[\bar{x}, \bar{z}_2]),$$

and

$$\psi'[\bar{x}, z] = \exists \bar{z}(f_{\theta'}(\bar{z}) = z \wedge \psi[\bar{x}, \bar{z}]).$$

Then the formula

$$\xi[z, \overline{a}/E_\theta] = \forall \overline{x}(\varphi'[\overline{x}, \overline{a}/E_\theta] \leftrightarrow \psi'[\overline{x}, z])$$

defines $\overline{b}/E_{\theta'}$.

1.12   Elimination of Imaginaries

Let M be an L-structure and $\varphi[\overline{x}, \overline{y}]$ an L-formula. If

$$\{(\overline{a}, \overline{b}) \in M^n \times M^n : M \models \varphi[\overline{a}, \overline{b}]\}$$

is an equivalence relation on M^n, we say that "$\varphi[\overline{x}, \overline{y}]$ *is an equivalence formula* on M." Further, each equivalence class $\overline{a}/E_\varphi, \overline{a} \in M^n$, is called an *imaginary element* of M.

We say that M has *elimination of imaginaries* if for every equivalence formula $\varphi[\overline{x}, \overline{y}]$ on M and each $\overline{a} \in M^n$, there is an L-formula $\psi[\overline{x}, \overline{z}]$ and a unique $\overline{b} \in M$ such that $\overline{a}/E_\varphi = \psi[M, \overline{b}]$. The notion of elimination of imaginaries was introduced by Poizat in [50].

In terms of M^{eq}, this is equivalently defined as follows.

Proposition 1.12.1 *An L-structure M has elimination of imaginaries if and only if for every equivalence formula $\theta[\overline{x}, \overline{y}]$ on M and every $\overline{a} \in M$, there is a $\overline{b} \in M$ such that $\overline{a}/E_\theta \in dcl(\overline{b})$ and $\overline{b} \in dcl(\overline{a}/E_\theta)$.*

Proof Suppose M has elimination of imaginaries, $\theta[\overline{x}, \overline{y}]$ an equivalence formula on M and $\overline{a} \in M$. Since M has elimination of imaginaries, there is an L-formula $\varphi[\overline{x}, \overline{z}]$ and a unique $\overline{b} \in M$ such that $\overline{a}/E_\theta = \varphi[M, \overline{b}]$. Now consider the L^{eq}-formula

$$\psi[z, \overline{z}] = \forall \overline{x}(f_\theta(\overline{x}) = z \leftrightarrow \varphi[\overline{x}, \overline{z}]),$$

where variable z is of sort S_θ. Then $\overline{a}/E_\theta$ is the unique imaginary c such that $M^{eq} \models \psi[c, \overline{b}]$. This implies that $\overline{a}/E_\theta \in dcl(\overline{b})$. Also, $\overline{b}$ is the unique $\overline{d} \in M$ such that $M^{eq} \models \psi[\overline{a}/E_\theta, \overline{d}]$. So, $\overline{b} \in dcl(\overline{a}/E_\theta)$.

We now prove if part of the result. Let $\overline{a}, \overline{b} \in M$, $\theta[\overline{x}, \overline{y}]$ an equivalence formula on M and $\psi[z, \overline{z}]$ and $\psi'[z, \overline{z}]$ be L^{eq}-formulas with z of sort S_θ such that $\overline{a}/E_\theta$ is the unique imaginary c satisfying $M^{eq} \models \psi[c, \overline{b}]$ and $\overline{b}$ the unique $\overline{d} \in M$ with $M^{eq} \models \psi'[\overline{a}/E_\theta, \overline{d}]$. Set

$$\varphi[\overline{x}, \overline{z}] = \psi[f_\theta(\overline{x}), \overline{z}] \wedge \psi'[f_\theta(\overline{x}), \overline{z}].$$

Then $\overline{b}$ is the unique $\overline{d}$ such that $\overline{a}/E_\theta = \varphi[M, \overline{d}]$. $\square$

Some authors define elimination of imaginaries in term of M^{eq}. However, we shall not take this approach.

If we can choose ψ independent of $\bar{a}$ also, then we say that M admits uniform elimination of imaginaries. Thus, we say that M admits *uniform elimination of imaginaries* if for every equivalence formula $\varphi[\bar{x}, \bar{y}]$ on M, there is an L-formula $\psi[\bar{x}, \bar{z}]$, $|\bar{x}| = n$ and $|\bar{z}| = m$, such that

$$M \models \forall \bar{y} \exists_{=1} \bar{z} \forall \bar{x} (\varphi[\bar{x}, \bar{y}] \leftrightarrow \psi[\bar{x}, \bar{z}]).$$

In this case, there is a $\emptyset$-definable function $F : M^n \to M^m$ such that $\bar{a}, \bar{b} \in M^n$ are equivalent if and only if $F(\bar{a}) = F(\bar{b})$. We can take F to be defined by the formula

$$\theta[\bar{y}, \bar{z}] = \forall \bar{x} (\varphi[\bar{x}, \bar{y}] \leftrightarrow \psi[\bar{x}, \bar{z}]).$$

Also, note that the converse is true, i.e. if for every equivalence formula $\varphi[\bar{x}, \bar{y}]$ on M, there is a $\emptyset$-definable function $F : M^n \to M^m$ such that $\bar{a}, \bar{b} \in M^n$ are equivalent if and only if $F(\bar{a}) = F(\bar{b})$, then M admits uniform elimination of imaginaries. For L-formula $\chi[\bar{x}, \bar{z}]$ defined by

$$\bar{z} = F(\bar{x}),$$

we have

$$M \models \forall \bar{y} \exists_{=1} \bar{z} \forall \bar{x} (\varphi[\bar{x}, \bar{y}] \leftrightarrow \chi[\bar{x}, \bar{z}]).$$

A theory T has *elimination of imaginaries* if each of its models has elimination of imaginaries.

Proposition 1.12.2 *Let T be an L-theory. The following statements are equivalent.*

1. *T has elimination of imaginaries.*
2. *For every $M \models T$ and for every definable $X \subset M^n$, there is an L-formula $\psi[\bar{x}, \bar{z}]$ and a unique $\bar{a} \in M$ such that $X = \psi[M, \bar{a}]$.*

Proof Clearly (2) implies (1). Assuming (1) we prove (2). Let $M \models T$, $\varphi[\bar{x}, \bar{y}]$ be an L-formula with $|\bar{y}| = m$, $\bar{a} \in M^m$ and $X = \varphi[M, \bar{a}]$. Consider

$$\theta[\bar{y}_1, \bar{y}_2] = \forall \bar{x} (\varphi[\bar{x}, \bar{y}_1] \leftrightarrow \varphi[\bar{x}, \bar{y}_2]).$$

Then θ is an equivalence formula on M. By (1) there is an L-formula $\psi'[\bar{y}_1, \bar{z}]$ and a unique $\bar{b} \in M$ such that $\bar{a}/\theta = \psi'[M, \bar{b}]$.

Now consider the L-formula

$$\psi[\bar{x}, \bar{z}] = \forall \bar{y}_1 (\psi'[\bar{y}_1, \bar{z}] \leftrightarrow \varphi[\bar{x}, \bar{y}_1]).$$

Then $\bar{b} \in M$ is the unique tuple such that $X = \psi[M, \bar{b}]$. $\square$

An L-structure M is called *uniformly 1-eliminable* if for every L-formula $\varphi[x, \bar{y}]$ there is an L-formula $\psi[x, \bar{z}]$ such that

$$M \models \forall \bar{y} \exists_{=1} \bar{z} \forall x (\varphi[x, \bar{y}] \leftrightarrow \psi[x, \bar{z}]).$$

Theorem 1.12.3 *If an L-structure M has definable Skolem functions and is uniformly 1-eliminable, then M has uniform elimination of imaginaries.*

Proof Take an equivalence L-formula $\varphi[\overline{x}, \overline{y}]$ on M. We shall show that there is a $\emptyset$-definable function $F = (F_1, \ldots, F_n) : M^n \to M^n$ such that

1. $\forall \overline{a} \in M^n M \models \varphi[F(\overline{a}), \overline{a}]$, and
2. $\forall \overline{a}, \overline{a'} \in M^n(M \models \varphi[\overline{a}, \overline{a'}] \Rightarrow F(\overline{a}) = F(\overline{a'}))$.

We shall define $F_1, \ldots, F_n$ inductively. To define F_1, consider

$$\varphi_1[x_1, \overline{y}] = \exists x_2 \ldots \exists x_n \varphi[\overline{x}, \overline{y}].$$

By uniform 1-elimination, there exists an L-formula $\psi_1[x_1, \overline{z}]$ such that

$$M \models \forall \overline{y} \exists_{=1} \overline{z} \forall x_1 (\varphi_1[x_1, \overline{y}] \leftrightarrow \psi_1[x_1, \overline{z}]).$$

Since M has definable Skolem functions, corresponding to ψ_1, there is a formula $\xi_1[x_1, \overline{z}]$ such that

$$M \models \forall \overline{z} (\exists_{=1} x_1 \xi_1[x_1, \overline{z}] \wedge (\exists x_1 \psi_1[x_1, \overline{z}] \to \forall x_1(\xi_1[x_1, \overline{z}] \to \psi_1[x_1, \overline{z}]))).$$

Thus, we have a function $F_1(\overline{y})$ defined by

$$x_1 = F_1(\overline{y}) \leftrightarrow \exists \overline{z}(\forall x(\varphi_1[x, \overline{y}] \leftrightarrow \psi_1[x, \overline{z}]) \wedge \xi_1[x_1, \overline{z}]).$$

Note that

$$M \models \varphi[\overline{y}, \overline{y'}] \to F_1(\overline{y}) = F_1(\overline{y'}).$$

Assume $F_1, \ldots, F_{i-1}$ have been defined. Consider the formula

$$\varphi_i[x_i, \overline{y}] = \exists x_{i+1}, \ldots \exists x_n \varphi[F_1(\overline{y}), \ldots, F_{i-1}(\overline{y}), x_i, x_{i+1}, \ldots, x_n, \overline{y}].$$

Now by uniform 1-elimination, get $\psi_i[x_i, \overline{z}]$ corresponding to φ_i and then, using definable Skolem functions on ψ_i get $\xi_i[x_i, \overline{z}]$ as above. This gives us an $\emptyset$-definable function $F_i(\overline{y})$ as above. $\qquad\qquad\square$

Now let T be an L-theory. An L-formula $\varphi[\overline{x}, \overline{y}]$ is called an *equivalence formula* (in T) if

$$T \models \text{``}\varphi \text{ is an equivalence relation''},$$

i.e. it is an equivalence formula on each model M of T. The theory T is said to admit *elimination of imaginaries* if for every equivalence formula $\varphi[\overline{x}, \overline{y}]$, there exists a formula $\psi[\overline{x}, \overline{z}]$ such that

$$T \models \forall \overline{y} \exists_{=1} \overline{z} \forall \overline{x} (\varphi[\overline{x}, \overline{y}] \leftrightarrow \psi[\overline{x}, \overline{z}]).$$

In this case, every model of T has uniform elimination of imaginaries.

Example 1.12.4 Peano arithmetic (PA) has uniform elimination of imaginaries.

More examples will be given later.

An L-structure M is said to have *semi-uniform elimination of imaginaries* if for every equivalence formula $\varphi[\overline{x}, \overline{y}]$ in M, there is a finite sequence of formulas $\{\varphi_i[\overline{x}, \overline{z}] : i < m\}$ such that

$$M \models \forall \overline{y} \exists_{=1} i < m \exists_{=1} \overline{z} \forall \overline{x} (\varphi[\overline{x}, \overline{y}] \leftrightarrow \varphi_i[\overline{x}, \overline{z}]).$$

Proposition 1.12.5 *Let M be an L-structure and there exist variable free terms t, s such that $M \models t \neq s$. Then if M has semi-uniform elimination of imaginaries, it has uniform elimination of imaginaries.*

Proof Let $\varphi[\overline{x}, \overline{y}]$ and $\{\varphi_i[\overline{x}, \overline{z}] : i < m\}$ be as in the definition above. Define

$$\psi[\overline{x}, \overline{z}, w_0, \ldots, w_{m-1}] = \vee_{i < m} (\varphi_i[\overline{x}, \overline{z}] \wedge (w_i = s) \wedge \wedge_{j \neq i} (w_j = t)).$$

Then

$$M \models \forall \overline{y} \exists_{=1} \overline{z}, w_0, \ldots, w_{m-1} \forall \overline{x} (\varphi[\overline{x}, \overline{y}] \leftrightarrow \psi[\overline{x}, \overline{z}, w_0, \ldots, w_{m-1}]).$$

$\square$

A theory T has *semi-uniform elimination of imaginaries* if for every equivalence formula $\varphi[\overline{x}, \overline{y}]$, there is a finite sequence of formulas $\{\varphi_i[\overline{x}, \overline{z}] : i < m\}$ such that

$$T \models \forall \overline{y} \exists_{=1} i < m \exists_{=1} \overline{z} \forall \overline{x} (\varphi[\overline{x}, \overline{y}] \leftrightarrow \varphi_i[\overline{x}, \overline{z}]).$$

Remark 1.12.6 If there exist two variable free terms t, s such that $T \models t \neq s$ and T has semi-uniform elimination of imaginaries, then T has elimination of imaginaries.

Exercise 1.12.7 Let M, $N \models T$ be isomorphic. Show that M^{eq} and N^{eq} are isomorphic models of T^{eq}.

Exercise 1.12.8 Let T be a first-order theory and $\mathcal{M} \models T^{eq}$. Show that there is a model M of T such that $\mathcal{M} = M^{eq}$.

Chapter 2
Basic Introductory Results

Abstract The goal of this chapter is to present basic introductory techniques of model theory. The main results presented are Łoś fundamental lemma on ultra-product of structures, compactness theorem and quantifier elimination. These are cornerstones of model theory. A large number of applications given in this chapter bear testimony to the importance of these results. We also introduce the notion of independence and dimension in minimal sets. Finally we give several applications of the results proved in this section in algebra and geometry. A large number of examples and exercises are given as we go along.

2.1 Ultraproduct of Structures

In this section, we introduce ultraproduct of models. It is a notion of the product of structures and a basic technique of constructing new models from old ones. It made its first appearance in Skolem [57]. The fundamental lemma was proved by Łoś in [36]. Since then ultraproduct has become a basic tool in model theory.

Let L be a first-order language and $\mathcal{F}$ a filter on a non-empty set I. Suppose for each $i \in I$ we are given an L-structure M_i of L. Set

$$M = \times_{i \in I} M_i.$$

For $\alpha, \beta \in M$, define

$$\alpha \sim \beta \Leftrightarrow \{i \in I : \alpha(i) = \beta(i)\} \in \mathcal{F}.$$

Since $I \in \mathcal{F}$, $\sim$ is reflexive. Clearly, it is symmetric. Since $\mathcal{F}$ is closed under finite intersections and supersets, $\sim$ is transitive. Thus, $\sim$ is an equivalence relation on $\times_i M_i$. For $\alpha \in M$, $[\alpha]$ will denote the $\sim$-equivalence class containing α. We set

$$M(\mathcal{F}) = M/\sim = \{[\alpha] : \alpha \in M\}.$$

© Springer Nature Singapore Pte Ltd. 2017

H. Sarbadhikari and S.M. Srivastava, *A Course on Basic Model Theory*,
DOI 10.1007/978-981-10-5098-5_2

45

We interpret the nonlogical symbols of L as follows:

1. If c is a constant symbol, $c^{M(\mathcal{F})} = [\alpha]$, where $\alpha(i) = c^{M_i}$, $i \in I$.
2. If p is an n-ary relation symbol,

$$p^{M(\mathcal{F})}([\alpha_1], \ldots, [\alpha_n]) \Leftrightarrow \{i \in I : p^{M_i}(\alpha_1(i), \ldots, \alpha_n(i))\} \in \mathcal{F}.$$

3. If f is an n-ary function symbol, we define

$$[\beta] = f^{M(\mathcal{F})}([\alpha_1], \ldots, [\alpha_n]),$$

where

$$\beta(i) = f^{M_i}(\alpha_1(i), \ldots, \alpha_n(i)), \ i \in I.$$

We need to show that $p^{M(\mathcal{F})}$ and $f^{M(\mathcal{F})}$ are well defined. Suppose $\alpha_j \sim \beta_j$, $1 \leq j \leq n$. Since $\mathcal{F}$ is closed under finite intersections, there is an $X \in \mathcal{F}$ such that $\alpha_j(i) = \beta_j(i)$ for all $1 \leq j \leq n$ and all $i \in X$. This implies the well-definedness of $p^{M(\mathcal{F})}$ and $f^{M(\mathcal{F})}$.

Proposition 2.1.1 *For every term $t[\overline{x}]$ and every $\alpha_0, \ldots, \alpha_{n-1}, \beta \in M$,*

$$t^{M(\mathcal{F})}[[\alpha_0], \ldots, [\alpha_{n-1}]] = [\beta] \Leftrightarrow \{i \in I : t^{M_i}[\alpha_0(i), \ldots, \alpha_{n-1}(i)] = \beta(i)\} \in \mathcal{F}.$$

Proof The result is proved easily by induction on the length of t. The details are left for the reader as an easy exercise. $\qquad\square$

Proposition 2.1.2 *For every atomic formula $\varphi[\overline{x}]$ and every $\overline{\alpha} \in M$,*

$$M(\mathcal{F}) \models \varphi[[\alpha_0], \ldots, [\alpha_{n-1}]] \Leftrightarrow \{i \in I : M_i \models \varphi[\alpha_0(i), \ldots, \alpha_{n-1}(i)]\} \in \mathcal{F}. \quad (*)$$

Proof Let $t[\overline{x}], s[\overline{x}]$ be terms and $\alpha_0, \ldots, \alpha_{n-1} \in M$. Define

$$\beta(i) = t^{M_i}[\alpha_0(i), \ldots, \alpha_{n-1}(i)], \ i \in I$$

and

$$\gamma(i) = s^{M_i}[\alpha_0(i), \ldots, \alpha_{n-1}(i)], \ i \in I.$$

By the last Proposition 2.1.1,

$$t^{M(\mathcal{F})}[[\alpha_0], \ldots, [\alpha_{n-1}]] = [\beta]$$

and

$$s^{M(\mathcal{F})}[[\alpha_0], \ldots, [\alpha_{n-1}]] = [\gamma].$$

Thus, $(*)$ holds for $t[\overline{x}] = s[\overline{x}]$ and $\alpha_0, \ldots, \alpha_{n-1}$.

Let $\varphi[\overline{x}]$ be an atomic formula $p[t_1[\overline{x}], \dots, t_m[\overline{x}]]$ and every $\alpha_0, \dots, \alpha_{n-1} \in M$. Set

$$\beta_j(i) = t_j^{M_i}[\alpha_0(i), \dots, \alpha_{n-1}(i)], \ i \in I, 1 \leq j \leq m.$$

Then

$$M(\mathcal{F}) \models p[t_1^{M(\mathcal{F})}([\alpha_0], \dots, [\alpha_{n-1}]), \dots, t_m^{M(\mathcal{F})}([\alpha_0], \cdots, [\alpha_{n-1}])]$$
$$\Leftrightarrow M(\mathcal{F}) \models p[[\beta_1], \dots, [\beta_m]]]$$
$$\Leftrightarrow \{i \in I : M_i \models p[t_1^{M_i}[\alpha_0(i), \dots, \alpha_{n-1}(i)], \dots, t_m^{M_i}[\alpha_0(i), \dots, \alpha_{n-1}(i)]]\} \in \mathcal{F}$$

The first equivalence holds by the last Proposition 2.1.1 and the last equivalence holds by definition. $\qquad \square$

In a fundamental contribution to model theory Łoś showed that $(*)$ holds for every formula if $\mathcal{F}$ is an ultrafilter on I.

Theorem 2.1.3 (Łoś Fundamental Lemma) *Let $\mathcal{U}$ be an ultrafilter on I, $\varphi[\overline{x}]$ an L-formula and $[\alpha_0], \dots, [\alpha_{n-1}] \in M(\mathcal{U})$. Then*

$$M(\mathcal{U}) \models \varphi[[\alpha_0], \dots, [\alpha_{n-1}]] \Leftrightarrow \{i \in I : M_i \models \varphi[\alpha_0(i), \dots, \alpha_{n-1}(i)]\} \in \mathcal{U}.$$
$$(**)$$

Proof For atomic φ, $(**)$ follows from the last Proposition 2.1.2. Suppose φ satisfies $(**)$ and ψ is the formula $\neg\varphi$. Take $[\alpha_0], \dots, [\alpha_{n-1}] \in M(\mathcal{U})$. Then

$$M(\mathcal{U}) \models \psi[[\alpha_0], \dots, [\alpha_{n-1}]] \Leftrightarrow M(\mathcal{U}) \not\models \varphi[[\alpha_0], \dots, [\alpha_{n-1}]]$$
$$\Leftrightarrow \{i \in I : M_i \models \varphi[\overline{\alpha}(i)]\} \notin \mathcal{U}$$
$$\Leftrightarrow \{i \in I : M_i \models \psi[\overline{\alpha}(i)]\} \in \mathcal{U},$$

where $\overline{\alpha}(i) = (\alpha_0(i), \dots, \alpha_{n-1}(i))$. The second equivalence holds because φ satisfies $(**)$ whereas the third equivalence holds because $\mathcal{U}$ is an ultrafilter. Similarly we show that if φ and ψ satisfy $(**)$, so does $\varphi \vee \psi$.

Now assume that $(**)$ holds for $\psi[x_0, x_1, \dots, x_n]$, $n \geq 0$ and all $(\alpha_0, \dots, \alpha_n) \in M^{n+1}$. Consider $\varphi = \exists x_0 \psi$. Take any $\alpha_1, \dots, \alpha_n \in M$ such that

$$M(\mathcal{U}) \models \varphi[[\alpha_1], \dots, [\alpha_n]].$$

Then there exists $[\alpha_0] \in M(\mathcal{U})$ such that

$$M(\mathcal{U}) \models \psi[[\alpha_0], \dots, [\alpha_n]].$$

By our hypothesis,

$$\{i \in I : M_i \models \psi[\alpha_0(i), \dots, \alpha_n(i)]\} \in \mathcal{U}.$$

This clearly implies that

$$\{i \in I : M_i \models \varphi[\alpha_1(i), \ldots, \alpha_n(i)]\} \in \mathcal{U}.$$

To prove the converse, assume that the set

$$U = \{i \in I : M_i \models \varphi[\alpha_1(i), \ldots, \alpha_n(i)]\} \in \mathcal{U}.$$

So, for each $i \in U$ there exists an $\alpha_0(i) \in M_i$ such that

$$M_i \models \psi[\alpha_0(i), \ldots, \alpha_n(i)].$$

Take any extension α_0 of $i \rightarrow \alpha_0(i)$, $i \in U$, to I. Then by our assumption

$$M(\mathcal{U}) \models \psi[[\alpha_0], \ldots, [\alpha_n]].$$

Thus,

$$M(\mathcal{U}) \models \varphi[[\alpha_1], \ldots, [\alpha_n]].$$

The result is thus seen by induction on the rank of φ.   □

If $\mathcal{U}$ is an ultrafilter on I, the structure $M(\mathcal{U})$ is called the *ultraproduct* of M_i's. If each $M_i = M$, it is denoted by $M^{\mathcal{U}}$ and is called an *ultrapower* of M.

Let $\{M_i : i \in I\}$ and $\{N_i : i \in I\}$ be families of sets and $\mathcal{U}$ an ultrafilter on I. Let $g_i, h_i : M_i \rightarrow N_i$, $i \in I$, be arbitrary maps. Define

$$\{g_i : i \in I\} \sim_{\mathcal{U}} \{h_i : i \in I\} \Leftrightarrow \{i \in I : g_i = h_i\} \in \mathcal{U}.$$

It is easy to see that $\sim_{\mathcal{U}}$ is an equivalence relation.

Fix $\{g_i : i \in I\} \sim_{\mathcal{U}} \{h_i : i \in I\}$ and $\bar{a} = (a_i) \sim (a_i') = \bar{a}'$. Then $(g_i(a_i)) \sim (h_i(a_i'))$. Hence, we have a well-defined map

$$(\Pi_i g_i)^{\mathcal{U}}([(a_i)]) = [(g_i(a_i))].$$

We make a series of simple observations whose proofs are left to the reader as a simple exercise.

1. If $\{i \in I : g_i$ is onto$\} \in \mathcal{U}$, then $(\Pi_i g_i)^{\mathcal{U}}$ is onto.
2. If $\{i \in I : g_i$ is one-to-one$\} \in \mathcal{U}$, then $(\Pi_i g_i)^{\mathcal{U}}$ is one-to-one.

Next assume that each M_i and each N_i, $i \in I$, are L-structures.

3. If $\{i \in I : g_i$ is a homomorphism$\} \in \mathcal{U}$, then $(\Pi_i g_i)^{\mathcal{U}}$ is a homomorphism. It follows that if $\{i \in I : g_i$ is an embedding (isomorphism)$\} \in \mathcal{U}$, then $(\Pi_i g_i)^{\mathcal{U}}$ is an embedding (isomorphism).
4. Using Łoś theorem (Theorem 2.1.3), it is easy to see that if $\{i \in I : g_i$ is elementary$\} \in \mathcal{U}$, then $(\Pi_i g_i)^{\mathcal{U}}$ is elementary.

Corollary 2.1.4 *Let T be an L-theory and $\{M_i : i \in I\}$ a family of models of T. Then for every ultrafilter $\mathcal{U}$ on I, the ultraproduct $M(\mathcal{U})$ is a model of T.*

Remark 2.1.5 Ultraproduct gives a new notion of product in the category of models of T; in particular, in any category of algebraic structures such as groups, rings, fields, etc. Further, by choosing the ultrafilter $\mathcal{U}$ suitably, one gets a model $M(\mathcal{U})$ with some desired properties.

Since $M(\mathcal{U})$ is in a sense a limit of $\{M_i : i \in I\}$, in general, no reasonable converse of the corollary exists. However, if T has only finitely many axioms, a converse of the corollary is true.

Proposition 2.1.6 *Let $\{M_i : i \in I\}$ be a family of L-structures and $\mathcal{U}$ an ultrafilter on I. Suppose an L-theory T has finitely many axioms only and $M(\mathcal{U}) \models T$. Then $\{i \in I : M_i \models T\} \in \mathcal{U}$.*

Proof Let $\varphi_1, \ldots, \varphi_n$ be all the axioms of T. Since $M(\mathcal{U}) \models T$, for each $1 \leq k \leq n$, the set $A_k = \{i \in I : M_i \models \varphi_k\} \in \mathcal{U}$. Then $A = \cap_{1 \leq k \leq n} A_k \in \mathcal{U}$ and for every $i \in A$, $M_i \models T$.   □

Corollary 2.1.7 *Let $\{\mathbb{K}_i : i \in I\}$ be a family of rings, $\mathcal{U}$ an ultrafilter on I and $p > 0$ a prime. Then the ultraproduct $\mathbb{K}(\mathcal{U})$ is a field of characteristic p if and only if $\{i \in I : \mathbb{K}_i$ is a field of characteristic $p\}$ is in $\mathcal{U}$.*

Example 2.1.8 For each prime $p > 0$, let $\mathbb{K}_p$ be a field of characteristic p and $\mathcal{U}$ a free ultrafilter on the set of all primes. Then the ultraproduct $\mathbb{K}(\mathcal{U})$ is a field of characteristic 0. To see this, let P denote the set of all primes. Fix a prime p. Since $\mathcal{U}$ is free, $\{q \in P : q > p\} \in \mathcal{U}$. Since $\mathbb{K}_q \models \underline{p} \neq 0$ for every $q > p$, $char(\mathbb{K}(\mathcal{U})) \neq p$ by Łoś Theorem 2.1.3. Our claim follows.

Proposition 2.1.9 *A class $\mathcal{C}$ of L-structures is elementary if and only if $\mathcal{C}$ is closed under elementary equivalences and ultraproducts.*

Proof The only if part is clear from Łoś theorem (Theorem 2.1.3). So, assume that $\mathcal{C}$ is closed under elementary equivalences and ultraproducts and $T = Th(\mathcal{C})$. We now show that $\mathcal{C}$ is precisely the class of all models of T. Clearly, if $M \in \mathcal{C}$, $M \models T$.

Now assume that $M \models T$. Let I denote the set of all non-empty finite subsets of $Th(M)$. Note that for each $i \in I$ there is a $M_i \in \mathcal{C}$ such that $M_i \models \wedge i$. If not, then $\neg(\wedge i) \in T$. But then both $\wedge i$ and $\neg(\wedge i)$ are true in M which is a contradiction. For each sentence $\varphi \in Th(M)$, set

$$A_\varphi = \{i \in I : \varphi \in i\}.$$

Given $\varphi_1, \ldots, \varphi_k$,

$$\{\varphi_1, \ldots, \varphi_k\} \in \wedge_{j=1}^{k} A_{\varphi_j}.$$

This implies that there is an ultrafilter $\mathcal{U}$ containing each A_φ, $\varphi \in Th(M)$. Set

$$N = \times_{i \in I} M_i / \mathcal{U}.$$

By our hypothesis $N \in C$. Our proof will be complete if we show that M is elementarily equivalent to N. But for any $\varphi \in Th(M)$,

$$A_\varphi \subset \{i \in I : M_i \models \varphi\}.$$

Thus, $Th(M) \subset Th(N)$. This implies that these two sets are equal, i.e. M and N are elementarily equivalent. $\square$

This result can be easily used to show that various classes of structures are not elementary. To illustrate this let L have no nonlogical symbol. So any non-empty set is an L-structure. Let C be the class of all finite sets. For $k > 0$, let $X_k = \{0, \ldots, k - 1\}$. Take any free ultrafilter $\mathcal{U}$ on the set of all positive integers. Now consider $X = \times_k X_k / \mathcal{U}$. For any positive integer m, let $\alpha_m \in \times_k X_k$ be a sequence which is eventually m. Then for $m \neq n$, $[\alpha_m] \neq [\alpha_n]$. Thus, X is infinite. Hence, C is not closed under ultraproducts. So, C is not elementary.

We saw earlier that the class of all fields of positive characteristic is not closed under ultraproducts. Hence, the class of all fields of positive characteristic is not elementary.

Exercise 2.1.10 Let $\mathcal{U}$ be an ultrafilter on I with $\cap \mathcal{U} = \{j\}$. Suppose $\{M_i : i \in I\}$ is a family of L-structures. Show that $M(\mathcal{U})$ is isomorphic to M_j.

Exercise 2.1.11 Let M be an L-structure and $\mathcal{U}$ an ultrafilter on I. Define the inclusion map $j : M \to M^\mathcal{U}$ by

$$j(x) = [c_x], x \in M,$$

where $c_x : I \to M$ is the constant map $c_x(i) = x, i \in I$. Show that j is an elementary embedding.

2.2 Compactness Theorem

In this section, we prove the compactness theorem for first-order theories. Because of its great importance, we also give several variants of this theorem. This was first proved for countable theories by Gödel in [15]. For general theories, it was independently proved by Mal'tsev in [39, 40] and by Henkin in [18].

Theorem 2.2.1 (Compactness theorem) *An L-theory T has a model if and only if each finite $T' \subset T$ has a model.*

Proof If part: For each finite $i \subset T$, let M_i be a model of i. Set $I = \{i : i \subset T \text{ finite}\}$. For each sentence φ, set

$$B_\varphi = \{i \in I : \varphi \in i\}.$$

Let $\varphi_1, \ldots, \varphi_n \in T$. $\{\varphi_1, \ldots, \varphi_n\} \in \cap_{i=1}^n B_{\varphi_i}$. Thus, the family $\{B_\varphi : \varphi \in T\}$ has finite intersection property. Hence, it is contained in an ultrafilter $\mathcal{U}$.

We claim that $M(\mathcal{U}) \models T$. Let $\varphi \in T$. Then for every $i \in B_\varphi$, $M_i \models \varphi$. Hence, by Łoś theorem, $M(\mathcal{U}) \models \varphi$.

The only if part is entirely trivial. This completes the proof of the compactness theorem. $\square$

We give an alternative proof of compactness theorem. This is essentially the semantic version of the syntactical proof given originally. This proof gives yet another technique of building models which will be used later also.

Let T be a finitely satisfiable set of L-sentences. The following observation is trivially seen.

Fact. For any L-sentence φ, at least one of $T \cup \{\varphi\}$ and $T \cup \{\neg\varphi\}$ is finitely satisfiable.

A finitely satisfiable set T of L-sentences will be called complete if for every sentence φ, φ or $\neg\varphi$ is in T. Using Zorn's lemma, it is immediately seen that.

Theorem 2.2.2 (Lindenbaum Theorem) *Every finitely satisfiable set of L-sentences is contained in a complete set of finitely satisfiable L-sentences.*

We leave the detail for the reader as a simple exercise.

A set of finitely satisfiable L-sentences T will be called *Henkin* if whenever a closed sentence of the form $\exists x \varphi \in T$, there is a constant symbol c such that $\varphi_x[c] \in T$. Note that if T is complete and finitely satisfiable, then the sentence $\exists x(x = x) \in T$. Otherwise, $\neg\exists x(x = x) \in T$, contradicting that T is finitely satisfiable. This, in particular, implies that L has constant symbols.

The main idea of the proof is the following.

Theorem 2.2.3 *Every complete, Henkin set of finitely satisfiable L-sentences T has a model.*

Proof Let M' denote the set of all variable-free L-terms. By the above remark, $M' \neq \emptyset$. If t and s are variable-free terms, define

$$t \sim s \text{ if } t = s \in T.$$

Using finite satisfiability and completeness of T, it can be easily proved that $\sim$ is an equivalence relation on M'. For instance, if t_1, t_2, t_3 are variable-free L-terms and $t_1 \sim t_2$ and $t_2 \sim t_3$ hold, then $t_1 \sim t_3$ must hold. For otherwise, by completeness of T, $t_1 = t_2, t_2 = t_3, t_1 \neq t_3 \in T$. This contradicts the finite satisfiability of T.

Let $M = M'/\sim$, the set of all $\sim$-equivalence classes. For any variable-free term t, let $[t]$ denote the equivalence class containing t. For any constant symbol c, take $c^M = [c]$. Let f be a n-ary function symbol, R a n-ary relation symbol and $[t_1], \ldots, [t_n] \in M$. Define

$$f^M([t_1], \ldots, [t_n]) = [f(t_1, \ldots, t_n)]$$

and
$$R^M([t_1], \ldots, [t_n]) \Leftrightarrow R[t_1, \ldots, t_n] \in T.$$

Using completeness and finite satisfiability of T it is easy to see that these are well defined. Thus, we have defined an L-structure M.

By induction on the rank of φ, we now show that for every L_M-sentence φ,

$$M \models \varphi \Leftrightarrow \varphi \in T. \tag{*}$$

(*) is true for all atomic φ by the definition of M. Suppose (*) holds for φ and $\psi = \neg\varphi$. Then

$$M \models \psi \Leftrightarrow M \not\models \varphi \Leftrightarrow \varphi \notin T \Leftrightarrow \psi \in T.$$

The second equivalence holds by the induction hypothesis and the third equivalence holds because T is complete.

Next, we assume that (*) holds for φ and ψ and $\xi = \varphi \vee \psi$. Suppose $M \models \xi$. Then $M \models \varphi$ or $M \models \psi$. Without any loss of generality, assume that $M \models \varphi$. Then by induction hypothesis, $\varphi \in T$. Hence, by the completeness of T, $\xi \in T$. Conversely, let $\xi \in T$. Then by the completeness of T, $\varphi \in T$ or $\psi \in T$. Hence, by induction hypothesis, $M \models \varphi$ or $M \models \psi$. In either case, $M \models \xi$.

Finally, let (*) holds for all L_M-sentences of length less than the length of $\exists x\varphi[x]$ which is assumed to be closed. Suppose $M \models \exists x\varphi$. Then there exists $[t] \in M$ such that $M \models \varphi[[t]]$. So, by induction hypothesis, $\varphi[[t]] \in T$. Hence, by completeness of T, $\exists x\varphi[x] \in T$. Now assume that $\exists x\varphi[x] \in T$. Since T is Henkin, there is a constant c such that $\varphi_x[c] \in T$. So, by induction hypothesis, $M \models \varphi_x[c]$. Thus, $M \models \exists x\varphi[x]$. $\qquad \square$

The model of T obtained in the last proposition is called the canonical model of T. To complete the proof of compactness theorem, we need one more result.

Proposition 2.2.4 *Let T be a finitely satisfiable set of L-sentences. Then there is an extension L_∞ of L obtained by adding new constant symbols only and a finitely satisfiable, Henkin set of L_∞-sentences T_∞ that contains T.*

Proof Set $L_0 = L$ and $T_0 = T$. Suppose L_n and a finitely satisfiable set of L_n-sentences T_n have been defined. For each L_n-sentence of the form $\exists x\varphi[x]$ which is not an L_m-sentence for any $m < n$, we add a new constant symbol $c_{\exists x\varphi}$ to L_n and the sentence $\exists x\varphi[x] \rightarrow \varphi_x[c_{\exists x\varphi}]$ to T_n. Call the resulting language L_{n+1} and resulting set of L_{n+1}-sentences T_{n+1}. It is straightforward to check that T_{n+1} is finitely satisfiable.

We put $L_\infty = \cup_n L_n$ and $T_\infty = \cup_n T_n$. These satisfy the conclusions of the proposition. $\qquad \square$

Proof of the compactness theorem. Let T be a finitely satisfiable set of L-sentences. Then we obtain L_∞ and T_∞ as in the last proposition. By Lindenbaum Theorem 2.2.2, there is a complete finitely satisfiable set of L_∞-sentences T' containing T_∞. Then T' is Henkin. The canonical model of T' is a model of T. $\qquad \square$

Exercise 2.2.5 Let L be a first-order language and $\mathcal{T}$ the set of all complete L-theories. This set of exercises defines a topology on $\mathcal{T}$ making it into a compact, Hausdorff, zero-dimensional space. For each L-sentence φ, set

$$B_\varphi = \{T \in \mathcal{T} : \varphi \in T\}$$

and $\mathcal{B} = \{B_\varphi : \varphi \text{ an } L - \text{sentence}\}$. Show the following.

1. $\mathcal{B}$ is closed under finite intersection and complementation. Thus, it is a base of a zero-dimensional, topology τ on $\mathcal{T}$.
2. Show that $(\mathcal{T}, \tau)$ is a compact, Hausdorff topological space.
3. Show that $(\mathcal{T}, \tau)$ is metrizable if the language L is countable.

There are some variants of compactness theorem which are quite useful.

Theorem 2.2.6 *For any sentence φ, $T \models \varphi$ if and only if $T' \models \varphi$ for some finite $T' \subset T$.*

Proof The if part is clear. For only if part, suppose for no finite T', $T' \models \varphi$. This implies that every finite part of $T'' = T \cup \{\neg\varphi\}$ has a model. Hence, by compactness theorem, T'' has a model, say M. But then $M \models T$ and $M \not\models \varphi$. So, $T \not\models \varphi$. □

Let L be a first-order language and Φ a set of formulas of L. Let $v_0, v_1, \ldots$ be all the variables (finitely or countably many), v_i's distinct, that has a free occurrence in a $\varphi \in \Phi$. We say that Φ is *satisfiable* if there is a structure M for L and $a_0, a_1, \ldots \in M$ such that for all $\varphi[v_0, \ldots, v_{n-1}] \in \Phi$, $M \models \varphi[\bar{a}]$. We say that Φ is finitely satisfiable if every finite $\Phi' \subset \Phi$ is satisfiable.

Proposition 2.2.7 *Every finitely satisfiable Φ is satisfiable.*

Proof Introduce in L a new constant c_i corresponding to each v_i that has a free occurrence in Φ and call the resulting language L'. Now consider

$$\Phi' = \{\varphi[\bar{c}] : \varphi[\bar{v}] \in \Phi\}.$$

Note that Φ is satisfiable if and only if Φ' has a model. By the compactness Theorem 2.2.1, it is sufficient to prove that each finite part of Φ' has a model. This follows because Φ is finitely satisfiable. □

Proposition 2.2.8 *Let M be an L-structure and Φ a set of L-formulas such that every finite $\Phi' \subset \Phi$ is satisfiable in an elementary extension of M. Then there is an elementary extension N of M in which Φ is satisfiable.*

Proof Consider $\Psi = \Phi \cup Diag_{el}(M)$. By our hypothesis, Ψ is finitely satisfiable. Hence, there is an L-structure N in which Ψ is satisfiable. Since $N \models Diag_{el}(M)$, N is an elementary extension of M. The proof is complete. □

Remark 2.2.9 Let L be a first-order language with uncountably many variables and Φ a set of L-formulas. In this case also the notion of finite satisfiability and satisfiability for Φ makes sense. Further, the last two propositions are seen to be true.

2.3 Some Consequences of Compactness Theorem

Proposition 2.3.1 *Let L be the language with constants* 0, 1, *binary function symbols* + *and* · *and a binary relation symbol* <. *Let* $\mathbb{N}$ *denote the standard model of natural numbers. There is a structure M for L elementarily equivalent to the standard model* $\mathbb{N}$ *and having an element b such that for every natural number n, n* < *b*.

Proof Introduce a new constant symbol c to $L_{\mathbb{N}}$. For each natural number m, let A_m be the formula $\underline{m} < c$. Now consider the theory

$$N' = Diag_{el}(\mathbb{N}) \cup \{A_m : m \in \mathbb{N}\}.$$

Since every finite set of natural numbers has an upper bound in $\mathbb{N}$, $\mathbb{N}$ is a model of each finite part of N'. Hence, by the compactness theorem, N' has a model M. This model has the required properties with $b = c_M$. $\qquad\square$

Proposition 2.3.2 *There is a non-Archimedean ordered field* $^*\mathbb{R}$ *elementarily equivalent to the ordered field* $\mathbb{R}$.

Proof Let L denote the language of the theory of ordered fields. Add a new constant symbol c to $L_{\mathbb{R}}$. For natural numbers n, let A_n be the formula $\underline{n} < c$ and consider

$$T = Diag_{el}(\mathbb{R}) \cup \{A_n : n \in \mathbb{N}\}.$$

Since the real line $\mathbb{R}$ is a model of each finite $T' \subset T$, by the compactness theorem, T has a model. Any model $^*\mathbb{R}$ of T does the job. $\qquad\square$

Proposition 2.3.3 *The class of all well-ordered sets is not elementary.*

Proof If possible, suppose there is a first-order theory T whose models are precisely well-ordered sets. Add to T a sequence $\{c_n\}$ of distinct and new constants and set $T' = T \cup \{c_{n+1} < c_n : n \in \omega\}$. Then, T' is finitely satisfiable. Hence, by compactness theorem, T' has a model, say M. But then $\{c_n^M\}$ is a non-empty subset of M with no least element. This is a contradiction. $\qquad\square$

Proposition 2.3.4 *The class of all fields of characteristic* 0 *is not finitely axiomatizable.*

Proof Let T be the theory of fields and φ_n denote the sentence $\underline{n} \neq 0$, $n > 1$. If possible, suppose ψ is a sentence in the language of rings such that $M \models \psi$ if and only if M is a field of characteristic 0. So, $T[\{\varphi_n : n > 1\}] \models \psi$. By compactness theorem, there is a positive integer N such that

$$T[\wedge_{i=2}^{N}\varphi_i] \models \psi.$$

Let $p > N$ be prime. It follows that $\mathbb{F}_p \models \psi$, a contradiction. $\qquad\square$

Exercise 2.3.5 Show that the class of all algebraically closed fields is not finitely axiomatizable.

(Hint: Use Proposition B.1.4)

Exercise 2.3.6 Show that the class of all archimedean ordered fields is not elementary.

Exercise 2.3.7 A graph (V, E) is called connected if for every $x \neq y \in V$, there exist $x_0, \ldots, x_n \in V$ such that $x_0 = x$, $x_n = y$ and for all $i < n$, $E[x_i, x_{i+1}]$. Show that the class of all connected graphs is not elementary.

Exercise 2.3.8 Show that the class of all torsion-free groups is not finitely axiomatizable.

Exercise 2.3.9 Show that a class $\mathcal{C}$ of L-structures is finitely axiomatizable if and only if both $\mathcal{C}$ and its complement are elementary.

Exercise 2.3.10 Let $\mathcal{F}$ denote the class of all finite fields. Call a field $\mathbb{F}$ *pseudofinite* if it is infinite and a model of $Th(\mathcal{F})$. Show that the class of all pseudofinite fields is elementary and non-empty.

Using compactness theorem we now show that every field is a subfield of an algebraically closed field. By easy algebra arguments, this will imply the existence of the algebraic closure of each field.

We shall use a standard fact from algebra. Let $\mathbb{F}$ be a field and $f(X) \in \mathbb{F}[X]$ an irreducible polynomial. Let (f) denote the ideal in $\mathbb{F}[X]$ generated by f. Then $\mathbb{F}[X]/(f)$ is a field extension of $\mathbb{F}$ in which f has a root. It then follows that given finitely many polynomials $f_1, \ldots, f_n \in \mathbb{F}[X]$ there is a field extension $\mathbb{K}$ of $\mathbb{F}$ in which each of $f_1, \ldots, f_n$ has a root.

Proposition 2.3.11 *Let $\mathbb{F}$ be a field. Then there is a field extension $\mathbb{K}$ of $\mathbb{F}$ such that every polynomial $f(X) \in \mathbb{F}[X]$ has a root in $\mathbb{K}$.*

Proof Let T denote the theory of fields in the language L of rings with identity. For each polynomial $f(X) \in \mathbb{F}[X]$ introduce a new constant symbol c_f to $L_{\mathbb{F}}$. Let φ_f be the sentence $f(c_f) = 0$ of $L_{\mathbb{F}} \cup \{c_f : f \in \mathbb{F}[X]\}$. By the above observation, each finite subset of the theory

$$T' = T \cup Diag(\mathbb{F}) \cup \{\varphi_f : f \in \mathbb{F}[X]\}$$

has a model. Hence, by compactness theorem, T' has a model, say $\mathbb{K}$. Such a $\mathbb{K}$ does our job. □

Proposition 2.3.12 *Every field is a subfield of an algebraically closed field.*

Proof Let $\mathbb{F}_0$ be a field. By repeatedly applying the last Proposition 2.3.11, we get a chain of fields $\mathbb{F}_0 \sqsubseteq \mathbb{F}_1 \sqsubseteq \mathbb{F}_2 \sqsubseteq \ldots$ such that for each n, every polynomial $f(X) \in \mathbb{F}_n[X]$ has a root in $\mathbb{F}_{n+1}$. Now take $\mathbb{K} = \cup_n \mathbb{F}_n$. □

2.4 Preservation Results

In this section, we use compactness theorem and prove several so-called preservation results.

Proposition 2.4.1 *Let N be a substructure of an L-structure M. Then N is existentially closed in M if and only if there is an extension M' of M in which N is elementarily embedded.*

Proof 'If part' is easy and was left as an exercise in Chap. 1. Assume then N is existentially closed in M. Take $T = Diag_{el}(N) \cup Diag(M)$. Sufficient to prove that T has a model. If not, then by the compactness theorem, there is an open L-formula $\varphi[\overline{x}]$ and $\overline{a} \in M$ such that $M \models \varphi[\overline{a}]$ and $Diag_{el}(N) \not\models \varphi[\overline{a}]$. Hence, there exists an elementary extension N' of N such that $\varphi[\overline{a}]$ is not satisfiable in N'. This implies that $N' \not\models \exists \overline{x}\varphi[\overline{x}]$. Hence, $N \not\models \exists \overline{x}\varphi[\overline{x}]$. This contradicts that N is existentially closed in M. $\square$

Proposition 2.4.2 *1. Let T be a first-order theory. Then $M \models T_\forall$ if and only if it is a substructure of a model N of T.*
   2. A theory T is universal if and only if every substructure of a model of T is a model of T.

Proof (1): 'If part' was given as an exercise in Chap. 1. Conversely, let $M \models T_\forall$. Set $T' = T \cup Diag(M)$. It is sufficient to show that T' is consistent. If not, then by compactness theorem, there is a finite set $\Gamma \subset Diag(M)$ such that $T[\Gamma]$ has no model. Let $\varphi_1[\overline{x}], \ldots, \varphi_n[\overline{x}]$ be open formulas and $\overline{c} \in M$ such that $\Gamma = \{\varphi_1[\overline{c}], \ldots, \varphi_n[\overline{c}]\}$. It now follows that $T[\exists \overline{x} \wedge_{i=1}^{n} \varphi_i[\overline{x}]]$ has no model. So, $T \models \forall \overline{x} \neg \wedge_{i=1}^{n} \varphi_i[\overline{x}]$. In other words, $\forall \overline{x} \neg \wedge_{i=1}^{n} \varphi_i[\overline{x}] \in T_\forall$. So, $M \models \forall \overline{x} \neg \wedge_{i=1}^{n} \varphi_i[\overline{x}]$. This contradicts that $\wedge_{i=1}^{n} \varphi_i[\overline{c}] \in Diag(M)$.

   (2) follows from (1) because T is universal if and only if T and $T_\forall$ have the same class of models. $\square$

Proposition 2.4.3 *Let T be a theory and $\varphi[\overline{x}]$ a formula. The following are equivalent:*

   1. *There is a universal formula $\psi[\overline{x}]$ such that $T \models \forall \overline{x}(\varphi[\overline{x}] \leftrightarrow \psi[\overline{x}])$.*
   2. *Whenever $M, N \models T$, $N \sqsubseteq M$ and $\overline{a} \in N$, $M \models \varphi[\overline{a}] \Rightarrow N \models \varphi[\overline{a}]$.*

Proof (1) implies (2) is easy and was given as an exercise in Chap. 1. So, assume (2). Add new constants $\overline{c}$ to the language of T and consider the theories, $T_1 = T[\varphi[\overline{c}]]$ and $T_2 = T[\neg\varphi[\overline{c}]]$. Then (2) says that no substructure of a model of T_1 can be a model of T_2. But substructures of models of T_1 are precisely models of $(T_1)_\forall$. Thus, by (2), $(T_1)_\forall \cup T_2$ is inconsistent. Since a finite conjunction of universal sentences is tautologically equivalent to a universal sentence, by compactness theorem, we get a $\psi[\overline{c}] \in (T_1)_\forall$ such that $T_2[\psi[\overline{c}]]$ has no model. It follows that

$$T[\varphi[\overline{c}]] \models \psi[\overline{c}] \ \& \ T[\neg\varphi[\overline{c}]] \models \neg\psi[\overline{c}].$$

Hence,

$$T \models \forall \overline{x}(\varphi[\overline{x}] \leftrightarrow \psi[\overline{x}]).$$

□

Proposition 2.4.4 $M \models T_{\forall\exists}$ *if and only if there is a* $N \models T$ *such that* M *is an existentially closed substructure of* N.

Proof 'If part' is easy and was given as an exercise in Chap. 1. For the converse, let $M \models T_{\forall\exists}$ and T' be the set of all universal L_M-sentences true in M.

Sufficient to show that $T \cup T'$ has a model, say N: Then $N \models T$. Since T' contains the atomic diagram of M, M has an embedding in N. Let φ be an existential L_M-sentence true in N. If possible suppose φ is not true in M. Then $\neg\varphi$, a universal L_M-sentence, is true in M. But then $N \models \neg\varphi$ which is a contradiction.

If possible, suppose $T \cup T'$ is inconsistent. By compactness theorem, there exist universal L_M-sentences $\varphi_1, \ldots, \varphi_k$ true in M such that $T \models \neg \wedge_{i=1}^{k} \varphi_i$. Since $\neg \wedge_{i=1}^{k} \varphi_i$ is equivalent to a closed existential formula, it belongs to $T_{\forall\exists}$. So, $M \models \neg \wedge_{i=1}^{k} \varphi_i$. Hence, $M \models \neg\varphi_i$ for some $1 \leq i \leq k$. This contradicts that $M \models \varphi_i$. □

A model M of a theory T is called an *existentially closed model* of T if M is existentially closed in every extension $N \sqsupseteq M$ which is a model of T.

Corollary 2.4.5 *Let* T *be a* $\forall\exists$ *theory and* $T' = T_\forall$. *Then every existentially closed model of* T' *is a model of* T.

Proof Let M be an existentially closed model of $T' = T_\forall$. By Proposition 2.4.2, there is an extension N of M that models T. Let $\forall\overline{x}\exists\overline{y}\varphi[\overline{y}, \overline{x}]$, φ open, be in $T_{\forall\exists}$. Take any $\overline{a} \in M$. Then $N \models \exists\overline{y}\varphi[\overline{y}, \overline{a}]$. Note that $M, N \models T'$. Since M is an existentially closed model of T', $M \models \exists\overline{y}\varphi[\overline{y}, \overline{a}]$. □

Corollary 2.4.6 *A theory* T *is* $\forall\exists$ *if and only if* T *is inductive.*

Proof 'Only if' part is easy and was proved in Proposition 1.5.12. So, assume that the class of models of T is closed under unions of chains. Let $M_0 \models T_{\forall\exists}$. We shall find an elementary extension M_∞ of M_0 which is a model of T. This will prove that $M_0 \models T$ and the proof will be complete.

Applying Propositions 2.4.4 and 2.4.1 alternatively, we have

$$M_0 \sqsubseteq N_0 \sqsubseteq M_1 \sqsubseteq N_1 \sqsubseteq M_2 \sqsubseteq \cdots$$

such that for each k, M_k is existentially closed in N_k, $N_k \models T$ and M_{k+1} is an elementary extension of M_k. Set $N_\infty = \cup_k N_k$ and $M_\infty = \cup_k M_k$. By our hypothesis, $N_\infty \models T$. But $M_\infty = N_\infty$. So, $M_\infty \models T$. Further, $M_0 \preceq M_\infty$. □

2.5 Extensions of Partial Elementary Maps

In this section using compactness theorem, we prove results on the extensions of partial elementary maps.

Proposition 2.5.1 *Let M, N be L-structures, $A \subset M$, $f : A \to N$ a partial elementary map and $a \in M$. Then there is an elementary extension N' of N and a partial elementary map $g : A \cup \{a\} \to N'$ that extends f. Moreover, if L, A and N are countable, we can choose N' to be countable.*

Proof Suppose $\overline{a} \in A$ and $\varphi[x, \overline{x}]$, an L-formula, is such that $M \models \varphi[a, \overline{a}]$. Then $M \models \exists x \varphi[x, \overline{a}]$. Since f is partial elementary, $N \models \exists x \varphi[x, f(\overline{a})]$. From this it is entirely routine to see that every finite subset of

$$T = Diag_{el}(N) \cup \{\varphi[x, f(\overline{a})] : \overline{a} \in A \land M \models \varphi[a, \overline{a}]\}$$

is finitely satisfiable in N. Hence, by compactness theorem, it is satisfiable. Therefore, there is an elementary extension N' of N and a $b \in N'$ such that $N' \models \varphi[b, f(\overline{a})]$ whenever $M \models \varphi[a, \overline{a}]$. Now take $g = f \cup \{(a, b)\}$.

In case L, A and N are countable, T is countable. Therefore, a countable model N' of T exists. $\square$

Applying this result repeatedly, by transfinite induction, we also have the following result.

Proposition 2.5.2 *Let M, N_0 be L-structures, $A \subset M$ and $f_0 : A \to N_0$ partial elementary. Then there exists an elementary extension N_∞ of N_0 such that f_0 can be extended to an elementary embedding $f_\infty : M \to N_\infty$. Moreover, if L, M and N_0 are countable, we can choose N_∞ to be countable.*

Proof Fix an enumeration $\{a_\alpha : \alpha < |M|\}$ of M. By transfinite induction, for each $\alpha < |M|$, we shall get an L-structure N_α and a partial elementary map $f_\alpha : A \cup \{a_\beta : \beta < \alpha\} \to N_\alpha$ satisfying the following conditions:

1. $N_{\alpha+1}$ is an elementary extension of N_α, $N_\alpha = \cup_{\beta < \alpha} N_\beta$ if α is a limit ordinal.
2. $f_{\alpha+1}$ extends f_α and $f_\alpha = \cup_{\beta < \alpha} f_\beta$ if α limit.

Suppose f_α, N_α satisfying the desired properties have been defined. If $a_\alpha \in domain(f_\alpha)$, we set $N_{\alpha+1} = N_\alpha$ and $f_{\alpha+1} = f_\alpha$. Otherwise, by the last Proposition 2.5.1, there is an elementary extension $N_{\alpha+1}$ of N_α and a partial elementary map

$$f_{\alpha+1} : A \cup \{a_\beta : \beta \leq \alpha\} \to N_{\alpha+1}$$

extending f_α. Finally take $N_\infty = \cup_{\alpha < |M|} N_\alpha$ and $f_\infty = \cup_\alpha f_\alpha$.

In case L, M and N_0 are countable, enumerate $M = \{a_n\}$ and proceed as above but choose at each stage N_n countable. $\square$

Proposition 2.5.3 *Let M, N be L-structures and $A \subset M$, $B \subset N$. Suppose f : $A \to B$ is a partial elementary map. Then f has a partial elementary extension f' : $acl(A) \to acl(B)$. Moreover, if f is surjective, we can choose f' to be surjective also.*

Proof By Zorn's lemma, there is a maximal elementary extension $f' : A' \to B'$ of f with $A' \subset acl(A)$ and $B' \subset acl(B)$. Note that $acl(A') = acl(A)$. If possible, suppose there exists $a \in acl(A') \setminus A'$. Get an L-formula $\varphi[x, \overline{y}]$, $\overline{a} \in A'$ and $n \geq 1$ such that

$$M \models \varphi[a, \overline{a}] \wedge \exists_{=n} x \varphi[x, \overline{a}].$$

Choose φ and $\overline{a}$ such that n is minimal possible. Since $f' : A' \to B'$ is partial elementary,

$$N \models \exists_{=n} x \varphi[x, f'(\overline{a})].$$

Clearly, there exists $b \in acl(B)$ such that $N \models \varphi[b, f'(\overline{a})]$.

We claim that $f' \cup \{(a, b)\}$ is partial elementary. (This will complete the proof of the first part of the result.) Let $\psi[x, \overline{y}]$ be an L-formula, $\overline{b} \in A'$ such that $M \models \psi[a, \overline{b}]$. By the minimality of n,

$$M \models \forall x (\varphi[x, \overline{a}] \to \psi[x, \overline{b}]).$$

Hence,

$$N \models \forall x (\varphi[x, f'(\overline{a})] \to \psi[x, f'(\overline{b})]).$$

Thus, $N \models \psi[b, f'(\overline{b})]$.

Now assume that f is surjective. Then $acl(B) = acl(B')$. Let $b \in acl(B)$. Since f is surjective, there exist an L-formula $\varphi[x, \overline{y}]$, an $\overline{a} \in A$ and a $n \geq 1$ such that

$$N \models \varphi[b, f(\overline{a})] \wedge \exists_{=n} x \varphi[x, f(\overline{a})].$$

Then $M \models \exists_{=n} x \varphi[x, \overline{a}]$. Let $a_1, \ldots, a_n$ be all $a \in acl(A)$ such that $M \models \varphi[a, \overline{a}]$. Since f' is defined on $acl(A)$, $b = f'(a_i)$ for some i. $\qquad\square$

2.6 Upward Löwenheim–Skolem Theorem

In Theorem 1.7.6 we proved Downward Löwenheim–Skolem Theorem which can be viewed as a method for building models of smaller cardinalities. In this section, we present a technique for building large models. First such result was proved by Tarski in 1928 who showed that every first-order theory with an infinite model has an uncountable model. The so-called Upward Löwenheim–Skolem theorem (Theorem 2.6.3) appeared in a paper by Tarski and Vaught in [63].

Proposition 2.6.1 *If a theory T has arbitrarily large finite models, it has an infinite model.*

Proof Let $\{c_n : n \in \mathbb{N}\}$ be a sequence of distinct symbols not appearing in L. Let T' be the extension of T obtained by adding each c_n as a new constant symbol and for each $m < n$, let the formula $c_n \neq c_m$ be an axiom.

Since T has arbitrarily large finite models, each finite $T'' \subset T'$ has a model. Hence, by the compactness theorem, T' has a model. Clearly, any model of T' is infinite and a model of T. □

Theorem 2.6.2 *Let κ be an infinite cardinal and T a consistent κ-theory. Assume that T has an infinite model M. Then T has a model of cardinality κ.*

Proof Fix a set $\{c_\alpha : \alpha < \kappa\}$ of cardinality κ of distinct symbols not appearing in L. Let L' be the extension of L obtained by adding each c_α as a constant symbol. Set $\Gamma = \{c_\alpha \neq c_\beta : \alpha < \beta < \kappa\}$ and consider the theory $T' = T[\Gamma]$ with language L'.

We claim that T' is finitely satisfiable. To see this, fix a finite subset Γ' of Γ. Let $c_{\alpha_1}, \ldots, c_{\alpha_k}$ be all the new constants that appear in a formula in Γ'. Since M is infinite, there exist distinct elements $b_1, \ldots, b_k$ of M. Interpret c_{α_i} by b_i, $1 \leq i \leq k$. Thus we get a model of $T[\Gamma']$. Hence, by the compactness theorem, T' has a model. Now note that any model of T' is of cardinality at least κ and a model of T.

Fix a model M of T'. By downward Löwenheim–Skolem Theorem 1.7.6, M has an elementary substructure N of cardinality at most κ. Evidently $|M| = \kappa$. □

Theorem 2.6.3 (Upward Löwenheim–Skolem theorem) *Let κ be an infinite cardinal and L a κ-language. Then every infinite structure N of L of cardinality at most κ has an elementary extension M of cardinality κ.*

Proof Note that elementary diagram $Diag_{el}(N)$ of N is a consistent κ-theory. Further, N is an infinite model of $Diag_{el}(N)$. Hence, $Diag_{el}(N)$ has a model M of cardinality κ by the last theorem. Since $M \models Diag_{el}(N)$, M is an elementary extension of N. □

Exercise 2.6.4 Show that there are structures of arbitrarily large infinite cardinality elementarily equivalent to $\mathbb{N} \models PA$.

2.7 Some Complete Theories

The following theorem was independently proved by Łoś in [37] and Vaught in [65].

Theorem 2.7.1 (Vaught's Categoricity Theorem) *Let κ be an infinite cardinal and T a consistent κ-theory all of whose models are infinite. If T is κ-categorical, T is complete.*

Proof Suppose a sentence φ is not decidable in T. The theories $T_1 = T[\varphi]$ and $T_2 = T[\neg\varphi]$ are consistent. Since T has no finite models, both T_1 and T_2 have infinite models. So, T_1 and T_2 have models M_1 and M_2 respectively of cardinality κ by Theorem 2.6.2. Hence, by the hypothesis of the theorem, they are isomorphic. But φ is true in M_1 and false in M_2 contradicting that T is κ-categorical. Hence, T is complete. $\square$

Example 2.7.2 The theory T of infinite sets is κ-categorical for every infinite cardinal κ, Hence, it is complete.

We saw in Chap. 1 that DLO is $\aleph_0$-categorical and DAG and $ACF(p)$, $p = 0$ or prime, are κ-categorical for all uncountable κ. Further, these three are countable theories with all models infinite. Hence,

Example 2.7.3 DLO, DAG and $ACF(p)$, $p = 0$ or prime, are complete theories. In particular, any two models of these theories are elementarily equivalent.

Exercise 2.7.4 Let G be an infinite group and T the theory of free G-spaces. Show that T is complete.

Exercise 2.7.5 Show that the theory of random graphs is complete.

2.8 Amalgamation

We continue with applications of compactness theorem and give quite handy conditions under which two structures have a common elementary extension.

Proposition 2.8.1 *Let A and B be elementarily equivalent L-structures. Then there is an elementary extension C of A such that there is an elementary embedding $g : B \to C$.*

Proof Let B' be an L-structure, $f : B \to B'$ an isomorphism and $A \cap B' = \emptyset$. Take

$$T = Diag_{el}(A) \cup Diag_{el}(B').$$

Let $\psi_1[\overline{b}], \ldots, \psi_n[\overline{b}] \in Diag_{el}(B')$. Then $B' \models \exists \overline{y} \wedge_{j=1}^{n} \psi_j[\overline{y}]$. Since A and B' are elementarily equivalent, $A \models \exists \overline{y} \wedge_{j=1}^{n} \psi_j[\overline{y}]$. Thus, A is a model of each finite part of T. Hence, by compactness theorem, T has a model, say C. Take $g = i \circ f$. Then $g : B \to C$ is elementary and C an elementary extension of A. $\square$

The following theorem is due to Abraham Robinson ([51], Theorem 4.2.2).

Theorem 2.8.2 (Elementary Amalgamation Theorem) *Let A and B be L-structures and $\overline{a} \in A$, $\overline{b} \in B$ be such that $(A, \overline{a})$ is elementarily equivalent to $(B, \overline{b})$. Let $\langle \overline{a} \rangle_A$ be the substructure of A generated by $\overline{a}$ and $f : \langle \overline{a} \rangle_A \to B$ the embedding such that $f(\overline{a}) = \overline{b}$. Then there is an elementary extension C of A and an elementary embedding $g : B \to C$ such that $g(f(\overline{a})) = g(\overline{b}) = \overline{a}$.*

Proof Replacing B by an isomorphic copy if necessary, without loss of generality, we assume that $A \cap B = \emptyset$. Set

$$T = Diag_{el}(A) \cup \{\varphi[\bar{a}, \bar{c}] : \varphi[\bar{b}, \bar{c}] \in Diag_{el}(B) \wedge \bar{b} \cap \bar{c} = \emptyset\}.$$

Note that since $(A, \bar{a})$ and $(B, \bar{b})$ are elementarily equivalent, $A \models \exists \bar{y} \varphi[\bar{a}, \bar{y}]$, whenever $\varphi[\bar{b}, \bar{c}] \in Diag_{el}(B)$ and $\bar{b} \cap \bar{c} = \emptyset$. Now, it is fairly routine to see that A models every finite part of T. Hence, by compactness theorem, T has a model.

Let $C \models T$. Then C is an elementary extension of A. Define $g : B \to C$ be $g(\bar{b}) = \bar{a}$ and $g(b) = b^C$, if $c \notin \bar{b}$. Then g is an elementary embedding of B into C. □

Let A, B, C and D be L-structures such that A is a common elementary substructure of B and C and B and C are elementary substructures of D. We call D a *heir-coheir amalgamation* of B and C over A or a *coheir-heir amalgamation* of C and B over A if for all L-formulas $\varphi[\bar{x}, \bar{y}]$, whenever $\bar{b} \in B, \bar{c} \in C$ and $D \models \varphi[\bar{c}, \bar{b}]$, there is an $\bar{a} \in A$ such that $B \models \varphi[\bar{a}, \bar{b}]$.

The following theorem is due to Lascar and Poizat [35].

Theorem 2.8.3 *Let A, B, C be L-structures with A a common elementary substructure of B and C. Then, there is a common elementary extension D of B and C which is a heir-coheir amalgamation of B and C over A.*

Proof Replacing B by an isomorphic copy if necessary, without loss of generality, we assume that $B \cap C = A$. Let T' be the theory

$$\{\neg \varphi[\bar{c}, \bar{b}] : \bar{b} \in B \wedge \bar{c} \in C \wedge \forall \bar{a} \in A(B \models \neg \varphi[\bar{a}, \bar{b}])\},$$

and

$$T = Diag_{el}(B) \cup Diag_{el}(C) \cup T'.$$

Clearly, it is sufficient to show that T has a model. This will follow if we show that B models every finite part of $Diag_{el}(C) \cup T'$.

Let $\bar{a} \in A, \bar{b} \in B, \bar{c} \in C \backslash A, \neg \varphi_1[\bar{a}, \bar{c}, \bar{b}], \ldots, \neg \varphi_k[\bar{a}, \bar{c}, \bar{b}] \in T'$ and $\psi[\bar{a}, \bar{c}] \in Diag_{el}(C)$. So, for all $\bar{a}', \bar{a}'' \in A$, $B \models \neg \varphi_i[\bar{a}', \bar{a}'', \bar{b}], 1 \leq i \leq k$.

Now, $C \models \psi[\bar{a}, \bar{c}]$ implies that $C \models \exists \bar{y} \psi[\bar{a}, \bar{y}]$. Hence, $A \models \exists \bar{y} \psi[\bar{a}, \bar{y}]$. So, there exists $\bar{a}'' \in A$ such that $A \models \psi[\bar{a}, \bar{a}'']$. Thus, $B \models \psi[\bar{a}, \bar{a}'']$. Clearly, $B \models \neg \varphi_i[\bar{a}, \bar{a}'', \bar{b}], 1 \leq i \leq k$. □

Remark 2.8.4 By interchanging the role of B and C in the above proof, we get a coheir-heir amalgamation of B and C over A. We shall see later that in a stable theory, every heir-coheir amalgam is a coheir-heir amalgam.

Remark 2.8.5 Let D be a heir-coheir amalgamation of B and C over A. Suppose $b \in B, c \in C$ and $D \models b = c$. Then there exists $a \in A$ such that $B \models b = a$. So, the overlap of B and C in D remains A. Such amalgamations are called *strong*.

2.9 Quantifier Elimination

In this section, we introduce yet another important technique in model theory, namely quantifier elimination. This was introduced and systematically studied by Tarski [62]. Results and examples that follow are due to him.

Let T be an L-theory. We say that T has *quantifier elimination* if for every L-formula $\varphi[\overline{x}]$ there is an open L-formula $\psi[\overline{x}]$ such that

$$T \models \forall \overline{x}(\varphi[\overline{x}] \leftrightarrow \psi[\overline{x}]).$$

Example 2.9.1 Let φ be a sentence decidable in T and the language of T have a constant symbol, say c. Then $T \models \varphi \leftrightarrow c = c$ if $T \models \varphi$, else $T \models \varphi \leftrightarrow c \neq c$.

In the rest of this section, we present some necessary and sufficient conditions for T to have quantifier elimination. Some examples of theories having quantifier elimination are given in the next section.

Proposition 2.9.2 *A theory T has quantifier elimination if and only if for every open formula $\varphi[x, \overline{y}]$, there is an open formula $\psi[\overline{y}]$ such that*

$$T \models \forall \overline{y}((\exists x \varphi[x, \overline{y}]) \leftrightarrow \psi[\overline{y}]).$$

Proof Since only if part of the result is clear, we need to prove if part only. By induction on the rank of formulas, we prove that for every formula $\varphi[\overline{x}]$ of L there is an open formula $\psi[\overline{x}]$ such that

$$T \models \forall \overline{x}(\varphi[\overline{x}] \leftrightarrow \psi[\overline{x}]). \tag{$*$}$$

$(*)$ is clearly true for open φ. It is easy to prove that if $(*)$ is true for φ, it is true for $\neg \varphi$. If $(*)$ holds for $\varphi = \varphi_1$ and $\varphi = \varphi_2$, it holds for $\varphi_1 \vee \varphi_2$.

To complete the proof, assume that $(*)$ holds for $\varphi[x, \overline{y}]$. Get an open formula $\eta[x, \overline{y}]$ such that

$$T \models \forall x \forall \overline{y}(\varphi[x, \overline{y}] \leftrightarrow \eta[x, \overline{y}]).$$

This implies that

$$T \models \forall \overline{y}((\exists x \varphi[x, \overline{y}]) \leftrightarrow \exists x \eta[x, \overline{y}]).$$

By our hypothesis, there is an open formula $\psi[\overline{y}]$ such that

$$T \models \forall \overline{y}((\exists x \eta[x, \overline{y}]) \leftrightarrow \psi[\overline{y}]).$$

Now it is clear that

$$T \models \forall \overline{y}((\exists x \varphi[x, \overline{y}]) \leftrightarrow \psi[\overline{y}]).$$

Our proof is complete.   □

Theorem 2.9.3 *Let T be a theory with a constant symbol c and $\varphi[\overline{x}]$ a formula of T. The following are equivalent:*

(1) *There is an open formula $\psi[\overline{x}]$ such that*

$$T \models \forall \overline{x}(\varphi[\overline{x}] \leftrightarrow \psi[\overline{x}]). \qquad (*)$$

(2) *For any two models $M, N \models T$, for any common substructure A of M, N and for any $\overline{a} \in A$,*
$$M \models \varphi[\overline{a}] \Leftrightarrow N \models \varphi[\overline{a}].$$

(3) *For any two models $M, N \models T$, for any common finitely generated substructure A of M, N and for any $\overline{a} \in A$,*

$$M \models \varphi[\overline{a}] \Leftrightarrow N \models \varphi[\overline{a}].$$

Proof (1) implies (2): Take M, N, A and $\overline{a}$ as in (2). By (1), there is an open formula $\psi[\overline{x}]$ such that $T \models \forall \overline{x}(\varphi(\overline{x}) \leftrightarrow \psi(\overline{x}))$. So,

$$M \models \varphi(\overline{a}) \Leftrightarrow M \models \psi[\overline{a}]$$

and

$$N \models \varphi(\overline{a}) \Leftrightarrow N \models \psi[\overline{a}].$$

But A being a common substructure of M and N, since $\overline{a} \in A$ and ψ is open,

$$M \models \psi(\overline{a}) \Leftrightarrow A \models \psi(\overline{a}) \Leftrightarrow N \models \psi(\overline{a}).$$

Hence,

$$M \models \varphi(\overline{a}) \Leftrightarrow N \models \varphi(\overline{a}).$$

(3) is a special case of (2).

(3) implies (1): Assume that $\varphi[\overline{x}]$ satisfies (3). When a closed formula φ satisfies (3), φ is either true in all models or in none. Now note that $T \models \varphi \leftrightarrow c = c$ if $T \models \varphi$. Otherwise $T \models \neg\varphi$ when $T \models \varphi \leftrightarrow c \neq c$. The same argument works when $\varphi[\overline{x}]$ is not closed but decidable in T, i.e. $\forall \overline{x}\varphi[\overline{x}]$ is decidable in T.

It remains to prove the result in case both $T[\varphi[\overline{x}]]$ and $T[\neg\varphi[\overline{x}]]$ are satisfiable. Introduce new constants $\overline{c}$ to the language to get a new language, say L'. Let T' be the new theory whose language is L' but no new nonlogical axiom. Consider

$$\Gamma = \{\psi[\overline{c}] : T' \models \varphi[\overline{x}] \rightarrow \psi[\overline{x}], \psi \text{ open}\}.$$

We first see that it is sufficient to prove that

$$T'[\Gamma] \models \varphi[\overline{c}]. \qquad (*)$$

Then by compactness theorem, there exist $\psi_0[\overline{c}], \ldots, \psi_{n-1}[\overline{c}] \in \Gamma$ such that

$$T' \models \wedge_{i<n} \psi_i[\overline{c}] \rightarrow \varphi[\overline{c}].$$

Since $\overline{c}$ are new constants, it follows that

$$T \models \forall \overline{x}(\varphi[\overline{x}] \leftrightarrow \wedge_{i<n} \psi_i[\overline{x}])$$

and $\wedge_{i<n} \psi_i[\overline{x}]$ is open.

We prove $(*)$ by contradiction. So, assume that

$$T'[\Gamma] \not\models \varphi[\overline{c}].$$

Let

$$M \models T'[\Gamma] \cup \{\neg\varphi[\overline{c}]\}.$$

Let A be the substructure of M generated by $\overline{c}^M$. So A is finitely generated. Now consider

$$\Delta = T \cup Diag(A) \cup \{\varphi[\overline{c}]\}.$$

We claim that Δ has a model. If not, then by compactness theorem, there exist $\psi_1[\overline{c}], \ldots, \psi_n[\overline{c}] \in Diag(A)$ such that

$$T' \models \wedge_{i=1}^{n} \psi_i[\overline{c}] \rightarrow \neg\varphi[\overline{c}].$$

Since $\overline{c}$ are new constants,

$$T \models \wedge_{i=1}^{n} \psi_i[\overline{x}] \rightarrow \neg\varphi[\overline{x}].$$

Set $\psi[\overline{x}] = \neg \wedge_{i=1}^{n} \psi_i[\overline{x}]$. Note that ψ is open. We have,

$$T \models \varphi[\overline{x}] \rightarrow \psi[\overline{x}].$$

Thus, $\psi[\overline{c}] \in \Gamma$. Hence, $M \models \psi[\overline{c}]$. Since ψ is open and $\overline{c}^M \in A$, $A \models \psi[\overline{c}]$, contradicting that $\psi_1[\overline{c}], \ldots, \psi_n[\overline{c}] \in Diag(A)$.

Now take a model $N \models \Delta$. By the Atomic diagram Theorem 1.5.13, A is a substructure of N. But $M \models \neg\varphi[\overline{c}]$ and $N \models \varphi[\overline{c}]$. This contradicts (3) and proves $(*)$.
$\square$

Since every open formula is equivalent to an open formula in disjunctive normal form (DNF), we now easily see that

Proposition 2.9.4 *Let T be a theory with a constant. The following are equivalent:*

(1) T has quantifier elimination.

(2) *For every conjunction of literals $\varphi[x, \overline{y}]$, for any two models $M, N \models T$, for every common substructure A of M, N and for every $\overline{a} \in A$, if there is a $b \in M$ such that $M \models \varphi[b, \overline{a}]$, there is a $c \in N$ such that $N \models \varphi[c, \overline{a}]$.*

The simple proof of this result is left to the reader as a simple exercise.

Let T be an L theory, $M, N \models T$, $A \subset M$ and $B \subset N$. A map $f : A \to B$ is called a *partial isomorphism* if f is onto and for every atomic L-formula $\varphi[\overline{x}]$ and every $\overline{a} \in A$,

$$M \models \varphi[\overline{a}] \Leftrightarrow N \models \varphi[f(\overline{a})]. \tag{$*$}$$

It is easy to see that every partial isomorphism $f : A \to B$ is a bijection and for every open L-formula $\varphi[\overline{x}]$ and every $\overline{a} \in A$,

$$M \models \varphi[\overline{a}] \Leftrightarrow N \models \varphi[f(\overline{a})].$$

If, moreover, $(*)$ is satisfied for every formula $\varphi[\overline{x}]$ and every $\overline{a} \in A$, we call f a *partial elementary*. In the next chapter, we shall study partial elementary maps in detail.

Theorem 2.9.5 *A theory T has quantifier elimination if and only if for every pair of models M, N of T every finite partial isomorphism $M \ni \overline{a} \to \overline{b} \in N$ is partial elementary.*

Proof The only if part of the result being clear, we prove the if part only. Take an L-formula $\varphi[\overline{x}]$, $\overline{x} = (x_0, \ldots, x_{n-1})$.

$$\Gamma[\overline{x}] = \{\psi[\overline{x}] : \psi[\overline{x}] \text{ an open } L - \text{formula \& } T \models \forall \overline{x}(\varphi[\overline{x}] \to \psi[\overline{x}])\}.$$

Add new constants $c_0, \ldots, c_{n-1}$ and consider

$$\Gamma[\overline{c}] = \{\psi[\overline{c}] : \psi[\overline{x}] \in \Gamma[\overline{x}]\}.$$

Claim. $T[\Gamma[\overline{c}]] \models \varphi[\overline{c}]$.

Assuming the claim, we complete the proof first. Since $\Gamma[\overline{c}]$ is closed under conjunctions, by compactness theorem, there is a $\psi[\overline{c}] \in \Gamma[\overline{c}]$ such that $T[\psi[\overline{c}]] \models \varphi[\overline{c}]$. It follows that

$$T \models \forall \overline{x}(\psi[\overline{x}] \to \varphi[\overline{x}]).$$

But we already have

$$T \models \forall \overline{x}(\varphi[\overline{x}] \to \psi[\overline{x}]).$$

Hence,

$$T \models \forall \overline{x}(\varphi[\overline{x}] \leftrightarrow \psi[\overline{x}]).$$

Proof of the claim. Suppose the claim does not hold. Then there exists a $M \models T[\Gamma[\overline{c}]] \cup \{\neg\varphi[\overline{c}]\}$. Let $a_i = c_i^M$, $i < n$. Set

$$p[\overline{x}] = \{\xi[\overline{x}] : \xi[\overline{x}] \text{ an open } L - \text{formula } \& M \models \xi[\overline{a}]\}.$$

Then $T \cup p[\overline{x}] \cup \{\varphi[\overline{x}]\}$ is satisfiable: If not, then it is not finitely satisfiable. Hence, there is a formula $\xi[\overline{x}] \in p[\overline{x}]$ such that

$$T \models \forall \overline{x}(\varphi[\overline{x}] \to \neg\xi[\overline{x}]).$$

This forces $\neg\xi[\overline{x}] \in \Gamma[\overline{x}]$ which is a contradiction.

Thus, there exist $N \models T$ and $\overline{b} \in N$ such that $N \models \varphi[\overline{b}]$ and for every open formula $\xi[\overline{x}]$,

$$M \models \xi[\overline{a}] \Leftrightarrow N \models \xi[\overline{b}].$$

Since $M \models \neg\varphi[\overline{a}]$, we have arrived at a contradiction. $\square$

We close this section by giving an application of partial elementary maps to quantifier elimination. Let M be a model of a theory T and $A \sqsubseteq M$. We say that M is *prime over A* or that M is a *prime model extension* of A if for every model N of T and every partial elementary map $h : A \to N$, there is an elementary embedding $g : M \to N$ such that $h = g|A$. We say that T has *algebraically prime models* if every model A of $T_\forall$ has an extension $M \models T$ such that M is prime over A. Recall that $A \models T_\forall$ if and only if it has an extension to a model of T (Proposition 2.4.2).

Example 2.9.6 Consider the theory ACF of algebraically closed fields. Let D be an integral domain and $\mathbb{F}$ the algebraic closure of the fraction field of D. We know that given any $\mathbb{K} \models ACF$ and a partial elementary map $h : D \to \mathbb{K}$ (an embedding, in particular), there is an embedding $g : \mathbb{F} \to \mathbb{K}$ such that $h = g|D$. Since ACF has quantifier elimination, g is elementary.

Example 2.9.7 Consider the theory $RCOF$ of real closed fields. Let D be an ordered integral domain and $\mathbb{F}$ the real closure of the ordered fraction field of D. We know that given any $\mathbb{K} \models RCF$ and an elementary map $h : D \to \mathbb{K}$, there is an embedding $g : \mathbb{F} \to \mathbb{K}$ such that $h = g|D$. Since $RCOF$ has quantifier elimination, g is elementary.

Example 2.9.8 Consider the theory DLO of dense linearly ordered sets with no end points. Let $(A, <)$ be a linearly ordered sets. We define a dense linearly ordered set A^* as follows: If A has a least element, say x, add a copy of $\mathbb{Q}$ with the usual order to the left of x, if A has a greatest element, say y, add a copy of $\mathbb{Q}$ with the usual order to the right of y and if $x < y$ are two elements of A with no element in between, add a copy of $\mathbb{Q}$ with the usual order between x and y. There is a canonical inclusion map $f : A \hookrightarrow A^*$. Now given any $B \models DLO$ and a partial elementary map $h : A \to B$, it is easy to define an embedding $g : A^* \to B$ such that $h = g \circ f$. Since DLO has quantifier elimination, g is elementary.

We leave the proof of following theorem for readers as an exercise:

Theorem 2.9.9 *Let T be a theory such that*

1. *T has algebraically prime models, and*
2. *for any two $M, N \models T$ with $M \sqsubseteq N$, for any conjunction of literals $\varphi[x, \overline{y}]$ and for every $\overline{a} \in M$,*

$$N \models \exists x \varphi[x, \overline{a}] \Rightarrow M \models \exists x \varphi[x, \overline{a}].$$

Then T has quantifier elimination.

Exercise 2.9.10 Show that the theory T of vector spaces over a fixed field has quantifier elimination.

2.10 Examples of Quantifier Elimination

In the following examples, we use Proposition 2.9.4 without mentioning it.

Example 2.10.1 The theory DLO of dense linear orders without end points has quantifier elimination.

Proof Let $\varphi[x, \overline{y}]$ be a conjunction of literals. For instance, suppose

$$\varphi[x, \overline{y}] = y_1 < \cdots < y_{i-1} < x < y_i < \cdots < y_n.$$

Suppose $M, N \models DLO$, A is a common substructure of $M, N, \overline{a} \in A$ and there is a $b \in M$ satisfying

$$a_1 < \cdots < a_{i-1} < b < a_i < \cdots < a_n.$$

This, in particular, implies that

$$a_1 < \cdots < a_{i-1} < a_i < \cdots < a_n.$$

Since $N \models DLO$, there is a $c \in N$ such that

$$a_1 < \cdots < a_{i-1} < c < a_i < \cdots < a_n.$$

Cases when $\varphi[x, \overline{y}]$ is "$x < y_1 < \cdots < y_n$" or "$y_1 < \cdots < y_n < x$" are dealt with similarly because N has no end points.   □

Example 2.10.2 The theory DAG of torsion-free divisible abelian groups has quantifier elimination.

Proof We take $G_1, G_2 \models DAG$, a common subgroup $H \subset G_1, G_2$. Let $\varphi[x, \overline{y}]$ be a conjunction of literals. Suppose $\overline{a} \in H$. Replacing H by its divisible hull considered as a common subgroup of both G_1 and G_2, we further assume that H too is divisible. Now $\varphi[x, \overline{y}]$, being a conjunction of literals, it can be assumed to be of the form

$$\wedge_{i=0}^{k-1} \sum_{j=1}^{m_i}(n_{ij}y_j + n_i x = 0) \wedge \wedge_{p=0}^{l-1} \sum_{j=1}^{r_p}(n'_{pj}y_j + n'_p x \neq 0)). \qquad (*)$$

Assume that there is a $b \in G_1$ such that

$$G_1 \models \varphi[b, \overline{a}].$$

We need to show that there is a $c \in G_2$ such that

$$G_2 \models \varphi[c, \overline{a}].$$

Since H is a substructure of G_2, it is sufficient to show that there is such a c in H.
If any $n_i \neq 0$, as H is divisible,

$$b = -\frac{\sum_{j=1}^{m_i} n_{ij}a_j}{n_i} \in H$$

and we are done. So, assume that all $n_i = 0$. Then b disappears from the equalities appearing in $(*)$. Since H is infinite, we can certainly find a $c \in H$ satisfying all inequalities in $(*)$. $\qquad\square$

Example 2.10.3 The theory $ODAG$ of ordered divisible abelian groups has quantifier elimination.

Proof As in the above case, we take ordered divisible abelian groups G_1 and G_2, a common subgroup H, a conjunction of literals $\varphi[x, \overline{y}]$ and an $\overline{a} \in H$. Assume that there is a $b \in G_1$, such that $G_1 \models \varphi[b, \overline{a}]$. Again, as in the last example, it is sufficient to show that if H' is the ordered divisible hull of H, there is a $c \in H'$ such that $H' \models \varphi[c, \overline{a}]$. Towards showing this, note that we can assume that $\varphi[x, \overline{y}]$ is of the form

$$\wedge_{i=0}^{k-1} \sum_{j=1}^{m_i}(n_{ij}y_j + n_i x = 0) \wedge \wedge_{p=0}^{l-1}(\sum_{j=1}^{r_p} n'_{pj}y_j < n'_p x).$$

Since H' is order-dense, arguing as in the last example, we get a required $c \in H'$. $\square$

Example 2.10.4 Let $\mathbb{K}$ be a field. Then the theory T of infinite vector spaces over $\mathbb{K}$ has quantifier elimination.

Proof Let $V_1, V_2 \models T$ and V be a common subspace of V_1 and V_2. Let $\varphi[x]$ be an open L_V-formula and there exists an $a \in V_1$ such that $V_1 \models \varphi[a]$. We need to show that $V_2 \models \exists x \varphi[x]$.

If $a \in V$, since V is a substructure of V_1 and V_2 and φ open, $V_2 \models \varphi[a]$. Next assume that $a \notin V$. If $V = V_2$, since V_2 is infinite, it has a proper elementary extension, say V_2'. If $V_2' \models \exists x \varphi[x]$, $V_2 \models \exists x \varphi[x]$. Hence, without any loss of generality, we assume that $V \neq V_2$. Let $b \in V_2 \backslash V$. Set $L_1 = span(V \cup \{a\})$ and $L_2 = span(V \cup \{b\})$. There is a linear isomorphism $f : L_1 \to L_2$ fixing V pointwise and $f(a) = b$. This implies that $L_2 \models \varphi[b]$. Since φ is open, $V_2 \models \varphi[b]$. □

Example 2.10.5 The theory ACF of algebraically closed fields has quantifier elimination.

Proof Note that a substructure of a field is an integral domain. Also, recall that if D is an integral domain, its quotient field embeds into every field in which D is embedded. Therefore, as in the last two cases, we only need to show that whenever $\mathbb{F} \subset \mathbb{K}$ are algebraically closed fields, $\varphi[x, \overline{y}]$ a conjunction of literals and $\overline{a} \in \mathbb{F}$, if there is a $b \in \mathbb{K}$ such that $\mathbb{K} \models \varphi[b, \overline{a}]$, there is a $c \in \mathbb{F}$ such that $\mathbb{F} \models \varphi[c, \overline{a}]$. Now note that we can take $\varphi[x, \overline{a}]$ in the form

$$\wedge_{i=0}^{k-1}(P_i(x) = 0) \wedge \wedge_{j=0}^{l-1}(Q_j(x) \neq 0),$$

$P_i[X]$'s and $Q_j[X]$'s are polynomials over the smallest subfield of $\mathbb{F}$ generated by $\overline{a}$. If $k \geq 1$, $b \in \mathbb{F}$ because it is algebraically closed. Otherwise, since $\mathbb{F}$ is infinite, it certainly has a c which is not a root of any $Q_j[X]$ which works for us. □

It is interesting to ask if the converse of Proposition 1.9.17 is true? We shall come back to this question later.

Corollary 2.10.6 *Let $\mathbb{K}$ be an algebraically closed field and $A \subset \mathbb{K}$. Then $a \in acl(A)$ if and only if a is algebraic in usual algebra sense over the subfield k generated by A.*

Proof Let $a \in acl(A)$. By quantifier elimination and the fact that every open formula is equivalent to a formula in disjunctive normal form, there exist polynomial terms $p_i(x, \overline{y})$, $i < n$, $q_j(x, \overline{y})$, $j < m$, and $\overline{a} \in A$ such that

$$\wedge_i p_i(a, \overline{a}) = 0 \wedge \wedge_j q_j(a, \overline{a}) \neq 0,$$

and that this equation has only finitely many solutions. But then $n > 0$. Hence a is algebraic over k. If part is straight forward. □

Exercise 2.10.7 Let $G \models DAG$ and $A \subset G$. Show that $acl(A) = dcl(A)$ and it equals the smallest divisible subgroup of G generated by A.

Exercise 2.10.8 Let V be an infinite vector space over a field $\mathbb{K}$ and $A \subset V$. Show that $acl(A) = dcl(A)$ and it equals the vector subspace of V generated by A.

Example 2.10.9 The theory $RCOF$ of real closed fields has quantifier elimination.

Proof As in the cases of say $ODAG$ and ACF etc. we only need to show that if $\varphi[x, \overline{y}]$ is a conjunction of literals, $\mathbb{F} \subset \mathbb{K} \models RCOF$ and $\overline{a} \in \mathbb{F}$, then

$$\mathbb{K} \models \exists x \varphi[x, \overline{a}] \Rightarrow \mathbb{F} \models \exists x \varphi[x, \overline{a}].$$

We can assume that $\varphi[x, \overline{y}]$ is of the form

$$\wedge_{i=1}^{n}(p_i(x, \overline{y}) = 0) \wedge \wedge_{j=1}^{m}(q_j(x, \overline{y}) > 0),$$

with p_i, q_j being terms.

Choose a $b \in \mathbb{K}$ such that

$$\mathbb{K} \models \varphi[b, \overline{a}].$$

If any of the equality term is present, since $\mathbb{F}$ has no proper real algebraic extension (Theorem B.3.10), $b \in \mathbb{F}$.

So, assume no p_i is present. Since $\mathbb{F}$ has no proper real algebraic extension, roots of q_j's, if any, belong to $\mathbb{F}$. If a q_j has no root in the field and since $q_j(b, \overline{a}) > 0$, by Weierstrass Nullstellensatz (Theorem B.3.9), $q_j(c, \overline{a}) > 0$ for all $c \in \mathbb{F}$. By considering finitely many roots of all q_j's (all of which belong to $\mathbb{F}$), we find a non-empty open interval I in $\mathbb{K}$ with end points in $\mathbb{F}$ such that $b \in I$ and $q_j(x, \overline{a}) > 0$ for all $x \in I$ and for all $1 \leq j \leq m$. Using the order-denseness of $\mathbb{F}$, we have a $b \in \mathbb{F}$ that lies in I. This b witnesses $\mathbb{F} \models \varphi[b, \overline{a}]$. □

Exercise 2.10.10 Show that the theory of random graphs has quantifier elimination and it is complete.

Exercise 2.10.11 Let $\mathbb{K}$ be a field. Show that the theory of infinite vector spaces over $\mathbb{K}$ is complete.

2.11 Strongly Minimal and O-Minimal Theories

As a consequence of the fact that ACF has quantifier elimination, we get

Proposition 2.11.1 *Let $\mathbb{F}$ be an algebraically closed field. Then $\mathbb{F}$ is infinite and $D \subset \mathbb{F}$ is definable if and only if D is either finite or cofinite in $\mathbb{F}$.*

Proof Note that a subset D of $\mathbb{F}$ is defined by an atomic formula if and only if it is the set of all roots of a polynomial in $\mathbb{F}$. Hence, such a set $D \subset \mathbb{F}$ is finite. Boolean algebra of subsets of $\mathbb{F}$ generated by all finite sets consists of all finite and cofinite sets. These are precisely sets defined by open formulas. Our claim is followed by Example 2.10.5. □

The same argument shows the following.

Proposition 2.11.2 *Let G be a torsion-free divisible abelian group. Then G is infinite and $D \subset G$ is definable if and only if D is either finite or cofinite in G.*

Corollary 2.11.3 $\mathbb{R}$ *is not a definable subset of the field* $\mathbb{C}$ *of complex numbers.*

Corollary 2.11.4 $\mathbb{Z}$ *and* $\mathbb{N}$ *are not definable subsets of the group* $\mathbb{Q}$ *of rational numbers.*

Remark 2.11.5 In a remarkable discovery, J. Robinson produced a formula $\varphi[x]$ in the language of rings such that for a rational number r,

$$\mathbb{Q} \models \varphi[r] \Leftrightarrow r \in \mathbb{N}.$$

(See [13, 52]).

Let M be an L-structure and $A \subset M^n$. We call A *minimal* if A is infinite and if for every L_M-formula $\varphi[\overline{x}]$ either $A \cap \varphi(M)$ or $A \backslash \varphi(M) = A \cap \neg\varphi(M)$ is finite. Thus, M is a minimal structure if and only if M is infinite and every definable subset of M is either finite or cofinite in M.

An L_M-formula $\varphi[\overline{x}]$ is called *minimal in* M if $\varphi(M)$ is minimal; it is called *strongly minimal in* M if φ is minimal in every elementary extension of M.

A theory T is called *strongly minimal* if every $M \models T$ is minimal. It follows that if T is strongly minimal, every model of T is strongly minimal. Whatever may be the language L, clearly all finite subsets and their complements in an L-structure M is definable. Thus definable subsets of models of a strongly minimal theory have simplest possible structure. This notion was introduced by Marsh [42]. Its importance was shown by Baldwin and Lachlan to give a simpler proof of Morley categoricity theorem [5].

Example 2.11.6 ACF and DAG are strongly minimal.

Remark 2.11.7 Consider the theory RCF of real closed fields (without order relation). The field of real numbers $\mathbb{R}$ is a model of it. We also have

$$x \geq 0 \Leftrightarrow \exists y(x = y \cdot y).$$

This shows that the real closed field $\mathbb{R}$ is not minimal. Hence, RCF does not admit quantifier elimination.

Exercise 2.11.8 Show that the theory T of vector spaces over a fixed field is strongly minimal.

Proposition 2.11.9 *Let M be an L-structure and $\varphi[\overline{x}]$ an L-formula. The following conditions are equivalent:*

1. φ *is strongly minimal in* M.
2. φ *is minimal in every structure* N *which is elementarily equivalent to* M.

Proof Since every elementary extension of M is elementarily equivalent to M, clearly (2) implies (1).

Now assume (1) and let N be elementarily equivalent to M. By Proposition 2.8.1, there exists an elementary extension A of M and an elementary embedding $g :$ $N \to A$. Let $\psi[\overline{x}, \overline{y}]$ be an L-formula and $\overline{a} \in N$. By (a), either $\varphi(A) \cap \psi(A, g(\overline{a}))$ or $\varphi(A) \cap \neg\psi(A, g(\overline{a}))$ is finite. Since g is elementary, either $\varphi(N) \cap \psi(N, \overline{a})$ or $\varphi(N) \cap \neg\psi(N, \overline{a})$ is finite. Thus, (1) implies (2). $\qquad \square$

Let $(X, <)$ be a linearly ordered set. An interval in X is a subset I of X such that whenever $x \leq y$ are in I and $x \leq z \leq y, z \in I$.

Here is a very important class of theories. Let T be a theory whose language has a binary relation symbol $<$ such that for every $M \models T$, $<^M$ is a linear order on M. We call T *O-minimal* if for every $M \models T$, $D \subset M$ is definable if and only if D is a finite union of intervals. Here 'O' stands for order. This concept was defined by Pillay and Steinhorn in [48, 49]. Today O-minimality is a major tool in geometry.

Since the theories DLO, $ODAG$ and $RCOF$ have quantifier elimination, we have the following example.

Example 2.11.10 Theories DLO, $ODAG$ and $RCOF$ are O-minimal.

Example 2.11.11 The theory $ODAG$ of ordered abelian groups has definable Skolem functions. To see this, let $\varphi[\overline{x}, y]$ be a formula. By quantifier elimination, we know that "$\{y : \varphi[\overline{x}, y]\}$ is a finite union of intervals and singletons." We define $\psi[\overline{x}, y]$ as the disjunction of following formulas:

$$(\forall z \neg\varphi[\overline{x}, z] \vee \forall z \varphi[\overline{x}, z]) \wedge y = 0,$$

$$\exists z (\forall u < z \varphi[\overline{x}, u] \wedge \forall w > z \exists v < w \neg\varphi[\overline{x}, v] \wedge y = z - 1),$$

$$\exists z (\forall u > z \varphi[\overline{x}, u] \wedge \forall w < z \exists v > w \neg\varphi[\overline{x}, v] \wedge y = z + 1),$$

$$\exists z_1, z_2 (\forall u < z_1 \neg\varphi[\overline{x}, u] \wedge ((\forall z_1 < u < z_2 \varphi[\overline{x}, u]) \vee (z_1 = z_2 \wedge \varphi[\overline{x}, z_1]))$$

$$\wedge \forall v > z_2 \exists z_2 \leq u < v \neg\varphi[\overline{x}, u] \wedge y = \frac{z_1 + z_2}{2}).$$

Then

$$ODAG \models \forall \overline{x} (\exists_{=1} y \psi[\overline{x}, y] \wedge (\exists y \varphi[\overline{x}, y] \to \forall y (\psi[\overline{x}, y] \to \varphi[\overline{x}, y]))).$$

Further, $\psi[\overline{x}, y]$ defines a function F whose graph is the set defined by $\psi[\overline{x}, y]$. We also have

$$ODAG \models \forall \overline{x} \forall \overline{x'} (\forall y (\varphi[\overline{x}, y] \leftrightarrow \varphi[\overline{x'}, y]) \to F(\overline{x}) = F(\overline{x'})).$$

In this sense, we call F invariant.

Example 2.11.12 We extend the idea contained in the last Example further. Let $\varphi[\overline{x}, \overline{y}]$ be a formula of $ODAG$. By induction on the arity n of $\overline{y}$, we show that there exists an invariant definable function $\overline{y} = F(\overline{x})$ such that

$$ODAG \models \forall \overline{x}(\exists \overline{y}\varphi[\overline{x}, \overline{y}] \to \varphi[\overline{x}, F(\overline{x})]).$$

For $n = 1$, this is done above.

For inductive step, take a formula $\varphi[\overline{x}, y_1, \dots, y_{n+1}]$. By induction hypothesis, there exists an invariant, definable Skolem function $f(\overline{x}, y_1)$ such that

$$ODAG \models \forall y_1 \forall \overline{x}(\exists y_2 \dots \exists y_{n+1}\varphi[\overline{x}, \overline{y}] \to \varphi[\overline{x}, y_1, f(\overline{x}, y_1)]).$$

By case $n = 1$, there exists an invariant definable Skolem function $g(\overline{x})$ such that

$$ODAG \models \forall \overline{x}(\exists y_1 \varphi[\overline{x}, y_1, f(\overline{x}, y_1)] \to \varphi[\overline{x}, g(\overline{x}), f(\overline{x}, g(\overline{x}))]).$$

Now take

$$F(\overline{x}) = (g(\overline{x}), f(\overline{x}, g(\overline{x}))).$$

Then $F(\overline{x})$ is an invariant function such that

$$ODAG \models \forall \overline{x}(\exists \overline{y}\varphi[\overline{x}, \overline{y}] \to \varphi[\overline{x}, F(\overline{x})]).$$

Further note that if $\varphi[\overline{x}, \overline{y}]$ is an equivalence formula, then $F(\overline{x})$ is a definable section of φ.

Example 2.11.13 Exactly the same arguments as in the last two examples show that $RCOF$ has definable Skolem functions. Further, since we can introduce $<$ in an extension by definition of RCF, we see that RCF too has definable Skolem functions.

Example 2.11.14 By Theorem 1.12.3 it follows that $ODAG$, $RCOF$ and RCF admit uniform elimination of imaginaries.

The theory of algebraically closed fields ACF also admits uniform elimination of imaginaries. However, it requires considerable work. This will be proved in Sect. 4.3.

2.12 Independence and Dimension in Minimal Sets

In this section, we generalise the notions of independence and basis to models of strongly minimal theories.

Theorem 2.12.1 (Exchange Lemma) *Let M be an L-structure, $A \subset M$ and X an L_A-definable minimal set. Let $a, b \in X$ be such that $b \in acl(A \cup \{a\}) \setminus acl(A)$. Then $a \in acl(A \cup \{b\})$.*

Proof If possible, suppose there exists $b \in acl(A \cup \{a\}) \setminus acl(A)$ such that $a \notin acl(A \cup \{b\})$. We shall arrive at a contradiction.

Since $b \in acl(A \cup \{a\})$, there exists an L_A-formula $\varphi[x, y]$ and $n \geq 1$ such that

$$M \models \varphi[b, a] \wedge \exists_{=n} x \varphi[x, a].$$

Since X is minimal A-definable and $a \in X \setminus acl(A \cup \{b\})$, there exists a finite set $Y \subset M$ such that for all $c \in X \setminus Y$,

$$M \models \varphi[b, c] \wedge \exists_{=n} x \varphi[x, c].$$

Let $\psi[y]$ be an L_A-formula that defines X and $|Y| = m$. We have

$$M \models \exists y_1 \ldots \exists y_m (\forall y ((\wedge_i (y \neq y_i) \wedge \psi[y]) \rightarrow (\varphi[b, y] \wedge \exists_{=n} x \varphi[x, y]))).$$

Since $b \notin acl(A)$, there exists an infinite set $Z \subset M$ such that for all $b' \in Z$,

$$M \models \exists y_1 \ldots \exists y_m (\forall y ((\wedge_i (y \neq y_i) \wedge \psi[y]) \rightarrow (\varphi[b', y] \wedge \exists_{=n} x \varphi[x, y]))).$$

Take distinct elements $b_0, \ldots, b_n \in Z$. Then there exists a $c \in X$ such that

$$M \models \wedge_{i=0}^n \varphi[b_i, c] \wedge \exists_{=n} x \varphi[x, c].$$

This is a contradiction. $\qquad\qquad\qquad\qquad\qquad\qquad\qquad\qquad\qquad\quad$ $\square$

We say that $A \subset M$ is *independent* if for every $a \in A$, $a \notin acl(A \setminus \{a\})$. If $C \subset M$, we say that A is *independent over* C if for every $a \in A$, $a \notin acl(C \cup (A \setminus \{a\}))$. This, in particular, implies that $A \cap C = \emptyset$. A subset B of A is called a *basis* of A if B is independent and $acl(B) = acl(A)$. Equivalently, B is a maximal independent subset of A.

Proposition 2.12.2 *Let X be an $\emptyset$-definable minimal subset of an L-structure M and A, B independent subsets of X with $A \subset acl(B)$. Then*

1. *Let $A_0 \subset A$, $B_0 \subset B$ and $A_0 \cup B_0$ a basis for $acl(B)$. Then for every $a \in A \setminus A_0$, there is a $b \in B_0$ such that $A_0 \cup \{a\} \cup (B_0 \setminus \{b\})$ is a basis of $acl(B)$.*
2. $|A| \leq |B|$.
3. *For every $Y \subset X$, any two bases of Y have the same cardinality.*

Proof Let $C \subset B_0$ be a set of minimum cardinality such that $a \in acl(A_0 \cup C)$. Since A is independent, $C \neq \emptyset$. Take a $b \in C$. Because C is of minimum possible cardinality,

$$a \in acl(A_0 \cup C) \setminus acl((A_0 \cup C) \setminus \{b\}).$$

Therefore, by exchange lemma (Theorem 2.12.1),

$$b \in acl((A_0 \cup \{a\}) \cup (C \setminus \{b\})).$$

Hence,

$$acl(B) = acl((A_0 \cup \{a\}) \cup (B_0 \setminus \{b\})).$$

We claim that $a \notin acl(A_0 \cup (B_0 \setminus \{b\}))$. For otherwise, $b \in acl(A_0 \cup (B_0 \setminus \{b\}))$ which contradicts that $A_0 \cup B_0$ is a basis of $acl(B)$. Using exchange lemma (Theorem 2.12.1) it is easy to see that $(A_0 \cup \{a\}) \cup (B_0 \setminus \{b\})$ is independent. Thus, $(A_0 \cup \{a\}) \cup (B_0 \setminus \{b\})$ is a basis of $acl(B)$. This proves (1).

First we prove (2) when B is finite. Let $|B| = n$. Set $A_0 = \emptyset$. Take any $a_1 \in A$. Get $b_1 \in B$ such that $\{a_1\} \cup (B \setminus \{b_1\})$ is a basis of $acl(B)$. Such an a_1 exists by (1). Suppose $1 \leq i < n$ and $a_1, \ldots, a_i \in A$ and $b_1, \ldots, b_i \in B$ be such that $\{a_1, \ldots, a_i\} \cup (B \setminus \{b_1, \ldots, b_i\})$ is a basis of $acl(B)$. If $A \neq \{a_1, \ldots, a_i\}$, take any $a_{i+1} \in A \setminus \{a_1, \cdots, a_i\}$. Get $b_{i+1} \in B \setminus \{b_1, \ldots, b_i\}$ such that $\{a_1, \ldots, a_{i+1}\} \cup (B \setminus \{b_1, \ldots, b_{i+1}\})$ is a basis of $acl(B)$. Such an a_{i+1} exists by (1). This process must stop in a maximum of n steps. Thus, $|A| \leq n$. If B is infinite

$$A = \cup \{A \cap acl(B_0) : B_0 \subset B \text{ finite}\}.$$

Hence, $|A| \leq |B|$. Thus, (2) is proved.

(3) is a direct corollary of (2).   □

Let M be an L-structure, X a $\emptyset$-definable minimal set in M and $A \subset X$. Then any two bases of A have the same cardinality which we call the *dimension* of A, denoted by $dim(A)$.

Proposition 2.12.3 *Let M and N be L-structures, $X \subset M$, $Y \subset N$ and $g : X \to Y$ partial elementary. Suppose $\psi[x]$ is an L-formula minimal in both M and N, $\{a_\alpha : \alpha < \kappa\}$ a sequence in $\psi(M)$ independent over X and $\{b_\alpha : \alpha < \kappa\}$ a sequence in $\psi(N)$ independent over Y. Then the extension $g : X \cup \{a_\alpha : \alpha < \kappa\} \to Y \cup \{b_\alpha : \alpha < \kappa\}$ of g (which we denote by g itself) defined by $g(a_\alpha) = b_\alpha$, $\alpha < \kappa$, is partial elementary.*

Proof Set $g_\beta = g|(X \cup \{a_\alpha : \alpha < \beta\})$, $\beta < \kappa$. Suffices to show that each g_β is partial elementary. This will follow if we show that whenever g_β is partial elementary, so is $g_{\beta+1}$.

Assume that $\beta < \kappa$ and g_β is partial elementary. Take an L-formula $\varphi[\overline{x}, \overline{y}, z]$, $\overline{a} \in \{a_\alpha : \alpha < \beta\}$ and $\overline{b} \in X$. Suppose

$$M \models \varphi[\overline{a}, \overline{b}, a_\beta].$$

Since $\psi(M)$ is minimal and $a_\beta \notin acl(X \cup \{a_\alpha : \alpha < \beta\})$, there exists a natural number m such that

$$M \models \exists_{=m} z(\psi[z] \wedge \neg\varphi[\overline{a}, \overline{b}, z]).$$

Since g_β is partial elementary, we have

$$N \models \exists_{=m} z(\psi[z] \land \neg\varphi[g_\beta(\overline{a}), g_\beta(\overline{b}), z]).$$

As $\psi(N)$ is minimal and $b_\beta \notin acl(Y \cup \{b_\alpha : \alpha < \beta\})$, we must have

$$N \models \varphi[g_\beta(\overline{a}), g_\beta(\overline{b}), b_\beta].$$

If $M \not\models \varphi[\overline{a}, \overline{b}, a_\beta]$, we repeat the above argument with $\neg\varphi$ to see that $N \not\models \varphi[g_\beta(\overline{a}), g_\beta(\overline{b}), b_\beta]$. Our proof is complete now. $\qquad\square$

Our next few exercises show that these notion of independence and basis generalise corresponding notions in vector spaces and fields.

Exercise 2.12.4 Let $\mathbb{K}$ be a field, V an infinite vector space over $\mathbb{K}$ and $A \subset V$. Show the following:

1. A is an independent set if and only if A is linearly independent.
2. A is a basis of V if and only if A is a basis of V in linear algebra sense.
3. $dim(V)$ equals the vector space dimension of V.

Exercise 2.12.5 Let $\mathbb{F}$ be an algebraically closed field and $A \subset \mathbb{F}$. Show the following:

1. A is an independent set if and only if A is algebraically independent.
2. A is a basis of $\mathbb{F}$ if and only if A is a transcendence basis of $\mathbb{F}$.
3. $dim(\mathbb{F})$ equals the transcendence degree of $\mathbb{F}$ over the prime field.

2.13 More Complete Theories

Quantifier elimination can be used to prove completeness of theories.

Proposition 2.13.1 *Let T have quantifier elimination and M an L-structure such that $T \cup Diag(M)$ is consistent. Then $T \cup Diag(M)$ is complete.*

Proof Let $M_1, M_2 \models T \cup Diag(M)$. Then $M \sqsubseteq M_1, M_2$ and $M_1, M_2 \models T$. Take a sentence φ. By quantifier elimination of T, there is an open sentence ψ such that $T \models \varphi \leftrightarrow \psi$. Now

$$M_1 \models \varphi \Leftrightarrow M_1 \models \psi \Leftrightarrow M \models \psi \Leftrightarrow M_2 \models \psi \Leftrightarrow M_2 \models \varphi.$$

This completes the proof. $\qquad\square$

An L-structure M is called a *prime structure* of an L-theory T if M is embeddable in every model of T.

Corollary 2.13.2 *If T has quantifier elimination and a prime structure M, then T is complete.*

Proof This follows from the fact that if $N \models T$, then $N \models T \cup Diag(M)$.   □

Now note the following:

1. $\mathbb{Q} \models DLO$ and it embeds into all models of DLO.
2. $\mathbb{Q} \models DAG$ and it embeds into all models of DAG.
3. $\mathbb{Q} \models ODAG$ and it embeds into all models of $ODAG$.
4. The field of all algebraic numbers is a model of $ACF(0)$ that embeds into all models of $ACF(0)$.
5. Let p be a prime and $\overline{\mathbb{F}}_p$ the algebraic closure of the field $\mathbb{F}_p$. Then $\overline{\mathbb{F}}_p$ is a model of $ACF(p)$ that embeds into all models of $ACF(p)$.
6. The field $\mathbb{R}_{alg}$ is a real closed field that embeds into all models of RCF.

Thus,

Theorem 2.13.3 *The theories DLO, DAG, ODAG, ACF(p), p = 0 or prime, and RCF are all complete. Hence, models of these theories are elementarily equivalent.*

A model M of a theory T is called a *prime model* of T if it is elementarily embeddable into every $N \models T$. If T has quantifier elimination, then every model of T which is a prime structure of T is a prime model of T. So, $DLO, DAG, ODAG, ACF(p), p = 0$ or prime, and RCF have prime models.

Remark 2.13.4 A word on decidability of theories and decidable structures: Suppose T is a theory with finitely many nonlogical symbols. Then Gödel coded each formula of T, a finite sequence of logical and nonlogical symbols, by a natural number. The theory T is called *axiomatised* if the set of codes of its axioms is computable. In a landmark result, Gödel showed that a complete, axiomatised theory is decidable. It follows that every model of such a T is decidable. Thus, we get many examples of classical structures such as $\mathbb{R}$ as a real closed field, $\mathbb{C}$, $\overline{\mathbb{F}}_p$, p a prime, etc. which are decidable. All these results are due to Tarski. Since this topic is beyond the scope of this book, we refer the reader to [59] for details.

2.14   Model Completeness

A theory T is called *model complete* if whenever $M, N \models T$ and N is a substructure of M, N is an elementary substructure of M. This notion was introduced and used, for instance, to prove Hilbert Nullstellensatz (Theorem 2.15.8) and give a model theoretic proof of Artin's theorem on Hilbert's seventeenth problem (Theorem 2.15.9) in [51].

Proposition 2.14.1 *If T is model complete and has a model which is a prime structure of T, then T is complete.*

Proposition 2.14.2 *If T has quantifier elimination, it is model complete.*

Proof Let $M, N \models T$ and M be a substructure of N. We need to show that the inclusion map $i : M \hookrightarrow N$ is an elementary embedding. Take a formula $\varphi[\overline{x}]$ and an $\overline{a} \in M$. By elimination of quantifiers, there is an open formula $\psi[\overline{x}]$ such that

$$T \models \forall \overline{x}(\varphi[\overline{x}] \leftrightarrow \psi[\overline{x}]).$$

So,

$$M \models \varphi[\overline{a}] \Leftrightarrow M \models \psi[\overline{a}],$$

$$N \models \varphi[\overline{a}] \Leftrightarrow N \models \psi[\overline{a}]$$

and since M is a substructure of N,

$$M \models \psi[\overline{a}] \Leftrightarrow N \models \psi[\overline{a}].$$

The result follows now. □

Corollary 2.14.3 *The theories DLO, DAG, $ODAG$, ACF, RCF and $RCOF$ are model complete.*

Proposition 2.14.4 *Let T be a model complete theory. Then*

1. *The class of all models of T is closed under unions of chains.*
2. *T is a $\forall \exists$ theory.*

Proof By model completeness, every chain of models of T is an elementary chain. Hence, their unions are models of T. By Corollary 2.4.6, (1) implies (2). □

Proposition 2.14.5 *An L-theory T is model complete if and only if for every model M of T, $T \cup Diag(M)$ is a complete theory.*

Proof Note that $T \cup Diag(M)$ is complete if and only if every model of $T \cup Diag(M)$ is elementarily equivalent to M. Further, every model of $T \cup Diag(M)$ is elementarily equivalent to M if and only if T is model complete. The result follows. □

Proposition 2.14.6 *Let T be a theory. The following statements are equivalent:*

1. *T is model complete.*
2. *For every $M, N \models T$ with $N \subseteq M$, for every formula $\varphi[\overline{x}]$ without parameters, for every $\overline{a} \in N$,*

$$M \models \varphi[\overline{a}] \Rightarrow N \models \varphi[\overline{a}].$$

3. *Every model of T is an existentially closed model of T.*
4. *Every existential formula is equivalent in T to a universal formula.*

5. *Every formula $\varphi[\overline{x}]$ (without parameters) is equivalent in T to a universal formula $\psi[\overline{x}]$ (without parameters).*
6. *Every formula $\varphi[\overline{x}]$ (without parameters) is equivalent in T to a existential formula $\xi[\overline{x}]$ (without parameters).*

Proof (1) implies (2) because for a model complete theory T, every submodel of a model of T is an elementary submodel. (3) is a special case of (2).

Now assume (3). Take $M, N \models T$ with $N \sqsubseteq M$. Let $\varphi[\overline{x}]$ be an existential formula, and $\overline{a} \in N$. By (3), N is existentially closed in M. Hence, $M \models \varphi[\overline{a}] \Rightarrow N \models \varphi[\overline{a}]$. Therefore, by Proposition 2.4.3, φ is equivalent to an universal formula.

Clearly, (4), (5) and (6) are equivalent. (5) and (6) together imply that T is model complete.   □

Let T be an L-theory. An L-theory T' is called a *model companion* of T if it satisfies the following three conditions:

1. T' is model complete.
2. Every model T has an extension which is a model of T'.
3. Every model T' has an extension which is a model of T.

Example 2.14.7   1. The theory of infinite sets is a model companion of the empty theory.
2. DLO is a model companion of the theory of linearly ordered sets.
3. DAG is a model companion of the theory of torsion-free abelian groups.
4. $ODAG$ is a model companion of the theory of ordered groups.
5. ACF is a model companion of the theory of integral domains.

Proposition 2.14.8 *A theory T can have at most one model companion.*

Proof Let T_0 and T_1 be model companions of T. Start with a model M_0 of T_0. Get an extension M of M_0 that models T. Then get a model N_0 of T_1 that extends M. There exists a model N of T that extends N_0, Now get a model M_1 of T_0 that extends N. Proceeding similarly, we get a chain of L-structures

$$M_0 \sqsubseteq N_0 \sqsubseteq M_1 \sqsubseteq N_1 \sqsubseteq \cdots$$

such that $\{M_k\}$ is a chain of models of T_0 and $\{N_k\}$ is a chain of models of T_1. But T_0 and T_1 are model complete. Hence these two chains are elementary. Let $M' = \cup_k M_k = \cup_k N_k$. Then M_0 is an elementary substructure of M' and $M' \models T_1$. Thus, every model of T_0 is a model of T_1. Likewise, every model of T_1 is a model of T_0.   □

Exercise 2.14.9 A linearly ordered set $(D, <)$ is called *discrete* if every element of D that is not the least element has an immediate predecessor and every element that is not the greatest element has an immediate successor. Show that the theory of discrete linear orders with no least element and no greatest element is not model complete. In Exercise 4.7.8 it is shown that this theory is complete.

2.15 Some Applications to Algebra and Geometry

Let $\mathbb{F}$ be a field. A set $C \subset \mathbb{F}^n$ is called constructible if and only if it belongs to the algebra of subsets of $\mathbb{F}^n$ generated by sets of the form $\{\bar{a} \in \mathbb{F}^n : f(\bar{a}) = 0\}$, $f \in \mathbb{F}[X_1, \ldots, X_n]$. Since ACF has quantifier elimination, we have the following result:

Proposition 2.15.1 *For every algebraically closed field $\mathbb{F}$, $C \subset \mathbb{F}^n$ is constructible if and only if it is definable.*

This is a generalisation of

Theorem 2.15.2 (Chevalley Projection Theorem) *If $\mathbb{F}$ is an algebraically closed field and $C \subset \mathbb{F}^{n+1}$ constructible, then its projection $\pi_{\mathbb{F}^n}(C) \subset \mathbb{F}^n$ is constructible.*

If $\mathbb{F}$ is a real closed ordered field, then $D \subset \mathbb{F}^n$ is definable if and only if it belongs to the algebra $\mathcal{A}_n$ of subsets of $\mathbb{F}^n$ generated by sets of the form $\{\bar{a} \in \mathbb{F}^n : p(\bar{a}) < 0\}$, where $p \in \mathbb{F}[X_1, \ldots, X_n]$. Geometers call sets in $\mathcal{A}_n$, $n \geq 1$, *semi-algebraic*. A function $f : \mathbb{F}^n \to \mathbb{F}^m$ is called *semi-algebraic* if its graph is semi-algebraic. So, semi-algebraic sets and functions in a real closed field are precisely those which are definable. This can be thought of as the counterpart of Chevalley's theorem in real case. We now have the following result of Tarski and Seidenberg.

Theorem 2.15.3 (Tarski–Seidenberg Theorem) *If $\mathbb{F}$ is a real closed field and $f : \mathbb{F}^n \to \mathbb{F}^m$, $C \subset \mathbb{F}^n$ and $D \subset \mathbb{F}^m$ semi-algebraic, then $f(C)$ and $f^{-1}(D)$ are semi-algebraic.*

Since RCF is complete, every model of RCF is elementarily equivalent to the ordered field of reals $\mathbb{R}$ or of real algebraic numbers $\mathbb{R}_{alg}$. Hence, $Th(\mathbb{R}) = Th(\mathbb{R}_{alg})$ is the set of all theorems of RCF. This is very useful in proving many theorems of RCF. We illustrate it by proving Rolle's theorem for real closed fields.

Let $\mathbb{F}$ be any field and $\sum_{i=0}^n a_i X^i \in \mathbb{F}[X]$. Then the *formal derivative* of f is the polynomial $f'(X) = \sum_{i=1}^n i a_i X^{i-1}$.

Theorem 2.15.4 (Rolle's Theorem for Real Closed Fields) *Let $\mathbb{F}$ be a real closed field, $a < b$ in $\mathbb{F}$ and $f \in \mathbb{F}[X]$ be such that $f(a) = f(b)$. Then there is $a < c < b$ such that $f'(c) = 0$.*

Proof For each $d \geq 1$, consider the sentence φ given by

$$\forall \bar{x} \forall x \forall y ((x < y \wedge \sum_{i=0}^d x_i x^i = \sum_{i=0}^d x_i y^i) \to \exists z (x < z < y \wedge \sum_{i=0}^{d-1} i x_i z^i = 0)).$$

By classical Rolle's theorem for $\mathbb{R}$, $\varphi \in Th(\mathbb{R})$. Since RCF is complete, it follows that $RCF \models \varphi$. □

Theorem 2.15.5 *Let φ be a sentence of the language of the theory of fields. The following statements are equivalent:*

(i) $\mathbb{C} \models \varphi$.

(ii) φ *is true in some algebraically closed field of characteristic 0.*

(iii) $ACF(0) \models \varphi$.

(iv) *There is an m such that for all prime $p > m$, $ACF(p) \models \varphi$.*

(v) *There is an m such that for all prime $p > m$, φ is true in some algebraically closed field of characteristic p.*

(vi) $ACF(p) \models \varphi$ *for infinitely many primes p.*

Proof Clearly (i) implies (ii). Since any two models of $ACF(0)$ are elementarily equivalent, (ii) implies (iii). Clearly (iii) implies (i).

Now assume (iii). Then by the compactness theorem, $T \models \varphi$, where T consists of some finitely many axioms of $ACF(0)$. Hence, there is an m such that for no prime $p > m$, $\underline{p} \neq 0$ belongs to T. Thus, $ACF(p) \models \varphi$ for all $p > m$. Thus, (iii) implies (iv).

Clearly (iv) implies (v). The statement (v) implies (iv) because each $ACF(p)$ is complete. (iv) clearly implies (vi).

We now show that (vi) implies (iii). Let $ACF(0) \not\models \varphi$. Since $ACF(0)$ is complete, it follows that $ACF(0) \models \neg\varphi$. Since (iii) implies (v), there is an m such that for all primes $p > m$, $ACF(p) \models \neg\varphi$. This completes the proof.   □

Let $p > 0$ be a prime and $\overline{\mathbb{F}}_p$ the algebraic closure of the field with p elements. It is a standard fact of algebra that every finitely generated subfield of $\overline{\mathbb{F}}_p$ is finite. Using this we easily get the following result.

Proposition 2.15.6 *Let $f_1, \ldots, f_n \in \overline{\mathbb{F}}_p[X_1, \ldots, X_n]$ be such that $\overline{f} = (f_1, \ldots, f_n) : \overline{\mathbb{F}}_p^n \to \overline{\mathbb{F}}_p^n$ is injective. Then $\overline{f}$ is surjective.*

Proof Assume that $\overline{f}$ is not surjective. Take any $\overline{b} \notin range(\overline{f})$. Let $\mathbb{K}$ be the smallest subfield of $\overline{\mathbb{F}}_p$ that contains $\overline{b}$ and coefficients of $f_1, \ldots, f_n$. As observed above $\mathbb{K}$ is finite. But then $\overline{f} : \mathbb{K}^n \to \mathbb{K}^n$ is one-to-one but not onto. This is a contradiction since $\mathbb{K}^n$ is finite.   □

Theorem 2.15.7 (Ax [1]) *Let $\mathbb{F}$ be an algebraically closed field and $f_1, \ldots, f_n \in \mathbb{F}[X_1, \ldots, X_n]$ be such that $\overline{f} = (f_1, \ldots, f_n) : \mathbb{F}^n \to \mathbb{F}^n$ is injective. Then $\overline{f}$ is surjective.*

Proof Let each f_i be of degree at most d. It is not hard to see that there is a sentence φ of the language of fields saying that if $f_1, \ldots, f_n$ are polynomials of degree at most d and if the map $f = (f_1, \ldots, f_n)$ is injective, it is surjective.

Let $\mathbb{F}$ be of characteristic p for some prime $p > 1$. By the last proposition $\overline{\mathbb{F}}_p \models \varphi$. Since any two models of $ACF(p)$ are elementarily equivalent, $\mathbb{F} \models \varphi$. As $\overline{\mathbb{F}}_p \models \varphi$ for all prime $p > 1$, by the above theorem, $ACF(0) \models \varphi$ also.   □

We now give some applications of model completeness.

Recall that for an ideal $I \subset \mathbb{K}[\overline{X}]$,

$$\sqrt{I} = \{f \in \mathbb{K}[\overline{X}] : f^n \in I \text{ for some } n \geq 1\}.$$

Then

$$\mathcal{V}(I) = \mathcal{V}(\sqrt{I}).$$

Theorem 2.15.8 (Hilbert Nullstellensatz) *Let $\mathbb{K}$ be an algebraically closed field and I an ideal in $\mathbb{K}[\overline{X}]$. Then*

$$\mathcal{I}(\mathcal{V}(I)) = \sqrt{I}.$$

Proof We clearly have $\sqrt{I} \subset \mathcal{I}(\mathcal{V}(I))$. If possible, suppose there is an $f \in \mathcal{I}(\mathcal{V}(I)) \setminus \sqrt{I}$. By prime decomposition theorem (Theorem B.2.4), there is a prime ideal $P \supset \sqrt{I}$ not containing f. Since P is a prime ideal in $\mathbb{K}[\overline{X}]$, $\mathbb{K}[\overline{X}]/P$ is an integral domain.

Let $\mathbb{F}$ be the algebraic closure of the quotient field of $\mathbb{K}[\overline{X}]/P$. By Hilbert's basis theorem (Theorem B.2.3), we fix a basis $g_1, \ldots, g_k \in \sqrt{I}$ generating $\sqrt{I}$. Note that each X_i can be regarded as an element of $\mathbb{K}[\overline{X}]$. Because $f \notin P$ and $g_1, \ldots, g_k \in \sqrt{I}$, we have

$$\mathbb{F} \models \wedge_{i=1}^{k} g_i([X_1], \ldots, [X_n]) = 0 \wedge f([X_1], \ldots, [X_n]) \neq 0.$$

In particular,

$$\mathbb{F} \models \exists \overline{y}(\wedge_{i=1}^{k} g_i(\overline{y}) = 0 \wedge f(\overline{y}) \neq 0).$$

By model completeness of RCF,

$$\mathbb{K} \models \exists \overline{y}(\wedge_{i=1}^{k} g_i(\overline{y}) = 0 \wedge f(\overline{y}) \neq 0).$$

This gives an $\overline{a} \in \mathbb{K}$ such that for all $1 \leq i \leq k$, $g_i(\overline{a}) = 0$ and $f(\overline{a}) \neq 0$. But if $g_i(\overline{a}) = 0$ for all $1 \leq i \leq k$, as $g_1, \ldots, g_k$ generate $\sqrt{I}$, $\overline{a} \in \mathcal{V}(\sqrt{I}) = \mathcal{V}(I)$. Since $f \in \mathcal{I}(\mathcal{V}(I))$, $f(\overline{a}) = 0$. This contradiction proves the result. $\square$

17th problem in Hilbert's famous list of 23 problems was

Hilbert's Seventeenth problem *Let $f \in \mathbb{R}(\overline{X})$ be a rational function such that for no $\overline{x} \in \mathbb{R}^n$, $f(\overline{x}) < 0$. Then is it true that f is a sum of squares of finitely many rational functions?*

This problem was answered in the affirmative by Artin. Abraham Robinson pointed out a strikingly beautiful proof of Artin's theorem using model completeness of $RCOF$. We refer the reader to the appendix in algebra and geometry for relevant definitions and results on real closed fields.

Theorem 2.15.9 *Let* $\mathbb{F}$ *be a real closed field and* $f \in \mathbb{F}(\overline{X}) = \mathbb{F}(X_1, \ldots, X_n)$ *a rational function over* $\mathbb{F}$ *in n variables such that for no* $\overline{x} \in \mathbb{F}^n$, $f(\overline{x}) < 0$. *Then* f *is a sum of squares of rational functions over* $\mathbb{F}$.

Proof By Proposition B.3.4, the field of rational functions $\mathbb{F}(\overline{X})$ is real. Suppose f is not a sum of squares. By Theorem B.3.7, there is a linear order $<$ on the field $\mathbb{F}(\overline{X})$ of rational functions over $\mathbb{F}$ making it into an ordered field such that $f < 0$.

Let $\mathbb{K}$ be the real closure of $\mathbb{F}(\overline{X})$ order compatible with $<$. Then

$$\mathbb{K} \models \exists \overline{x}(f(\overline{x}) < 0).$$

(Take $x_i = X_i \in \mathbb{K}$.) By model completeness of RCF,

$$\mathbb{F} \models \exists \overline{x}(f(\overline{x}) < 0).$$

But there is no $\overline{a} \in \mathbb{F}$ such that $f(\overline{a}) < 0$. Hence, f must be a sum of squares of rational functions over $\mathbb{F}$. $\qquad\qquad\qquad\qquad\qquad\qquad\qquad\qquad\square$

Chapter 3
Spaces of Types

Abstract In this chapter, we shall make a general study of types. This topic is quite important because most of the modern concepts and techniques of model theory are based on types. We introduce Stone topology on spaces of complete types. Omitting types theorem is an important result proved in this chapter. A systematic study of types was first made by Vaught in [66].

3.1 Realised Types

Let L be a first-order language and M an L-structure. For $\overline{a} \in M^n$, we define

$$tp^M(\overline{a}) = \{\varphi[\overline{x}] : M \models \varphi[\overline{a}], \varphi \text{ an } L - \text{formula}\},$$

and call it the *type of $\overline{a$ in M*. Note that if $M \preceq N$, $tp^M(\overline{a}) = tp^N(\overline{a})$. Also observe that for every formula $\varphi[\overline{x}]$, exactly one of $\varphi[\overline{x}]$ and $\neg\varphi[\overline{x}]$ belongs to $tp^M(\overline{a})$. So, $tp^M(\overline{a})$ may be considered to be the collection of everything that can be said about the tuple $\overline{a}$.

Next take any $A \subset M$. We define

$$tp^M(\overline{a}/A) = \{\varphi[\overline{x}] : M \models \varphi[\overline{a}], \varphi \text{ an } L_A - \text{formula}\},$$

and call it the *relative type of $\overline{a}$ in M over A*, or simply the *type of $\overline{a}$ in M over A*. Again note that if $A \subset M \preceq N$, then $tp^M(\overline{a}/A) = tp^N(\overline{a}/A)$.

Let M, N be L-structures, $A \subset M$ (including $A = \emptyset$) and $f : A \to N$ a map. Recall f is called partial elementary if for every L-formula $\varphi[\overline{x}]$ and for every $\overline{a} \in A$,

$$M \models \varphi[\overline{a}] \Leftrightarrow N \models \varphi[f(\overline{a})].$$

Note that if for some $A \subset M$ there is a partial elementary map $f : A \to N$, then $M \simeq N$. Further, the empty function from M into N is partial elementary if and only if M and N are elementarily equivalent.

© Springer Nature Singapore Pte Ltd. 2017
H. Sarbadhikari and S.M. Srivastava, *A Course on Basic Model Theory*,
DOI 10.1007/978-981-10-5098-5_3

Suppose $\bar{a} \in M^n$ and $\bar{b} \in N^n$. Then $tp^M(\bar{a}) = tp^N(\bar{b})$ if and only if the map $\bar{a} \to \bar{b}$ is partial elementary. Further, assume that $A \subset M$ and $\bar{a}, \bar{b} \in M^n$. Then $tp^M(\bar{a}/A) = tp^M(\bar{b}/A)$, if and only if the map $f : A \cup \{a_i : i < n\} \to M$, where $f|A$ is the identity map on A and $f(a_i) = b_i$, $i < n$, is partial elementary.

Proposition 3.1.1 *Let M be an L-structure and $\bar{a}, \bar{b} \in M^n$. Then $tp^M(\bar{a}) = tp^M(\bar{b})$, if and only if there is an elementary extension N of M and an automorphism $\alpha : N \to N$ such that $\alpha(\bar{a}) = \bar{b}$. Moreover, if L and M are countable, we can choose N to be countable.*

Proof The if part of the result is clear. So, we need to prove only the only if part.

Assume that $\bar{a}, \bar{b} \in M^n$ are such that $tp^M(\bar{a}) = tp^M(\bar{b})$. Set $M_0 = M$. By repeatedly using Proposition 2.5.2, we define an elementary chain

$$M_0 \preceq N_0 \preceq M_1 \preceq N_1 \preceq M_2 \preceq N_2 \preceq \cdots$$

and elementary embeddings $\alpha_k : M_k \to N_k$ satisfying the following conditions:

1. $\alpha_0(\bar{a}) = \bar{b}$.
2. For each k, α_{k+1} extends α_k.
3. For each k, $N_k \subset \alpha_{k+1}(M_{k+1})$.

Taking $N = \cup_k M_k = \cup_k N_k$ and $\alpha = \cup_k \alpha_k$, we get our result.

Since $tp^M(\bar{a}) = tp^M(\bar{b})$, $M_0 \ni \bar{a} \xrightarrow{\beta} \bar{b} \in M_0$ is partial elementary. Hence, by Proposition 2.5.2, there exists an elementary extension N_0 of M_0 and an elementary extension $\alpha_0 : M_0 \to N_0$ of β.

Now assume that $M_i, N_i, \alpha_i, i \leq k$ have been defined. Hence,

$$N_k \supset \alpha_k(M_k) \xrightarrow{\alpha_k^{-1}} M_k \subset N_k$$

is partial elementary. By Proposition 2.5.2, we get $M_{k+1} \succeq N_k$ and an elementary extension $\beta_k : N_k \to M_{k+1}$ of α_k^{-1}.

Then

$$M_{k+1} \supset \beta_k(N_k) \xrightarrow{\beta_k^{-1}} N_k \subset M_{k+1}$$

is partial elementary. By Proposition 2.12.3, there exist $N_{k+1} \succeq M_{k+1}$ and an elementary extension $\alpha_{k+1} : M_{k+1} \to N_{k+1}$ of β_k^{-1}. Clearly, α_{k+1} extends α_k. $\square$

Let α be an ordinal number and $\bar{a} = \{a_\beta : \beta < \alpha\}$ a sequence in M of length α. Then also, we can talk of $tp^M(\bar{a})$ by starting with a first-order language with a sequence of variables $\bar{x} = \{x_\beta : \beta < \alpha\}$ of length α. For a formula φ (with free variables), $\varphi[\bar{a}]$ will denote the L_M-formula obtained by replacing each free occurrence of x_β in φ by a_β. With these definitions, we define

$$tp^M(\bar{a}) = \{\varphi[\bar{x}] : M \models \varphi[\bar{a}], \varphi \text{ an } L - \text{formula}\}.$$

$tp^M(\overline{a}/A)$ in this case is defined as before.

We shall prove results for types of finite sequences only. But the reader should observe that many of the arguments done in the conventional first-order language goes through in this general set up also where the sequence of variables is of length α, α an arbitrary ordinal.

3.2 n-Types

Let M be an L-structure, $A \subset M$, $n \geq 1$, and $p = p(\overline{x})$ a set of L_A-formulas $\varphi[\overline{x}]$, where $\overline{x} = (x_0, \ldots, x_{n-1})$. We call p a n-type in M over A if for every finite set $\varphi_0, \ldots, \varphi_{k-1}$ of formulas in p there is a $\overline{a} \in M^n$ such that $M \models \wedge_{i<k}\varphi_i[\overline{a}]$. This is equivalent to saying that p is a n type in M over A if for every finite set $\varphi_0, \ldots, \varphi_{k-1}$ of formulas in p there is an elementary extension N of M and a $\overline{a} \in N^n$ such that $N \models \wedge_{i<k}\varphi_i[\overline{a}]$. If $A = \emptyset$, we call $p(\overline{x})$ just a n-type in M.

Clearly, for every $\overline{a} \in M^n$, $tp^M(\overline{a}/A)$, or any subset of $tp^M(\overline{a}/A)$, is a n-type in M over A. Further, if $N \succeq M$ and $\overline{a} \in N^n$, then $tp^N(\overline{a}/A)$ is a n-type in M over A. Later, we shall see that every n-type in M over A is a subset of $tp^N(\overline{a}/A)$ for some $N \succeq M$ and $\overline{a} \in N^n$.

We say that p is *realised* in M if $p \subset tp^M(\overline{a}/A)$ for some $\overline{a} \in M$. In this case, we say that $\overline{a}$ realises p and write $\overline{a} \models p$. The set of all realisations of p in M will be denoted by $p(M)$. If no $\overline{a}$ in M realises p, we say that M *omits* p.

Example 3.2.1 Consider the standard model $\mathbb{N}$ of Peano arithmetic. Let

$$p(x) = \{x > \underline{n} : n \in \omega\}.$$

Then $p(x)$ is a type in $\mathbb{N}$, which is omitted in $\mathbb{N}$.

Since every model of $Diag_{el}(M)$ is an elementary extension of M, note that p is a n-type in M over A if and only if $p \cup Diag_{el}(M)$ is finitely satisfiable. Now a straightforward application of compactness theorem gives us the following result.

Proposition 3.2.2 *Let M be an L-structure, $A \subset M$ and $p(\overline{x})$ a set of L_A-formulas $\varphi[\overline{x}]$, where $\overline{x} = (x_0, \ldots, x_{n-1})$. Then p is an n-type in M over A, if and only if there is a $N \succeq M$ and a $\overline{a} \in N^n$ such that $N \models \varphi[\overline{a}]$ for every $\varphi \in p$.*

In fact, we can say more.

Proposition 3.2.3 *Given any L-structure M, there is an elementary extension N of M such that for every $A \subset M$, every type $p(\overline{x})$ over A is realised in N.*

Proof For each n-type $p(\overline{x})$ (over some subset A of M), add a n-tuple of constant symbols $\overline{c}_p$ to L. Note that

$$\mathcal{P} = Diag_{el}(M) \cup \{\varphi[\overline{c}_p] : \varphi[\overline{x}] \in p\}$$

is finitely satisfiable. Hence, $\mathcal{P}$ has a model, say N. Then N is an elementary extension of M such that for every $p \in \mathcal{P}$, the interpretation of $\overline{c}_p$ realises p in N.   □

Remark 3.2.4 If $\kappa \geq \aleph_0$ and $|L|, |M| \leq \kappa$, then every n-type p in M over A is realised in an elementary extension N of M with $|N| \leq \kappa$. This can now be easily seen by downward Löwenheim–Skolem theorem.

An Observation. Let $p(\overline{x})$ be a n-type in M over A and $\varphi[\overline{x}]$ an L_A-formula such that none of φ and $\neg\varphi$ belongs to p. Then, at least one of $p \cup \{\varphi\}$ or $p \cup \{\neg\varphi\}$ is a n-type in M over A. If not, then there exist finite $\Phi_1, \Phi_2 \subset p$ such that neither $\Phi_1 \cup \{\varphi\}$ nor $\Phi_2 \cup \{\neg\varphi\}$ is satisfiable in M. But then $\Phi_1 \cup \Phi_2 \subset p$ is finite and not satisfiable in M. This contradicts that p is a type in M over A.

A n-type $p(\overline{x})$ in M over A is called a *complete n-type in M over A* if for every L_A-formula $\varphi[\overline{x}]$ either φ or $\neg\varphi$ is in p. Since p is finitely satisfiable, this is equivalent to saying that exactly one of φ and $\neg\varphi$ is in p. Using Zorn's lemma and the above observation, we see that every n-type in M over A is contained in a complete n-type in M over A. Further, complete n-types in M over A are precisely maximal n-types in M over A.

We let $S_n(M/A)$ denote the set of all complete n-types in M over A. If $A = \emptyset$, we shall write $S_n(M)$ instead of $S_n(M/\emptyset)$. Note that $p \in S_n(M/A)$ if and only if there is an elementary extension N of M and a n-tuple $\overline{a} \in N$ such that $p = tp^N(\overline{a}/A)$. Thus, each complete n-type in M is the collection of all statements that hold for some n-tuple in an elementary extension of M, though such a n-tuple may not exist in M. If α is an ordinal number, then $S_\alpha(M/A)$ will denote the set of all complete types in M over A in a sequence $\overline{x} = \{x_\beta : \beta < \alpha\}$ of variables of length α.

Here are some simple observations on complete n-types in M over A, which will be used in the sequel without specific mention. Let $p, q \in S_n(M/A)$.

(1) $p \subset q \Rightarrow p = q$: If possible, suppose there exists $\psi \in q \setminus p$. Since p is complete, $\neg\psi \in p \subset q$. Thus, both $\psi, \neg\psi \in q$ contradicting that q is finitely satisfiable.

(2) Let $\varphi[\overline{x}] \in p$, $\psi[\overline{x}]$ an L_A-formula and $M \models \forall \overline{x}(\varphi \to \psi)$. Then $\psi \in p$. Suppose not. Then $\neg\psi \in p$. Since $M \models \forall \overline{x}(\varphi \to \psi)$, $\{\varphi, \neg\psi\}$ is not satisfiable in M, a contradiction. It follows that if $M \models \forall \overline{x}(\varphi[\overline{x}] \leftrightarrow \psi[\overline{x}])$, then either both φ, ψ belong to p or none of these two belongs to p.

(3) $\varphi_1 \vee \cdots \vee \varphi_k \in p$ if and only if $\varphi_i \in p$ for some $1 \leq i \leq k$. If part follows from (2). If no $\varphi_i \in p$, then by the completeness of p, $\neg\varphi_1, \ldots, \neg\varphi_k \in p$. Since $\{\vee_{i=1}^k \varphi_i, \neg\varphi_1, \ldots, \neg\varphi_k\}$ is not satisfiable, we have $\varphi_1 \vee \cdots \vee \varphi_k \notin p$.

(4) $\varphi_1 \wedge \cdots \wedge \varphi_k \in p$ if and only if $\varphi_i \in p$ for all $1 \leq i \leq k$. This is easily seen as in (3).

Now, we define types in an L-theory T. Let T be an L-theory and $p = p(\overline{x})$ a set of L-formulas $\varphi[\overline{x}]$, where $\overline{x} = (x_0, \ldots, x_{n-1})$. We call p a n-type in T if $p \cup T$ is finitely satisfiable. By compactness theorem, this is equivalent to saying that there is a model M of T in which p is realised. p will be called a *complete n-type in T* if for every L-formula $\varphi[\overline{x}]$ either φ or $\neg\varphi$ is in p. $S_n(T)$ will denote the set of all complete n-types in T. Similar observations as above are easily seen to be true.

Exercise 3.2.5 Let T be a complete theory and $M \models T$. For every $n \geq 1$, show that $S_n(T) = S_n(M)$.

3.3   Stone Topology on the Space of Complete Types

Let M be an L-structure and $A \subset M$. For an L_A-formula $\varphi[\bar{x}], \bar{x} = (x_0, \ldots, x_{n-1})$, define

$$[\varphi] = \{p \in S_n(M/A) : \varphi \in p\}.$$

We then have

(∗) $[x_0 \neq x_0] = \emptyset$ and $[x_0 = x_0] = S_n(M/A)$.
(∗) $[\varphi] \cap [\psi] = [\varphi \wedge \psi]$.
(∗) $[\varphi] \cup [\psi] = [\varphi \vee \psi]$.
(∗) $S_n(M/A) \setminus [\varphi] = [\neg \varphi]$.

This shows that $\{[\varphi] : \varphi$ an $L_A -$ formula$\}$ is a base of a topology on $S_n(M/A)$ which is zero-dimensional, i.e. the topology has a clopen base. We equip $S_n(M/A)$ with this topology and call it the *Stone topology*.

$S_n(M/A)$ is Hausdorff: Let $p \neq q \in S_n(M/A)$. Then, there is a $\varphi \in p \setminus q$. Since q is complete, $\neg \varphi \in q$. Thus, $p \in [\varphi], q \in [\neg \varphi]$. Since $[\varphi] \cap [\neg \varphi] = \emptyset$, our contention follows.

$S_n(M/A)$ is compact: Let $\mathcal{F} = \{[\varphi] : \varphi$ an $L_A -$ formula$\}$ be a family of basic clopen sets with finite intersection property. Then $\gamma = \{\varphi : [\varphi] \in \mathcal{F}\}$ is a n-type in M over A: Let $\varphi_1, \ldots, \varphi_k \in \gamma$. By our hypothesis, there is a n-type p in M over A that contains each of $\varphi_1, \ldots, \varphi_k$. Hence, $\{\varphi_1, \ldots, \varphi_k\}$ is satisfiable in M. So, there is a complete n-type $p \supset \gamma$ in M over A. This implies that $p \in \cap \mathcal{F}$.

If L and A are countable, then $S_n(M/A)$ has a countable base. This implies that $S_n(M/A)$ is metrisable. The following theorem sums up the above observations.

Theorem 3.3.1 $S_n(M/A)$ *is a compact, zero-dimensional, Hausdorff space. Moreover, if L and A are countable, $S_n(M/A)$ is a compact, zero-dimensional metrisable space.*

We make a series of simple but useful observations on the topology on the space of complete types.

Lemma 3.3.2 *Let M be an L-structure and $A \subset B \subset M$. Define*

$$\cdot | A : S_n(M/B) \to S_n(M/A)$$

by

$$p|A = \{\varphi \in p : \varphi \text{ an } L_A - formula\}, \quad p \in S_n(M/B).$$

Then, $\cdot|A : S_n(M/B) \to S_n(M/A)$ *is a continuous surjection.*

Proof Let $q \in S_n(M/A)$. Then q is a type in M over B. Let $p \in S_1(M/B)$ contain q. Since q is complete (over A), $p|A = q$. This proves that the map $|A$ is surjective. It is also a continuous map: Let $\varphi[\overline{x}]$ be an L_A-formula. Then,

$$(|A)^{-1}([\varphi]) = \{p \in S_n(M/B) : \varphi \in S_n(M/B)\}.$$

$\square$

We call $p|A$ a *restriction* of p, p an *extension* of $p|A$.

Lemma 3.3.3 *Let* M, N *be* L-*structures,* $A \subset M$, $f : A \to N$ *partial elementary and* $B = f(A)$. *For* $p \in S_n(M/A)$, *define*

$$f(p) = \{\varphi[\overline{x}, f(\overline{a})] : \varphi[\overline{x}, \overline{y}] \text{ an } L - \text{formula } \& \varphi[\overline{x}, \overline{a}] \in p\}.$$

The map $p \to f(p)$ *from* $S_n(M/A)$ *to* $S_n(N/f(A))$ *is a homeomorphism.*

Proof Since f is one-to-one and f, f^{-1} are elementary, it is easily seen that $p \to f(p)$ is a bijection from $S_n(M/A)$ to $S_n(N/B)$.

This map is continuous: Take any L-formula $\varphi[\overline{x}, \overline{y}]$ and $\overline{b} \in B$. Then, there is a unique tuple $\overline{a} \in A$ such that $f(\overline{a}) = \overline{b}$. Now note that $f^{-1}([\varphi[\overline{x}, \overline{b}]]) = [\varphi[\overline{x}, \overline{a}]]$. By usual topology arguments or by reversing the above argument, we see the map $p \to f(p)$ from $S_n(M/A)$ to $S_n(N/B)$ is a homeomorphism. $\square$

Lemma 3.3.4 *Let* M *be an* L-*structure and* $A \subset M$. *For* $p(\overline{x}, \overline{y}) \in S_{n+m}(M/A)$, *define*

$$\pi(p) = \{\varphi[\overline{x}, \overline{a}] : \varphi[\overline{x}, \overline{a}] \in p(\overline{x}, \overline{y})\}.$$

Then, $\pi : S_{n+m}(M/A) \to S_n(M/A)$ *is a homeomorphism.*

Proof Arguing as in Lemma 3.3.2, we see that $\pi(p) \in S_n(M/A)$ and $\pi : S_{n+m(M/A)} \to S_n(M/A)$ is a continuous surjection.

The map π is open also. To see this take an L_A-formula $\psi[\overline{x}, \overline{y}]$. Let $p[\overline{x}, \overline{y}] \in S_{n+m}(M/A)$ contain $\psi[\overline{x}, \overline{y}]$. Then $\varphi[\overline{x}] = \exists \overline{y}\psi[\overline{x}, \overline{y}] \in p$. If not, then $\neg\varphi[\overline{x}] \in p$. But $\{\neg\varphi[\overline{x}], \psi[\overline{x}, \overline{y}]\}$ is not satisfiable which is a contradiction. Now take any $q(\overline{x}) \in S_n(M/A)$ containing $\varphi[\overline{x}]$. Let $N \succeq M$ realise q, say by $\overline{b}$. Since $N \models \varphi[\overline{b}]$, there is a $\overline{c} \in N$ such that $N \models \psi[\overline{b}, \overline{c}]$. This shows that $q \cup \{\psi[\overline{x}, \overline{y}]\}$ is a $(n + m)$-type in M over A. Hence, there is a $p(\overline{x}, \overline{y}) \in S_{n+m}(M/A)$ such that $\pi(p) = q$. Thus, we have proved that $\pi([\psi]) = [\varphi]$.

Next take an $\overline{a} \in M^n$. For each $p(\overline{x}, \overline{y}) \in S_{n+m}(M/A)$ with $\pi(p) = tp^M(\overline{a}/A)$ define

$$h(p) = \{\psi[\overline{a}, \overline{y}] : \psi[\overline{x}, \overline{y}] \in p\}.$$

For every $\psi[\overline{x}, \overline{y}] \in p$, $\exists \overline{y} \psi[\overline{x}, \overline{y}] \in p$. So, by our assumption, $M \models \exists \overline{y} \psi[\overline{a}, \overline{y}]$. Now it is easily seen that $h(p) \in S_m(M/A\overline{a})$. Conversely, for any $q \in S_m(M/A\overline{a})$, define

$$f(q) = \{\psi[\overline{x}, \overline{y}] : \psi[\overline{a}, \overline{y}] \in q\}.$$

Then, $f(q) \in S_{n+m}(M/A)$, $\pi(f(q)) = tp^M(\overline{a}/A)$ and $f = h^{-1}$. Also note that for any $L_{A\overline{a}}$-formula $\psi[\overline{a}, \overline{y}]$, $h^{-1}([\psi[\overline{a}, \overline{y}]]) = [\psi[\overline{x}, \overline{y}]]$. Hence, h is continuous, and so a homeomorphism. $\qquad \square$

Remark All these arguments hold for $S_\alpha(M/A)$ where α is an ordinal number. In particular, the Stone topology on $S_\alpha(M/A)$ is compact, Hausdorff, and zero-dimensional.

Let T be a first-order theory and $n \geq 1$. Likewise, we topologise $S_n(T)$. For an L-formula $\varphi[\overline{x}]$, $\overline{x} = (x_0, \ldots, x_{n-1})$, define

$$[\varphi] = \{p \in S_n(T) : \varphi \in p\}.$$

These form a base of a topology on $S_n(T)$ with respect to which $S_n(T)$ is a compact, Hausdorff, zero-dimensional space. Moreover, if L is countable, $S_n(T)$ is a compact. zero-dimensional, metric space.

Exercise 3.3.5 Let T be an L-theory. Show that $A \subset S_n(T)$ is clopen if and only if $A = [\varphi]$ for some L-formula φ. Show also that A is closed if and only if there is a set of L-formulas $\mathcal{F}$ such that

$$A = \{p \in S_n(T) : \mathcal{F} \subset p\}.$$

Exercise 3.3.6 Let T be an L-theory. For $p(\overline{x}, \overline{y}) \in S_{n+m}(T)$, where $\overline{x} = (x_0, \ldots, x_{n-1})$ and $\overline{y} = (y_0, \ldots, y_{m-1})$, define

$$\pi(p) = \{\varphi[\overline{x}] : \varphi[\overline{x}] \in p\}.$$

Show that $\pi(p) \in S_n(T)$ and the map $p \rightarrow \pi(p)$ from $S_{n+m}(T)$ to $S_n(T)$ is continuous, open, and surjective.

3.4 Isolated Types

Let M be an L-structure and $A \subset M$. A $p \in S_n(M/A)$ is called an *isolated type* if p is an isolated point of $S_n(M/A)$. So, p is isolated if and only if $\{p\} = [\varphi]$ for some L_A-formula $\varphi[\overline{x}]$.

Proposition 3.4.1 *Let $p \in S_n(M/A)$ and $\varphi[\overline{x}]$ an L_A-formula satisfiable in M. Then, the following two conditions are equivalent.*

1. $\{p\} = [\varphi]$.
2. For every L_A-formula $\psi[\overline{x}]$,

$$\psi[\overline{x}] \in p \Leftrightarrow M \models \forall \overline{x}(\varphi[\overline{x}] \rightarrow \psi[\overline{x}]).$$

3. For every L_A-formula $\psi[\overline{x}]$,

$$\psi[\overline{x}] \in p \Rightarrow M \models \forall \overline{x}(\varphi[\overline{x}] \rightarrow \psi[\overline{x}]).$$

Proof Assume (1) and let $\psi[\overline{x}]$ be an L_A-formula. Since p is a complete type, $\psi \in p$ whenever $M \models \forall \overline{x}(\varphi[\overline{x}] \rightarrow \psi[\overline{x}])$. Now assume that $M \not\models \forall \overline{x}(\varphi[\overline{x}] \rightarrow \psi[\overline{x}])$. Then $\{\varphi, \neg\psi\}$ is satisfiable in M. So, there is a complete n-type in M over A containing φ and $\neg\psi$. Since p is the only complete n-type in M over A containing φ, $\neg\psi \in p$. Hence, $\psi \notin p$.

Now let (2) be true and (1) false. Get a $q \in S_n(M/A)$ containing φ different from p. Suppose $\psi[\overline{x}]$ is an L_A-formula in $q \setminus p$. Since p is complete, $\neg\psi \in p$. Hence, by (2), $M \models \forall \overline{x}(\varphi[\overline{x}] \rightarrow \neg\psi[\overline{x}])$. But then $\{\varphi, \psi\}$ is not satisfiable in M. This contradicts that q is a type.

We now prove that (3) implies (2). Let

$$M \models \forall \overline{x}(\varphi[\overline{x}] \rightarrow \psi[\overline{x}]).$$

If possible, suppose $\psi[\overline{x}] \notin p$. Since p is complete, $\neg\psi[\overline{x}] \in p$. By (3),

$$M \models \forall \overline{x}(\varphi[\overline{x}] \rightarrow \neg\psi[\overline{x}]).$$

Since there exists an $\overline{a} \in M$ such that $M \models \varphi[\overline{a}]$, we have now a contradiction. □

We shall need this notion for incomplete types also. A n-type p in M over $A \subset M$ (not necessarily complete) is called *isolated*, if there is an L_A-formula $\varphi[\overline{x}]$ such that

$$\psi[\overline{x}] \in p \Rightarrow M \models \forall \overline{x}(\varphi[\overline{x}] \rightarrow \psi[\overline{x}]).$$

If $\{p\} = [\varphi]$, we say that φ *isolates* p. We make a series of preliminary observations first.

Remark 3.4.2 If φ and φ' isolate p, then $M \models \forall \overline{x}(\varphi[\overline{x}] \leftrightarrow \varphi'[\overline{x}])$

Remark 3.4.3 If $A \subset B \subset M, \overline{a} \in M$ and $\varphi[\overline{x}]$ is an L_A-formula that isolates $tp^M(\overline{a}/B)$, then φ isolates $tp^M(\overline{a}/A)$

Remark 3.4.4 If $\varphi[\overline{x}, \overline{y}]$ is an L_A-formula that isolates $tp^M(\overline{a}, \overline{b}/A)$, then $\varphi[\overline{x}, \overline{b}]$ isolates $tp^M(\overline{a}/A\overline{b})$.

Remark 3.4.5 If $tp^M(\overline{a}, \overline{b}/A)$ is isolated, then $tp^M(\overline{a}/A)$ is isolated. To see this let $\psi[\overline{x}, \overline{y}]$ be an L_A-formula that isolates $tp^M(\overline{a}, \overline{b}/A)$. Then $\exists \overline{y} \psi[\overline{x}, \overline{y}]$ isolates $tp^M(\overline{a}/A)$: Let $\varphi[\overline{x}]$ be an L_A-formula. Then

$$
\begin{aligned}
M \models \varphi[\overline{a}] &\Leftrightarrow M \models \forall \overline{x} \forall \overline{y}(\psi[\overline{x}, \overline{y}] \to \varphi[\overline{x}]) \\
&\Leftrightarrow M \models \forall \overline{x}(\forall \overline{y} \neg \psi[\overline{x}, \overline{y}] \vee \varphi[\overline{x}]) \\
&\Leftrightarrow M \models \forall \overline{x}((\exists \overline{y} \psi[\overline{x}, \overline{y}]) \to \varphi[\overline{x}])
\end{aligned}
$$

Proposition 3.4.6 *Let M be an L-structure and $A \subset B \subset M$. Suppose for every $\overline{b} \in B^m$, $tp^M(\overline{b}/A)$ is isolated. Then for every $\overline{a} \in M^n$, $tp^M(\overline{a}/B)$ is isolated implies that $tp^M(\overline{a}/A)$ is isolated.*

Proof Get an L-formula $\psi[\overline{x}, \overline{y}]$ and a $\overline{b} \in B$ such that $\psi[\overline{x}, \overline{b}]$ isolates $tp^M(\overline{a}/B)$. By the last remark, the proof will be complete if we show that $tp^M(\overline{a}, \overline{b}/A)$ is isolated.

Get an L_A-formula $\varphi[\overline{y}]$ that isolates $tp^M(\overline{b}/A)$. We show that $\psi[\overline{x}, \overline{y}] \wedge \varphi[\overline{y}]$ isolates $tp^M(\overline{a}, \overline{b}/A)$. Fix an L_A-formula $\eta[\overline{x}, \overline{y}]$. Then

$$
\begin{aligned}
M \models \eta[\overline{a}, \overline{b}] &\Leftrightarrow M \models \forall \overline{x}(\psi[\overline{x}, \overline{b}] \to \eta[\overline{x}, \overline{b}]) \\
&\Leftrightarrow M \models \forall \overline{y}(\varphi[\overline{y}] \to \forall \overline{x}(\psi[\overline{x}, \overline{y}] \to \eta[\overline{x}, \overline{y}])) \\
&\Leftrightarrow M \models \forall \overline{x} \forall \overline{y}((\psi[\overline{x}, \overline{y}] \wedge \varphi[\overline{y}]) \to \eta[\overline{x}, \overline{y}])
\end{aligned}
$$

This completes the proof. □

A variant of this result is the following, whose proof is left for the reader.

Remark 3.4.7 If $tp^M(\overline{b}/A)$ and $tp^M(\overline{a}/A\overline{b})$ are isolated, then $tp^M(\overline{a}, \overline{b}/A)$ is isolated.

Remark 3.4.8 If $a \in acl(A)$, then $tp^M(a/A)$ is isolated. Since $a \in acl(A)$, there is an L_A-formula $\varphi[x]$ and a $n \geq 1$ such that $M \models \varphi[a] \wedge \exists_{=n} x \varphi[x]$. We choose such a φ with n least possible. If there exists a $\psi[x] \in tp^M(a/A)$ such that $M \not\models \forall x(\varphi[x] \to \psi[x])$, then the number of $b \in M$ such that $M \models (\varphi \wedge \psi)[b]$ is less than n. This is a contradiction. Hence, φ isolates $tp^M(a/A)$.

Exercise 3.4.9 Let M be an L-structure, $A \subset M$ and $a \in acl(A)$. Show that $tp^M(a/A)$ is isolated.

Proposition 3.4.10 *Let M be an L-structure and $A \subset B \subset M \models T$ be such that for every $\overline{d} \in B$, $tp^M(\overline{d}/A)$ is isolated. Assume that $tp^M(\overline{a}/B)$ is isolated. Then, for every $\overline{b} \in B$, $tp^M(\overline{a}, \overline{b}/A)$ is isolated.*

Proof Let $\theta_1[\overline{x}, \overline{z}]$ be an L-formula and $\overline{c} \in B$ be such that $\theta_1[\overline{x}, \overline{c}]$ isolates $tp^M(\overline{a}/B)$. Take any $\overline{b} \in B$. Let $\theta_2[\overline{y}, \overline{z}]$ be an L_A-formula isolating $tp^M(\overline{b}, \overline{c}/A)$.

Now let $\varphi[\overline{x}, \overline{y}] \in tp^M(\overline{a}, \overline{b}/A)$. Then, $\varphi[\overline{x}, \overline{b}] \in tp^M(\overline{a}/B)$. Hence,

$$
M \models \forall \overline{x}(\theta_1[\overline{x}, \overline{c}] \to \varphi[\overline{x}, \overline{b}]),
$$

i.e.,

$$\forall \overline{x}(\theta_1[\overline{x}, \overline{z}] \to \varphi[\overline{x}, \overline{y}]) \in tp^M(\overline{b}, \overline{c}/A).$$

Therefore,

$$M \models \forall \overline{y} \forall \overline{z}(\theta_2[\overline{y}, \overline{z}] \to \forall \overline{x}(\theta_1[\overline{x}, \overline{z}] \to \varphi[\overline{x}, \overline{y}])).$$

By prenex operations, we now get

$$M \models \forall \overline{x} \forall \overline{y} \forall \overline{z}(\theta_2[\overline{y}, \overline{z}] \to \theta_1[\overline{x}, \overline{z}] \to \varphi[\overline{x}, \overline{y}]).$$

Therefore,

$$M \models \forall \overline{x} \forall \overline{y} \forall \overline{z}((\neg\theta_2[\overline{y}, \overline{z}] \vee \neg\theta_1[\overline{x}, \overline{z}]) \vee \varphi[\overline{x}, \overline{y}]).$$

Hence,

$$M \models \forall \overline{x} \forall \overline{y}(\forall \overline{z}(\neg\theta_2[\overline{y}, \overline{z}] \vee \neg\theta_1[\overline{x}, \overline{z}]) \vee \varphi[\overline{x}, \overline{y}]).$$

Thus,

$$M \models \forall \overline{x} \forall \overline{y}(\exists \overline{z}(\theta_2[\overline{y}, \overline{z}] \wedge \theta_1[\overline{x}, \overline{z}]) \to \varphi[\overline{x}, \overline{y}]). \qquad (*)$$

Conversely, let $\varphi[\overline{x}, \overline{y}]$ be an L_A-formula such that $(*)$ holds. Since

$$M \models \theta_2(\overline{b}, \overline{c}) \wedge \theta_1(\overline{a}, \overline{c}),$$

$\varphi[\overline{x}, \overline{y}] \in tp^M(\overline{a}, \overline{b}/A)$. Thus, the L_A-formula $\exists \overline{z}(\theta_2[\overline{y}, \overline{z}] \wedge \theta_1[\overline{x}, \overline{z}])$ isolates $tp^M(\overline{a}, \overline{b}/A)$. $\qquad \square$

Let T be an L-theory and $p \in S_n(T)$. We say that p is an *isolated type in* T if it is an isolated point of $S_n(T)$. So, $p \in S_n(T)$ is isolated if and only if there is an L-formula $\varphi[\overline{x}]$ such that $\{p\} = [\varphi]$. In this case, we say that φ *isolates* p. Again, we have the following theorem whose proof we leave for the reader.

Proposition 3.4.11 *Let* $p \in S_n(T)$ *and* $\varphi[\overline{x}]$ *be a consistent L-formula. Then, the following statements are equivalent.*

1. $\{p\} = [\varphi]$.
2. *For every L-formula* $\psi[\overline{x}]$,

$$\psi[\overline{x}] \in p \Leftrightarrow T \models \forall \overline{x}(\varphi[\overline{x}] \to \psi[\overline{x}]).$$

3. *For every L-formula* $\psi[\overline{x}]$,

$$\psi[\overline{x}] \in p \Rightarrow T \models \forall \overline{x}(\varphi[\overline{x}] \to \psi[\overline{x}]).$$

Example 3.4.12 If T is a complete theory, then every isolated type in T is realised in every model of T. To see this, take an isolated type $p(\overline{x})$ in T. Let $\varphi[\overline{x}]$ be an L-formula that isolates p. Now take any model M of T. Then $M \models \exists \overline{x} \varphi[\overline{x}]$. If not, then $M \models \neg \exists \overline{x} \varphi[\overline{x}]$. Since T is complete, this shows that $T \models \neg \exists \overline{x} \varphi[\overline{x}]$. But then $T \cup \{\exists \overline{x} \varphi[\overline{x}]\}$ is not satisfiable. This is a contradiction. Now let $\overline{a} \in M$ be such that $M \models \varphi[\overline{a}]$. Since φ isolates p, for every $\psi[\overline{x}] \in p$, $T \models \forall \overline{x}(\varphi[\overline{x}] \to \psi[\overline{x}])$. Hence, $M \models \psi[\overline{a}]$. This completes the proof.

Example 3.4.13 Let M be an L-structure, $A \subset M$ and $\overline{a} \in A^n$. Then, the formula $\bigwedge_{i<n} x_i = a_i$ isolates $tp(\overline{a}/A)$.

Example 3.4.14 For every $\overline{k} \in \mathbb{N}^n$, the formula $\bigwedge_{i<n} x_i = k_i$ isolates $tp^{\mathbb{N}}(\overline{k})$.

Example 3.4.15 Consider $\mathbb{R} \models RCOF$. Let $a < b$ be two real numbers. Then, there exist integers n, m with $n > 0$ such that $n \cdot a < m < n \cdot b$. It follows that the formula $\underline{n} \cdot x < \underline{m}$ is in $tp^{\mathbb{R}}(a)$ but not in $tp^{\mathbb{R}}(b)$. Thus $|S_1(\mathbb{R})| \geq \mathfrak{c}$.

Example 3.4.16 Consider the theory DLO. Take an isolated $p(\overline{x}) \in S_n(DLO)$. Since DLO is complete, $p(\overline{x})$ is realised in $\mathbb{Q}$, say by $\overline{r} = (r_0, \ldots, r_{n-1})$. So, $p = tp^{\mathbb{Q}}(\overline{r})$.

We now show that every realised type $tp^{\mathbb{Q}}(\overline{r})$ is isolated. For simplicity, assume that $r_0, r_1, \ldots, r_{n-1}$ are all distinct. Suppose π is the permutation of $0, 1, \ldots, n-1$ such that $r_{\pi(0)} < r_{\pi(1)} < \cdots < r_{\pi(n-1)}$ and

$$\psi[\overline{x}] = x_{\pi(0)} < x_{\pi(1)} < \cdots < x_{\pi(n-1)}.$$

Note that for $\overline{t}, \overline{s} \in \mathbb{Q}^n$, $\overline{t} \to \overline{s}$ is an order isomorphism if and only if there is an $\alpha \in Aut(\mathbb{Q})$ such that $\alpha(\overline{t}) = \overline{s}$. Using this it is easy to show that $\psi[\overline{x}]$ isolates $tp^{\mathbb{Q}}(\overline{r})$. We invite the reader to complete the proof.

Example 3.4.17 Consider $\mathbb{Q} \models DLO$. Let $p \in S_1(\mathbb{Q}/\mathbb{Q})$. Since p is a complete 1-type in $\mathbb{Q}$ over $\mathbb{Q}$, for each $a \in \mathbb{Q}$,

$$x < a \vee x = a \vee a < x \in p.$$

Hence, for each $a \in \mathbb{Q}$, exactly one of the formulas $x < a, x = a, a < x$ belongs to p.

Assume that for some $a \in \mathbb{Q}$, $x = a$ is in p. If possible, suppose there exists a $\psi[x] \in tp(a/\mathbb{Q})$ which is not in p. But then, by the completeness of p, $\neg \psi \in p$. This is a contradiction because $\{x = a, \neg \psi[x]\}$ is not satisfiable in $\mathbb{Q}$. Hence, by completeness, $p = tp(a/\mathbb{Q})$. We also see that in this case p is isolated by $x = a$.

Assume that $p \in S_1(\mathbb{Q}/\mathbb{Q})$ is not realised in $\mathbb{Q}$. Set

$$L_p = \{a \in \mathbb{Q} : a < x \in p\} \ \& \ U_p = \{b \in \mathbb{Q} : x < b \in p\}.$$

Then $L_p \cap U_p = \emptyset$, $L_p \cup U_p = \mathbb{Q}$, $a < b$ whenever $a \in L_p$ and $b \in U_p$, if $a \in L_p$ and $a' < a$ in $\mathbb{Q}$, $a' \in L_p$, whereas if $b \in U_p$ and $b < b'$, $b' \in U_p$. So, each $p \in S_1(\mathbb{Q}/\mathbb{Q})$ determines a cut (L_p, U_p) in $\mathbb{Q}$.

Conversely, suppose (L, U) is a cut in $\mathbb{Q}$. Then,

$$\{a < x : a \in L\} \cup \{x < b : b \in U\}$$

is finitely satisfiable in $\mathbb{Q}$. So, there is a complete 1-type containing all these formulas. Now let p, q be complete 1-types over $\mathbb{Q}$ such that

$$p, q \in \cap_{a \in L}[a < x] \cap \cap_{b \in U}[x < b].$$

This implies that p and q contain the same atomic $L_{\mathbb{Q}}$-formulas $\varphi[x]$. By induction on the rank of open formulas and completeness of p and q, it follows that p and q contain the same open $L_{\mathbb{Q}}$-formulas $\varphi[x]$. Since DLO has quantifier elimination, it follows that $p = q$. Thus, there is a natural one-to-one correspondence between $S_1(\mathbb{Q}/\mathbb{Q})$ and cuts in $\mathbb{Q}$.

This immediately implies that $|S_1(\mathbb{Q}/\mathbb{Q})| = 2^{\aleph_0}$.

Example 3.4.18 Let $\mathbb{K} \models ACF$ and $A \subset \mathbb{K}$. Let κ denote the prime field of $\mathbb{K}$, $\mathbb{A}$ the subfield generated by A and $n \geq 1$.

We claim that $p \to p|A$ is a bijection from $S_n(\mathbb{K}/A) \to S_n(\mathbb{K}/A)$. It is easy to see that this map is a surjection. (We have noted this in Sect. 3.3) Take $p \neq q \in S_n(\mathbb{K}/A)$. Then there is an L_A-formula $\varphi[\bar{x}, \bar{a}]$ such that $\varphi \in p$ and $\neg\varphi \in q$. Now note that there exist $b_1, \ldots, b_m \in A$ and for each a_i, an $f_i \in \kappa(X_1, \ldots X_m)$ such that $a_i = f_i(b_1, \ldots, b_m)$. It is fairly routine to get an L_A-formula $\psi'[\bar{x}]$ with parameters $b_1, \ldots, b_m$ such that $\psi' \in p$ and $\neg\psi' \in q$.

For a complete n-type p in $S_n(\mathbb{K}/\mathbb{A})$, define

$$I_p = \{f[\overline{X}] \in \mathbb{A}[\overline{X}] : f(\overline{x}) = 0 \in p\}.$$

Let $f(\overline{X}), g(\overline{X}) \in \mathbb{A}[\overline{X}]$. Then,

$$\mathbb{K} \models \forall \overline{x}((f(\overline{x}) = 0 \wedge g(\overline{x}) = 0) \to (f + g)(\overline{x}) = 0)$$

and

$$\mathbb{K} \models \forall \overline{x}(f(\overline{x}) = 0 \to (f \cdot g)(\overline{x}) = 0).$$

Since p is complete, these imply that I_p is an ideal on $\mathbb{A}[\overline{X}]$. Now suppose $f, g \in \mathbb{A}[\overline{X}]$ and $f \cdot g \in I_p$, i.e., $(f \cdot g)(\overline{x}) = 0 \in p$. This is equivalent to

$$f(\overline{x}) = 0 \vee g(\overline{x}) = 0 \in p.$$

Since p is complete, either $f(\overline{x}) = 0 \in p$ or $g(\overline{x}) = 0 \in p$. It follows that $f \in I_p$ or $g \in I_p$, i.e., I_p is a prime ideal.

Interestingly, every prime ideal J of $\mathbb{A}[\overline{X}]$ induces a complete n-type p over $\mathbb{A}$ such that $J = I_p$ and this correspondence between $S_n(\mathbb{K}/\mathbb{A})$ and the $Spec(\mathbb{A}[\overline{X}])$, the set of all prime ideals of $\mathbb{A}[\overline{X}]$, is a bijection.

By Proposition 9.2.8, for every prime ideal J in $\mathbb{A}[\overline{X}]$, there is a prime ideal I in $\mathbb{K}[\overline{X}]$ such that

$$J = I \cap \mathbb{A}[\overline{X}].$$

Then, $\mathbb{K}[\overline{X}]/I$ is an integral domain. Let $\mathbb{F}$ denote the algebraic closure of its quotient field. By model-completeness of ACF, $\mathbb{F}$ is an elementary extension of $\mathbb{K}$. Set $a_i = [X_i] \in \mathbb{F}$. Note that, for $f \in \mathbb{K}[\overline{X}]$,

$$f(\overline{a}) = 0 \Leftrightarrow f \in I.$$

So, if $p = tp(\overline{a}/\mathbb{A})$, $I_p = J$.

The quantifier elimination of ACF helps us to prove that this correspondence is one-to-one. Let $p, q \in S_n(\mathbb{K}/\mathbb{A})$ and $I_p = I_q$. Any open $L_\mathbb{A}$-formula is equivalent to a disjunction of formulas of the form

$$\wedge_{i=i}^m (f_i(\overline{x}) = 0) \wedge \wedge_{j=1}^n (g_j(\overline{x}) \neq 0),$$

$f_i, g_j \in \mathbb{A}[\overline{X}]$. Since $I_p = I_q$, it follows that both p and q contain the same open $L_\mathbb{A}$-formulas. Since ACF has quantifier elimination, it follows that p and q contain same formulas and so are equal.

We close this section by giving an application of isolated types.

Proposition 3.4.19 *Let T be a complete theory, $M, N \models T$ and φ a strongly minimal L-formula such that $dim(\varphi(M)) = dim(\varphi(N))$. Then, there is an elementary bijection $g : \varphi(M) \to \varphi(N)$.*

Proof Let A be a basis of $\varphi(M)$ and B that of $\varphi(N)$. Take any bijection $f : A \to B$, Then f is partial elementary by Proposition 2.12.3. By Zorn's lemma, there exists a maximal partial elementary extension $g : A' \to B'$ of f with $A' \subset \varphi(M)$ and $B' \subset \varphi(N)$. If possible, suppose there exists an $a \in \varphi(M) \setminus A'$. Since $a \in acl(A')$, $tp^M(a/A')$ is isolated by Remark 3.4.8. Let $\psi[x, \overline{a}]$ be an $L_{A'}$-formula isolating $tp^M(a/A')$. In particular, $M \models \exists x(\psi[x, \overline{a}] \wedge \varphi[x])$. So there exists a $b \in \varphi[N]$ such that $N \models \psi[b, g(\overline{a})]$. It is easily seen that $g \cup \{(a, b)\}$ is partial elementary, contradicting the maximality of g. Thus, $dom(g) = \varphi(M)$. Similarly, we show that $dom(g^{-1}) = \varphi(N)$. □

Applying this result to the formula $x = x$, we have the following important result as a corollary.

Corollary 3.4.20 *If T is a complete, strongly minimal theory and $M, N \models T$, then M and N are isomorphic if and only if $dim(M) = dim(N)$.*

Corollary 3.4.21 *Let T be a countable, complete, strongly minimal theory. Then, for all $\lambda > \aleph_0$, T is λ-categorical.*

Proof Let $M, N \models T$ be of cardinality $\lambda > \aleph_0$ and A, B be bases of M and N respectively. Then

$$|M| = |acl(A)| = \max\{|A|, \aleph_0\} = \lambda.$$

Hence, $|A| = \lambda$. By the same argument $|B| = \lambda$. The result follows from the last corollary. □

3.5 Algebraic Types

Let M be an L-structure and $A \subset M$. An L_A-formula $\varphi[x]$ is called *algebraic* if there exists a positive integer n such that

$$M \models \exists_{=n}\varphi[x].$$

The integer n is called the *degree* of φ and is denoted by $deg(\varphi)$. If no such n exists, φ is called *non-algebraic*.

A $p \in S_1^M(A)$ is called *algebraic* if it contains an algebraic formula. Otherwise p is called *non-algebraic*.

Proposition 3.5.1 *Let $p \in S_1^M(A)$ be an algebraic type. Then p is an isolated type. In particular, p is realised in M.*

Proof Get a $\varphi[x] \in p$ realised by minimum number of elements in M. Let $\psi \in p$. Then $\varphi \wedge \psi \in p$. So, $(\varphi \wedge \psi)(M) = \varphi(M)$. This implies that $M \models \forall x(\varphi[x] \rightarrow \psi[x])$. Hence, p is isolated by φ by Proposition 3.4.1. □

Proposition 3.5.2 *$p \in S_1^M(A)$ is algebraic if and only if p has only finitely many realisations in every elementary extension of M.*

Proof Let $\varphi[x] \in p$ and suppose there is an $n \geq 1$ such that $M \models \exists_{=n}x\varphi[x]$. Hence, $N \models \exists_{=n}x\varphi[x]$ whenever N is an elementary extension of M. Conversely assume that p is non-algebraic. Add constant symbols $\{c_i : i \in \omega\}$ and consider the theory

$$T = Diag_{el}(M) \cup \{c_i \neq c_j : i \neq j\} \cup \{\varphi[c_i] : \varphi \in p \wedge i \in \omega\}.$$

This is finitely satisfiable in M. Hence, it has a model N which is an elementary extension of M in which p has infinitely many realisations. □

Proposition 3.5.3 *Let $p \in S_1^M(A)$ be non-algebraic and $A \subset B$. Then, p has a non-algebraic extension $q \in S_1^M(B)$.*

Proof Consider

$$q = p \cup \{\neg\psi : \psi \text{ an algebraic } L_B - \text{formula}\}.$$

Then q is finitely satisfiable. Otherwise, there exist algebraic L_B-formula $\psi_1, \ldots, \psi_n$ and a $\varphi \in p$ such that

$$M \models \forall x (\varphi[x] \rightarrow (\psi_1[x] \vee \cdots \vee \psi_n[x])).$$

This implies that $\varphi \in p$ is algebraic, a contradiction. Extend q to a complete type over B, say r. Clearly, r is complete and contains no algebraic L_B-formula. $\qquad \square$

3.6 Omitting Types Theorems

Let T be a complete theory. In the last section we saw that if p is an isolated n-type in T, then there is no model M of T which omits p. Quite interestingly the converse of this fact is also true if T is a countable consistent theory. This is proved by imitating Henkin style proof of the compactness theorem that we presented earlier. The following theorem is due to Henkin [19] and Orey [45].

Theorem 3.6.1 (Omitting Types Theorem) *Let T be a countable consistent L-theory and p a non-isolated n-type in T. Then, there is a countable model M of T that omits p.*

Proof We add countably many distinct new constant symbols $\{c_k\}$ to the language of T and no new non-logical axiom. Denote the new language by L and the new theory by T also. Let $\{\psi_n\}$ be an enumeration of all closed L-formulas and $\{a_k\}$ an enumeration of all n-tuples of new constants c_m's.

We shall get a complete Henkin simple extension T^* of T and a countable model M of T^* such that

(a) Every element of M is the interpretation of some c_k.
(b) For every k, there is a formula $\psi[\overline{x}] \in p$ such that $M \not\models \psi[i_{a_k}]$.

It will then follow that M is a countable model of the original T that omits p.

To construct such a T^* we shall first define a sequence of closed L-formulas $\{\varphi_n\}$ such that

(c) For every k, $T[\varphi_k]$ is consistent.
(d) For $k < l$, $T \models \varphi_l \rightarrow \varphi_k$.
(e) For every k, if ψ_k is an existential sentence $\exists v \eta[v]$ such that $T[\varphi_{2k}] \models \psi_k$, then $T[\varphi_{2k+1}] \models \eta_v[c_m]$ for some m.
(f) For every k, there is a $\psi \in p$ such that $T[\varphi_{2k+2}] \models \neg\psi[a_k]$.

Take φ_0 to be any sentence consistent with T. Suppose φ_{2k} has been defined so that $T[\varphi_{2k}]$ is consistent.

If ψ_k is not an existential sentence, or if $T[\varphi_{2k}] \not\models \psi_k$, we take $\varphi_{2k+1} = \varphi_{2k}$. Otherwise, ψ_k is a closed existential formula, say $\exists v \eta[v]$, and $T[\varphi_{2k}] \models \psi_k$. Since only a finite number of the new constants c_l appear in φ_{2k} and ψ_k, take a constant symbol c_m that does not occur in φ_{2k} and ψ_k. Set $\varphi_{2k+1} = \varphi_{2k} \wedge \eta_v[c_m]$.

We need to show that $T[\varphi_{2k+1}]$ is consistent, i.e., $T[\{\varphi_{2k}, \eta_v[c_m]\}]$ has a model. To see this take a model N of $T[\varphi_{2k}]$. Then $N \models \psi_k$. So, there is a $b \in N$ such that $N \models \eta_v[b]$. Now interpret c_m by b in N. We definitely have $T[\varphi_{2k+1}] \models \eta_v[c_m]$.

Let $a_k = (c_{i_1}, \ldots, c_{i_n})$. Replace each occurrences of c_{i_j} in φ_{2k+1} by a new variable x_j and each $c_m \notin \{c_{i_j} : 1 \le j \le n\}$ occurring in φ_{2k+1} by a new variable y_m to get φ' and set $\varphi''[\overline{x}] = \exists \overline{y} \varphi'$. Because p is not isolated, there is a $\psi[\overline{x}] \in p$ such that

$$T \not\models \forall \overline{x}(\varphi'' \to \psi). \tag{$*$}$$

Set $\varphi_{2k+2} = \varphi_{2k+1} \wedge \neg\psi[a_k]$. We must show that $T[\varphi_{2k+2}]$ is consistent. By $(*)$, there is a model N of T and a $\overline{b} \in N$ such that

$$N \models \varphi''[\overline{b}] \wedge \neg\psi[\overline{b}].$$

Interpreting c_{i_j} by b_j, $1 \le j \le n$, we see that $N \models \varphi_{2k+2}$. This completes our construction.

Let T' be the theory $T[\{\varphi_k : k \ge 0\}]$. Then, T' is a countable consistent Henkin theory such that for every k there exists a $\psi \in p$ such that $T' \models \neg\psi[a_k]$. By Lindenbaum's theorem, T' has a complete simple extension T^*. Clearly, T^* is countable.

Let M be the canonical model of T^* obtained in the proof of Theorem 2.2.3. We claim that every element in M is an interpretation of some c_k. To see this, take a variable-free term $t = f t_1 \cdots t_l$ and consider $\psi = \exists x(x = t)$. Then $T' \models \psi$. So, ψ must have occurred at some stage (e) in our construction, giving us a c_l such that $T^* \models c_l = t$.   □

Proposition 3.6.2 *Let T be a countable consistent theory and $\{p_m\}$ a sequence of non-isolated n-types in T. Then, there is a countable model of T that omits each of p.*

This is proved by imitating the last proof with the following change. For each k of the form $2^q(2r+1)-1$ we ensure that there is a $\psi \in p_q$ such that $T[\varphi_{2k+2}] \models \neg\psi[a_r]$. This clearly can be done. The model M thus built will still be countable such that for each q and each r, there is a $\psi \in p_q$ such that $M \not\models \psi[a_r]$.

Example 3.6.3 Omitting types theorem is not necessarily true if T is not countable. Let T be the theory with uncountably many constant symbols and no other non-logical symbol. Let $C \cup D$ be the set of all constant symbols with C uncountable, D countably infinite, and $C \cap D = \emptyset$. The axioms of T are formulas $c \ne c'$, where c and c' are distinct constant symbols that belong to C.

Let $p(x) = \{x \ne d : d \in D\}$. Clearly $p(x)$ is a 1-type in T. Let $\varphi[x]$ be a formula consistent with T. Since only finitely many constants occur in φ, there is a $d \in D$ such that $\varphi[x] \wedge (x = d)$ is consistent with T. Thus, p is not isolated. As every model of T is uncountable, $p(x)$ is realised in every model of T.

3.7 Cardinalities of the Spaces of Complete Types

In this section, we prove two results on possible cardinalities of spaces of complete types. We need to fix some notation. $2^{<\mathbb{N}}$ will denote the set of all finite sequence of 0's and 1's including the empty sequence e. For $s, t \in 2^{<\mathbb{N}}$ and $\epsilon = 0, 1$, $|s|$ will denote the length of s, $s \prec t$ will mean that t extends s and $s\epsilon$ will stand for the concatenation of s and ϵ.

Lemma 3.7.1 *Let T be an L-theory and $\varphi[\overline{x}]$, $\overline{x} = (x_0, \ldots, x_{n-1})$, an L-formula such that $[\varphi] \neq \emptyset$ and contains no isolated type in T. Then, there exists a formula $\psi[\overline{x}]$ such that*

$$[\varphi \wedge \psi] \neq \emptyset \neq [\varphi \wedge \neg\psi].$$

Proof Suppose for every ψ, $[\varphi \wedge \psi] \neq \emptyset$ implies $[\varphi \wedge \neg\psi] = \emptyset$. Set

$$p = \{\psi : T \models \forall \overline{x}(\varphi \to \psi)\}.$$

Since $T \cup \{\varphi\}$ is satisfiable, $T \cup p$ is satisfiable. In particular, p is a n-type in T.

Also p is complete: Suppose not. Then there is a formula $\psi[\overline{x}]$ such that

$$T \not\models \forall \overline{x}(\varphi \to \psi) \ \& \ T \not\models \forall \overline{x}(\varphi \to \neg\psi).$$

This implies that both $T[\varphi \wedge \psi]$ and $T[\varphi \wedge \neg\psi]$ are satisfiable. Hence,

$$[\varphi \wedge \psi] \neq \emptyset \neq [\varphi \wedge \neg\psi],$$

a contradiction.

Since φ isolates p, we have a contradiction to the fact that $[\varphi]$ contains no isolated type. $\qquad \square$

Theorem 3.7.2 *Let T be an L-theory such that isolated types are not dense in $S_n(T)$. Then $|S_n(T)| \geq 2^{\aleph_0}$.*

Proof We are now going to use the last lemma and for each $s \in 2^{<\mathbb{N}}$ define an L-formula φ_s satisfying the following conditions:

1. If $s \prec t$, $T \models \varphi_t \to \varphi_s$.
2. $[\varphi_s] \neq \emptyset$ and contains no isolated types in T.
3. If $s \neq t$ and $|s| = |t|$, $T[\varphi_s] \models \neg\varphi_t$.

For empty sequence e, let φ_e be an L-formula such that $[\varphi_e] \neq \emptyset$ and contains no isolated types.

Suppose φ_s is defined so that $[\varphi_s] \neq \emptyset$ and contains no isolated types. Further, $t \prec s$ implies that $T \models \varphi_s \to \varphi_t$. By the last lemma, there is a formula ψ such that $[\varphi_s \wedge \psi]$ as well as $[\varphi_s \wedge \neg\psi]$ are non-empty. Set $\varphi_{s0} = \varphi_s \wedge \psi$ and $\varphi_{s1} = \varphi_s \wedge \neg\psi$.

Since $[\varphi_s]$ does not contain an isolated type, neither of $[\varphi_{s0}]$ and $[\varphi_{s1}]$ contains an isolated type. Our construction is complete.

For each $\alpha \in 2^{\mathbb{N}}$, $\{\varphi_{\alpha|n} : n \in \omega\}$ is a type in T. Choose and fix a $p_\alpha \in S_n(T)$. Suppose $\alpha \neq \beta \in 2^{\mathbb{N}}$. Get an n such that $\alpha|n \neq \beta|n$. Then $\varphi_{\alpha|n} \in p_\alpha \backslash p_\beta$. This completes the proof. $\square$

A variant of the above argument gives us the next result. We need a lemma first.

Lemma 3.7.3 *Let κ be an infinite cardinal, L a κ-language, M an L-structure, $A \subset M$ of cardinality at most κ and $\varphi[\overline{x}]$ an L_A-formula such that $|[\varphi[\overline{x}]]| > \kappa$. Then, there is an L_A-formula $\psi[\overline{x}]$ such that $|[\varphi \wedge \psi]| > \kappa$ as well as $|[\varphi \wedge \neg\psi]| > \kappa$.*

Proof Suppose not. Set

$$p = \{\psi : |[\varphi \wedge \psi]| > \kappa\}.$$

Note that $\varphi \in p$. Since $[\varphi] = [\varphi \wedge \psi] \cup [\varphi \wedge \neg\psi]$, for each ψ, either ψ or $\neg\psi$ belongs to p. So, by our assumption, for every ψ exactly one of ψ, $\neg\psi$ belongs to p.

We claim that $p \cup Th_A(M)$ is finitely satisfiable. If not, then there exists $\psi_1, \ldots, \psi_k$ in p such that $\{\wedge_{i=1}^{k} \psi_i\} \cup Th_A(M)$ is not satisfiable. In particular, $\wedge_{i=1}^{k} \psi_i \notin p$. Hence, $\vee_{i=1}^{k} \neg\psi_i \in p$. But this implies that $|[\varphi \wedge \neg\psi_i]| > \kappa$ for some $1 \leq i \leq k$, a contradiction.

Thus, p is a complete n-type and for every $\psi \notin p$, $|[\varphi \wedge \psi]| \leq \kappa$. Now note that

$$[\varphi] = \{p\} \cup \cup_{\psi \notin p}[\varphi \wedge \psi].$$

Since L and A are of cardinality at most κ, it follows that $|[\varphi]| \leq \kappa$. This contradiction proves our result. $\square$

Theorem 3.7.4 *Let T be a countable complete theory, $M \models T$ and $\kappa \geq \aleph_0$. Suppose there exists $A \subset M$ of cardinality κ such that $|S_n^M(A)| > \kappa$. Then, there exists a countable $A_0 \subset A$ such that $|S_n^M(A_0)| = 2^{\aleph_0}$.*

Proof Since $|A| = \kappa$ and T is countable, there are only κ-many L_A-formulas. So, there is an L_A-formula φ_e such that $|[\varphi_e]| > \kappa$.

For each $s \in 2^{<\mathbb{N}}$ will define an L_A-formula φ_s satisfying the following conditions:

1. If $s \prec t$, $T \models \varphi_t \rightarrow \varphi_s$.
2. $|[\varphi_s]| > \kappa$.
3. If $s \neq t$ and $|s| = |t|$, $T[\varphi_s] \models \neg\varphi_t$.

Suppose φ_s is defined so that $|[\varphi_s]| > \kappa$. By the last lemma, there is a formula ψ such that $|[\varphi_s \wedge \psi]|$ as well as $|[\varphi_s \wedge \neg\psi]|$ are greater than κ. Set $\varphi_{s0} = \varphi_s \wedge \psi$ and $\varphi_{s1} = \varphi_s \wedge \neg\psi$. Our construction is complete.

For each $\alpha \in 2^{\mathbb{N}}$, $\{\varphi_{\alpha|k} : k \in \omega\}$ is a n-type in M over A. Choose and fix a $p_\alpha \in \cap_n[\varphi_{\alpha|n}]$. Suppose $\alpha \neq \beta \in 2^{\mathbb{N}}$. Get an n such that $\alpha|n \neq \beta|n$. Then, $\varphi_{\alpha|n} \in p_\alpha \backslash p_\beta$.

Let A_0 be the set of all parameters from A that appear in φ_s's. Clearly, A_0 is countable and all the p_α's are complete n-types over A_0. In particular, $|S_n^M(A_0)| = 2^{\aleph_0}$. □

Corollary 3.7.5 *Let T be a countable, complete theory, with $M \models T$ and $A \subset M$ countable. Then, exactly one of the following holds:*

1. $|S_n^M(A)| \leq \aleph_0$.
2. $|S_n^M(A)| = 2^{\aleph_0}$.

By imitating these proofs, we get the following two results.

Lemma 3.7.6 *Let M be an L-structure and $A \subset M$. Suppose $\varphi[\overline{x}]$ is an L_A-formula such that $[\varphi] \neq \emptyset$ and contains no isolated type. Then, there is an L_A-formula ψ such that*

$$[\varphi \wedge \psi] \neq \emptyset \neq [\varphi \wedge \neg\psi].$$

Theorem 3.7.7 *Let T be a countable, complete theory, $M \models T$, and $A \subset M$. Suppose isolated types are not dense in $S_n^M(A)$. Then, there exists a countable $A_0 \subset A$ such that $|S_n^M(A_0)| = 2^{\aleph_0}$.*

Remark 3.7.8 Suppose $A \subset M$ is countable and $|S_n^M(A)| < 2^{\aleph_0}$. Then isolated types are dense in $S_n^M(A)$.

Chapter 4
Good Structures and Good Theories

Abstract In this chapter, we initiate a systematic study of important classes of structures and theories. Of particular importance are saturated structures and $\aleph_0$-categorical and stable theories. We also introduce Morley rank and Morley degrees and generalise the notion of independence in minimal sets given in Chap. 2 to forking independence.

4.1 Homogeneous Structures

Let M be an L-structure and κ an infinite cardinal. We call M κ-*homogeneous* if for all $A \subset M$ of cardinality less than κ, for all partial elementary map $f : A \to M$ and for all $a \in M$, there is a partial elementary map $g : A \cup \{a\} \to M$ extending f. This is the same as saying the following: For every sequences $\bar{a}, \bar{b} \in M$ of length $< \kappa$ with $tp^M(\bar{a}) = tp^M(\bar{b})$, for every $a \in M$ there is a $b \in M$ such that $tp^M(\bar{a}a) = tp^M(\bar{b}b)$. We call M *homogeneous* if it is $|M|$-homogeneous.

We call M κ-*strongly homogeneous* if for all $A \subset M$ of cardinality less than κ, every partial elementary map $f : A \to M$ extends to an automorphism of M. This is the same as saying the following: For every sequences $\bar{a}, \bar{b} \in M$ of length $< \kappa$, $tp^M(\bar{a}) = tp^M(\bar{b})$ if and only if there is an automorphism g of M such that $g(\bar{a}) = \bar{b}$. We call M *strongly homogeneous* if it is $|M|$-strongly homogeneous.

Example 4.1.1 The linearly ordered set of rationals $\mathbb{Q}$ is strongly homogeneous. This follows from Exercise 1.6.11.

Following the back and forth argument, we have the following theorem.

Proposition 4.1.2 *Every homogeneous L-structure is strongly homogeneous.*

Proof Let $A \subset M$ be of cardinality less than that of M and $f : A \to M$ partial elementary. Enumerate $M \setminus A = \{a_\alpha : \alpha < |M|\}$. Set $f_0 = f$. By transfinite induction, for each $\alpha < |M|$, we define a partial elementary map f_α from a subset of M into M such that

© Springer Nature Singapore Pte Ltd. 2017
H. Sarbadhikari and S.M. Srivastava, *A Course on Basic Model Theory*,
DOI 10.1007/978-981-10-5098-5_4

(∗) $|domain(f_\alpha)| < |M|$,
(∗) for $\beta < \alpha < |M|, f_\alpha$ extends f_β,
(∗) $f_\alpha = \cup_{\beta<\alpha} f_\beta$ if α is limit, and
(∗) a_α belongs to the domain as well as to the range of $f_{\alpha+1}$.

This will complete the proof, because $\cup_{\alpha<|M|} f_\alpha$ will be an automorphism of M extending f.

Suppose f_α has been defined. If a_α is in the domain of f_α, set $g = f_\alpha$. Otherwise, by homogeneity, there is a partial elementary extension g of f_α from $domain(f_\alpha) \cup \{a_\alpha\}$ to M.

If a_α is in the range of g, take $f_{\alpha+1} = g$. Otherwise, by homogeneity, there is a partial elementary map h extending g^{-1} from $range(g) \cup \{a_\alpha\}$ to M. Now take $f_{\alpha+1} = h^{-1}$.   □

Corollary 4.1.3 *Let M be an infinite homogeneous L-structure, $A \subset M$ with $|A| < |M|$ and $\overline{a}, \overline{b} \in M^n$. Then $tp^M(\overline{a}/A) = tp^M(\overline{b}/A)$ if and only if there is an $\alpha \in Aut(M)_A$ such that $\alpha(\overline{a}) = \overline{b}$. In particular, $a \in M$ has finitely many conjugates in M over A if and only if $tp^M(a/A)$ has only finitely many realisations in M.*

Proof Note that $|A \cup \{a_i : i < |\overline{a}|\}| < |M|\}$ and $f : A \cup \{a_i : i < |\overline{a}|\} \rightarrow M$ such that f is identity on A and $f(\overline{a}) = \overline{b}$ is partial elementary. Hence, the result follows from the last proposition.   □

Lemma 4.1.4 *Let M be an L-structure and $\overline{a}, \overline{b} \in M^n$ be such that $\overline{a} \rightarrow \overline{b}$ is partial elementary. Then for every $c \in M$, there is an elementary extension N of M and a $d \in N$ such that $(\overline{a}, c) \rightarrow (\overline{b}, d)$ is partial elementary. Further, if M is infinite and L countable, we can choose N so that $|N| = |M|$.*

Proof Consider the theory

$$T = Diag_{el}(M) \cup \{\varphi[x, \overline{b}] : M \models \varphi[c, \overline{a}]\}.$$

Since $\overline{a} \rightarrow \overline{b}$ is partial elementary, whenever $M \models \varphi_i[c, \overline{a}]$, $i < n$, $M \models \exists x \wedge_i \varphi_i[x, \overline{b}]$. Thus, T is finitely satisfiable. Hence, there is an L-structure N in which T is realised. Further, by downward Löwenheim–Skolem theorem, if L is countable and M infinite, we can choose N so that $|N| = |M|$. Such an N has all the desired properties.   □

Lemma 4.1.5 *Let M be an infinite L-structure with L countable. Then there exists an elementary extension N of M satisfying the following conditions:*

1. *$|N| = |M|$, and*
2. *whenever $\overline{a}, \overline{b} \in M$ with $\overline{a} \rightarrow \overline{b}$ partial elementary, for every $c \in M$ there is a $d \in N$ such that $(\overline{a}, c) \rightarrow (\overline{b}, d)$ is partial elementary.*

Proof Let $\{(\overline{a}_\alpha, \overline{b}_\alpha, c_\alpha) : \alpha < |M|\}$ be an enumeration of all finite tuples $(\overline{a}, \overline{b}, c)$ in M with $\overline{a} \rightarrow \overline{b}$ partial elementary. Applying the last lemma repeatedly, by transfinite induction, we can build an elementary chain $\{N_\alpha : \alpha < |M|\}$ of L-structures satisfying the following conditions:

(*) $N_0 = M$.
(*) $N_\alpha = \cup_{\beta < \alpha} N_\beta$ if α is limit.
(*) $\forall \alpha < |M|(|N_\alpha| = |M|)$, and
(*) $\forall \alpha < |M| \exists d_\alpha \in N_{\alpha+1}((\bar{a}_\alpha, \bar{b}_\alpha, c_\alpha) \to (\bar{a}_\alpha, \bar{b}_\alpha, d_\alpha)$ is partial elementary).

Finally, take $N = \cup_{\alpha < |M|} N_\alpha$. □

Proposition 4.1.6 *Let M be an infinite L-structure with L countable. Then there exists a $\aleph_0$-homogeneous elementary extension N of M such that $|N| = |M|$.*

Proof Set $N_0 = M$. By repeatedly applying the last lemma, we have an elementary chain $\{N_k : k \in \omega\}$ such that whenever $\bar{a}, \bar{b} \in N_k$ with $\bar{a} \to \bar{b}$ partial elementary, for every $c \in N_k$ there is a $d \in N_{k+1}$ such that $(\bar{a}, c) \to (\bar{b}, d)$ is partial elementary. Now take $N = \cup_k N_k$. □

Using the back-and-forth argument, we get the following surprising result.

Proposition 4.1.7 *Let M and N be countable homogeneous L-structures. The following conditions are equivalent.*

1. *M and N are isomorphic.*
2. *For every $k \geq 1$,*

$$\{tp^M(\bar{a}) : \bar{a} \in M^k\} = \{tp^N(\bar{b}) : \bar{b} \in N^k\}.$$

Proof Clearly (1) implies (2). Next assume (2). Then for every $n \in \omega$, $|M| = n$ if and only if $|N| = n$. Further, if M is finite, M is isomorphic to N. So, assume that $|M| = |N| = \aleph_0$. Fix enumerations $\{a_k\}$ and $\{b_k\}$ of M and N, respectively.

Set $a_0' = a_0$ and consider $tp^M(a_0')$. By our hypothesis, there is a $b \in N$ such that $tp^M(a_o') = tp^N(b)$. Let b_0' be the first such b in the above enumeration of N.

Now let b_1' be the first element in the enumeration of N different from b_0'. By our hypothesis, there exists $a, a' \in M$ such that $tp^M(a, a') = tp^N(b_0', b_1')$. In particular, $tp^M(a) = tp^N(b_0') = tp^M(a_0')$. So, $a \to a_0'$ is partial elementary. Since M is homogeneous, there is an $a'' \in M$ such that $(a, a') \to (a_0', a'')$ is partial elementary. Therefore, $tp^N(b_0', b_1') = tp^M(a, a') = tp^M(a_0', a'')$. Since $b_0' \neq b_1'$, $x \neq y$ is in $tp^N(b_0', b_1')$. This implies that $a_0' \neq a''$. We let a_1' denote the first such a'' in the enumeration of M.

Now let a_2' be the first element in the enumeration of M not belonging to $\{a_0', a_1'\}$. By our hypothesis, there exist $b, b', b'' \in N$ such that $tp^N(b, b', b'') = tp^M(a_0', a_1', a_2')$. In particular, $tp^N(b, b') = tp^M(a_0', a_1') = tp^N(b_0', b_1')$. So, $(b, b') \to (b_0', b_1')$ is partial elementary. Since N is homogeneous, there exists a $b''' \in N$ such that $(b, b', b'') \to (b_0', b_1', b''')$ is partial elementary. Hence, $tp^N(b_0', b_1', b''') = tp^N(b, b', b'') = tp^M(a_0', a_1', a_2')$. Since $a_2' \notin \{a_0', a_1'\}$, $b''' \notin \{b_0', b_1'\}$. Let b_2' be the first such b''' in the enumeration of N.

Continuing this back-and-forth method, we shall get enumerations $\{a'_k\}$ and $\{b'_k\}$ of M and N, respectively, such that for every k, $(a'_o, \ldots, a'_k) \rightarrow (b'_o, \ldots, b'_k)$ is partial elementary. Plainly, $a'_i \rightarrow b'_i$ defines an isomorphism from M to N. $\square$

We now extend this result for arbitrary cardinal numbers.

Proposition 4.1.8 *Let M and N be elementarily equivalent L-structures with N κ-homogeneous, where κ is an infinite cardinal. Suppose for every $\overline{a} \in M^n$ there exists $\overline{b} \in N^n$ such that $tp^M(\overline{a}) = tp^N(\overline{b})$. Then for every $A \subset M$ with $|A| \leq \kappa$, there is a partial elementary map $f_\infty : A \rightarrow N$.*

Proof First assume that A is finite, say $A = \{a_0, \ldots, a_n\}$. By our hypothesis, there exist $b_0, \ldots, b_n \in B$ such that $tp^M(\overline{a}) = tp^N(\overline{b})$. Plainly, $a_i \rightarrow b_i$, $i \leq n$, is an elementary map from A into N.

We complete the proof by induction on $|A|$. Let $\lambda \leq \kappa$ and the result be true for $A \subset M$ of cardinality less than λ. Take any $A = \{a_\alpha : \alpha < \lambda\} \subset M$ of cardinality λ. By induction on $\alpha < \lambda$, we define elementary maps $f_\alpha : \{a_\beta : \beta < \alpha\} \rightarrow N$ such that f_α extends f_β whenever $\beta < \alpha$. This will then complete the proof by taking $f_\infty = \cup_{\alpha < \lambda} f_\alpha$.

Suppose $\alpha < \lambda$ and f_β, $\beta < \alpha$, have been defined. If α is a limit ordinal, we define $f_\alpha = \cup_{\beta < \alpha} f_\beta$. Now suppose $\alpha = \beta + 1$ is a successor ordinal. By our assumption, there is a partial elementary map $f : \{a_\gamma : \gamma \leq \beta\} \rightarrow N$. Let $B = f(\{a_\gamma : \gamma < \beta\})$ and C the range of f_β. Note that every partial elementary map is injective. So, we have a partial elementary map $g = f_\beta \circ f^{-1} : B \rightarrow C$. Since N is κ-homogeneous, there is a partial elementary map $h : B \cup \{f(a_\beta)\} \rightarrow N$ extending g. Suppose $b = h(f(a_\beta))$. Take $f_\alpha = f_\beta \cup \{(a_\beta, b)\}$. $\square$

Proposition 4.1.9 *Let M and N be elementarily equivalent homogeneous L-structures such that $|M| = |N|$. The following conditions are equivalent.*

1. *M and N are isomorphic.*
2. *For every $k \geq 1$,*

$$\{tp^M(\overline{a}) : \overline{a} \in M^k\} = \{tp^N(\overline{b}) : \overline{b} \in N^k\}.$$

Proof We need to prove (2) implies (1) only. Further, we can assume that $|M| = |N| = \kappa > \aleph_0$. Fix enumerations $M = \{a_\alpha : \alpha < \kappa\}$ and $N = \{b_\alpha : \alpha < \kappa\}$. Set f_0 to be the empty function.

By induction, we define partial elementary maps f_α, $\alpha < \kappa$, as follows: We take f_0 to be the empty function. Assume f_α has been defined. Let a be the first element in the above enumeration of M that does not belong to the domain of f_α. Arguing as the last proposition we see that there is a partial elementary map, say $g = f_\alpha \cup \{(a, b)\}$ into N. Now let b be the first element in the above enumeration of N not in the range of g. Similarly there is a partial elementary map, say $h = g^{-1} \cup \{(b, a)\}$ into M. We take $f_{\alpha+1} = h^{-1}$. In the limit case, we take $f_\alpha = \cup_\beta f_\beta$.

Plainly, $f = \cup_{\alpha < \kappa} f_\alpha$ is an isomorphism from M onto N. $\square$

Exercise 4.1.10 Equip $\mathbb{Q} \times \mathbb{R}$ with lexicographic order. Show that it is not homogeneous.

Exercise 4.1.11 Let κ be a regular cardinal and $\{M_\alpha : \alpha < \kappa\}$ an elementary chain of κ-homogeneous L-structures. Show that $\cup_{\alpha<\kappa}M_\alpha$ is κ-homogeneous.

Exercise 4.1.12 Let M be a countable structure of a countable language L and $\bar{a}, \bar{b} \in M^n$. Show that $tp^M(\bar{a}) = tp^M(\bar{b})$ if and only if there is an elementary extension N of M and a $f \in Aut(N)$ such that $f(\bar{a}) = \bar{b}$.

Exercise 4.1.13 Show that every algebraically closed field is homogeneous.

4.2 Atomic Structures

We call an L-structure M *atomic* if for every $n \geq 1$ and every $\bar{a} \in M^n$, $tp^M(\bar{a})$ is isolated.

Example 4.2.1 Consider the linearly ordered set $\mathbb{Q}$. In Example 3.4.16 we showed that $tp^{\mathbb{Q}}(\bar{a})$ is isolated for every finite tuple $\bar{a}$ of rational numbers. Hence, $\mathbb{Q}$ is atomic.

Example 4.2.2 Consider the ordered field $\mathbb{R}$. In Example 3.4.15, we saw that $tp^{\mathbb{R}}(a) \neq tp^{\mathbb{R}}(b)$ whenever a and b are distinct real numbers. Since there are only countably many formulas, there are at most countably many isolated $tp^{\mathbb{R}}(a)$. Hence, the ordered field $\mathbb{R}$ is not atomic.

Theorem 4.2.3 *Every countable atomic L-structure M is homogeneous.*

Proof Let $\bar{a}, \bar{b} \in M^n$ such that the map $\bar{a} \to \bar{b}$ is partial elementary. Take an $a \neq a_i$, $i < n$. Since M is atomic, there is a formula $\varphi[x_0, \ldots, x_n]$ that isolates $tp^M(\bar{a}, a)$. In particular, $M \models \varphi[\bar{a}, a]$ implying $M \models \exists x \varphi[\bar{a}, x]$. Since $\bar{a} \to \bar{b}$ is partial elementary, $M \models \exists x \varphi[\bar{b}, x]$. Hence, there is a $b \in M$ such that

$$M \models \varphi[\bar{b}, b]. \tag{$*$}$$

The proof will be completed by showing that

$$tp^M(\bar{a}, a) = tp^M(\bar{b}, b).$$

Let $\psi[\bar{x}, x] \in tp^M(\bar{a}, a)$. Then

$$M \models \forall \bar{x} \forall x (\varphi[\bar{x}, x] \to \psi[\bar{x}, x]).$$

By $(*)$, $M \models \psi[\bar{b}, b]$, i.e. $\psi \in tp^M(\bar{b}, b)$. So, $tp^M(\bar{a}, a) \subset tp^M(\bar{b}, b)$. This implies that $tp^M(\bar{a}, a) = tp^M(\bar{b}, b)$. $\square$

Theorem 4.2.4 *Let T be a countable complete theory. Then a model M of T is prime if and only if it is countable and atomic.*

Proof Let $M \models T$ be prime. Since every countable consistent theory has a countable model (Corollary 1.7.7), every prime model of a countable theory is countable. Hence, M is countable.

Take any $\bar{a} \in M$. If possible, suppose $tp^M(\bar{a})$ is non-isolated. By omitting types theorem (Theorem 3.6.1), there is a model N of T omitting $tp^M(\bar{a})$. Since M is prime, there is an elementary embedding $\alpha : M \to N$. However, as $\alpha : M \to N$ elementary, $\alpha(\bar{a})$ realises $tp^M(\bar{a})$. This contradiction proves the only if part of the result.

To prove the converse, let M be a countable atomic model of T and $N \models T$. Fix an enumeration $\{a_k\}$ of M. Since M is atomic, for each $(a_0, \ldots, a_k)$ there is a formula $\varphi_k[x_0, \ldots, x_k]$ that isolates $tp^M(a_0, \ldots, a_k)$. By induction, for each k, we shall define a partial elementary map $\alpha_k : \{a_i : i < k\} \to N$ such that α_{k+1} extends α_k for each k. It will follow that $\alpha = \cup_k \alpha_k : M \to N$ is elementary.

Since T is complete, M and N are elementarily equivalent. So, the empty function from M to N is indeed elementary. Suppose $\alpha_k : \{a_i : i < k\} \to N$ has been defined and is partial elementary.

Since $M \models \varphi_k[a_0, \ldots, a_k], M \models \exists x \varphi_k[a_0, \ldots, a_{k-1}, x]$. Since α_k is partial elementary, we get $N \models \exists x \varphi_k[\alpha_k(a_0), \ldots, \alpha_k(a_{k-1}), x]$. This gives us a $b \in N$ such that $N \models \varphi_k[\alpha_k(a_0), \ldots, \alpha_k(a_{k-1}), b]$. We let $\alpha_{k+1} : \{a_i : i \le k\} \to N$ be the extension of α_k with $\alpha_{k+1}(a_k) = b$. We need to show that α_{k+1} is partial elementary. Because φ_k isolates $tp^M(a_0, \ldots, a_k)$, as we argued earlier,

$$tp^M(a_0, \ldots, a_k) = tp^N(\alpha_{k+1}(a_0), \ldots, \alpha_{k+1}(a_k)),$$

showing that α_{k+1} is partial elementary. $\square$

As a corollary we get,

Example 4.2.5 The field of all algebraic numbers, $\overline{\mathbb{F}}_p$, p a prime, the set of all real algebraic numbers both as a field as well as an ordered field and $\mathbb{Q}$ both as models of *DAG* and *ODAG* are atomic.

Theorem 4.2.6 *Let T be a countable complete L-theory and M and N prime models of T. Then M and N are isomorphic.*

Proof By the last theorem (Theorem 4.2.4), M and N are countable and atomic. Hence, they are homogeneous by Theorem 4.2.3. Also, each realised types in M and N is isolated. Then since T is complete, M and N realise the same complete types. To see this take an $\bar{a} \in M^n$ and an L-formula $\varphi[\bar{x}]$ that isolates $tp^M(\bar{a}) \in S_n(T)$. So, for every L-formula $\psi[\bar{x}]$,

$$M \models \psi[\bar{a}] \Leftrightarrow T \models \forall \bar{x}(\varphi[\bar{x}] \to \psi[\bar{x}]).$$

Since $M \models \exists \bar{x} \varphi[\bar{x}]$, as T is complete, there is a $\bar{b} \in N^n$ such that $N \models \varphi[\bar{b}]$. It follows that $tp^N(\bar{b}) = tp^M(\bar{a})$. Similarly, we prove that for every $\bar{b} \in N^n$, there is an $\bar{a} \in M^n$ such that $tp^M(\bar{a}) = tp^N(\bar{b})$. Hence, by Theorem 4.1.7, M and N are isomorphic. $\square$

Proposition 4.2.7 *Let T be a complete theory. If T has an atomic model, then isolated types are dense in $S_n(T)$.*

Proof Fix an atomic $M \models T$. Suppose $[\varphi[\bar{x}]] \neq \emptyset$. Then there is a model $N \models T$ such that $N \models \exists \bar{x} \varphi$. Since T is complete, $M \models \exists \bar{x} \varphi$. So, there exists $\bar{a} \in M$ such that $M \models \varphi[\bar{a}]$. Thus, $tp^M(\bar{a}) \in [\varphi]$. Since M is atomic, $tp^M(\bar{a})$ is isolated and our result is proved. □

Interestingly, the converse of this result is true for countable, complete theories. The proof is a Henkin type construction of models.

Theorem 4.2.8 *Let T be a countable complete theory such that for every $n \geq 1$, isolated types are dense in $S_n(T)$. Then T has a countable atomic model. In particular, T have a prime model.*

Proof We add an infinite sequence of distinct and new constants, say $c_0, c_1, c_2, \ldots$, to T but no new non-logical axiom and still call the theory T. Let $\{\varphi_n\}$ be an enumeration of all the sentences of T.

By induction on n, we shall now define a sequence of sentences $\{\psi_n\}$ such that $T[\{\psi_n\}]$ is a complete Henkin theory whose canonical structure is a countable atomic model of T.

We take $\psi_0 = \exists x(x = x)$. Suppose $n = 3m$ and $\psi_0, \ldots \psi_n$ have been defined.

If $T[\psi_n \wedge \varphi_m]$ is satisfiable, we take $\psi_{n+1} = \psi_n \wedge \varphi_m$, else set $\psi_{n+1} = \psi_n \wedge \neg \varphi_m$. So, $T[\psi_{n+1}]$ is consistent.

If φ_m is not a closed existential formula, we take $\psi_{n+2} = \psi_{n+1}$. Suppose φ_m is a closed existential formula, say $\exists x \varphi[x]$. If $T[\psi_{n+1}] \nvdash \varphi_m$, take $\psi_{n+2} = \psi_{n+1}$. Otherwise, we take the first new constant symbol c_k not occurring in $T[\psi_{n+1}]$ and set $\psi_{n+2} = \psi_{n+1} \wedge \varphi[c_k]$. It is easy to see that $T[\psi_{n+2}]$ is consistent.

Finally, let k be the first integer such that the constants occurring in $T[\psi_{n+2}]$ are among $c_0, \ldots, c_k$. By choosing a variant of ψ_{n+2}, if necessary, let $\psi[x_0, \ldots, x_k]$ be such that $\psi_{n+2} = \psi[\bar{c}]$. So, $[\psi] \neq \emptyset$. Let $p \in [\psi]$ be an isolated $(k+1)$-type, isolated by, say $\eta[\bar{x}]$. We set $\psi_{n+3} = \psi_{n+2} \wedge \eta[\bar{c}]$. Clearly, $T[\psi_{n+3}]$ is consistent.

It is fairly routine to check that $T[\{\psi_n\}]$ is a complete Henkin theory. Let M be its canonical model. We claim that M is an atomic model of T. Take $\bar{a} \in M$. Let k be the least integer such that all a_i's occur among $(c_0)_M, \ldots, (c_k)_M$ and there exists an $m = 3j + 2$ such that all the constants occurring in ψ_m occur among $c_0, \ldots, c_k$. By our construction, $tp^M[\bar{c}_M]$ is isolated. Hence, as proved before, $tp^M(\bar{a})$ is isolated. □

Corollary 4.2.9 *Let T be a countable complete theory such that for some $n \geq 1$, $|S_n(T)| < 2^{\aleph_0}$. Then T has a prime model.*

Proof Let $M \models T$. Since T is complete, $S_n(T) = S_n^M(\emptyset)$. Under our hypothesis, by Corollary 3.7.8, isolated points are dense in $S_n(T)$. The result now follows by the last theorem. □

4.3   Saturated Structures

Let M be an L-structure and κ an infinite cardinal. We call M κ-*saturated* if for every $A \subset M$ of cardinality less than κ, every type in $S_1(M/A)$ is realised in M. We call M *saturated* if it is $|M|$-saturated.

Proposition 4.3.1 *If M is κ-saturated, $\kappa \geq \aleph_0$, and $\overline{x}$ is a sequence of distinct variables of length $\alpha \leq \kappa$, then for every $A \subset M$ of cardinality less than κ, every complete type $p(\overline{x})$ over A is realised in M.*

Proof By transfinite induction, we define a sequence $\{a_\beta : \beta < \alpha\}$ in M such that for every $\beta < \alpha$, $\{a_\gamma : \gamma < \beta\}$ realises the type $p_\beta(\{x_\gamma : \gamma < \beta\})$ over A defined by

$$\{\exists\{x_\delta : \delta \geq \beta\}\varphi[\{x_\gamma : \gamma < \beta\}, \{x_\delta : \delta \geq \beta\}] : \varphi \in p(\overline{x})\}.$$

Suppose $\beta < \alpha$ and $\{a_\gamma : \gamma < \beta\}$ satisfying the above condition have been defined. Now consider the 1-type

$$q(x_\beta) = \{\exists\{x_\delta : \delta > \beta\}\varphi[\{a_\gamma : \gamma < \beta\}, x_\beta, \{x_\delta : \delta > \beta\}] : \varphi \in p(\overline{x})\}$$

over $A \cup \{a_\gamma : \gamma < \beta\}$. Since κ is infinite, $|A \cup \{a_\gamma : \gamma < \beta\}| < \kappa$. As M is κ-saturated, there is an $a_\beta \in M$ realising $q(x_\beta)$. This completes the construction and our proof.   □

Proposition 4.3.2 *Let M and N be L-structures with N κ-saturated, $\kappa \geq \aleph_0$, $A \subset M$ of cardinality less than κ, $f : A \to N$ partial elementary and $a \in M \setminus A$. Then there is a partial elementary map $g : A \cup \{a\} \to N$ that extends f.*

Proof Take any $\overline{a} \in A^n$ and a formula $\varphi[x, \overline{x}]$ of L such that $M \models \varphi[a, \overline{a}]$. Then $M \models \exists x \varphi[x, \overline{a}]$. Since f is partial elementary, $N \models \exists x \varphi[x, f(\overline{a})]$. From this it easily follows that

$$p = \{\varphi[x, f(\overline{a})] : \overline{a} \in A^n \wedge M \models \varphi[a, \overline{a}]\}$$

is finitely satisfiable in N, i.e. p is a 1-type over $f(A)$ which is of cardinality less than κ. Since N is κ-saturated, there is a $b \in N$ that realises it. This implies that $tp^M(a, \overline{a}) = tp^N(b, f(\overline{a}))$. So, $g = f \cup \{(a, b)\}$ is partial elementary. This proves our result.   □

Corollary 4.3.3 *Every κ-saturated L-structure is κ-homogeneous.*

Corollary 4.3.4 *Every saturated L-structure is strongly homogeneous.*

A model M of an L-theory T is called κ-*universal* if every model N of T of cardinality less that κ is elementarily embedded into M.

Proposition 4.3.5 *Let T be a complete theory. Then every κ-saturated model M of T is κ^+-universal.*

Proof Let $N \models T$ and $|N| \leq \kappa$. Fix an enumeration $\{a_\alpha : \alpha < \kappa\}$ of N. For $\alpha < \kappa$, set $A_\alpha = \{a_\beta : \beta < \alpha\}$.

Set f_0 to be the empty function. Since T is complete, M and N are elementarily equivalent. So, f_0 is partial elementary. Proceeding by transfinite induction and using the last proposition repeatedly, for each $\alpha < \kappa$, we get a partial elementary map $f_\alpha : A_\alpha \to M$ such that f_α extends f_β whenever $\beta < \alpha$ and $f_\alpha = \cup_{\beta<\alpha} f_\beta$; if α is a limit ordinal. Then $f = \cup_{\alpha<\kappa} f_\alpha : N \to M$ is an elementary embedding. $\square$

A converse of this result is true.

Proposition 4.3.6 *Let $\kappa \geq \aleph_0$. Every κ-homogeneous, κ^+-universal model M of a κ-theory T is κ-saturated. Moreover, if κ is uncountable and T countable, we can replace the condition κ^+-universality by κ-universality of M.*

Proof Let $A \subset M$ be of cardinality $< \kappa$ and $p \in S_1(M/A)$. Then, by downward Löwenheim–Skolem theorem, there is an infinite elementary substructure N of M containing A of cardinality $\leq \kappa$. Moreover, if κ is uncountable and T countable, we get N of cardinality $< \kappa$. Since $p \in S_1(N/A)$, there is an elementary extension N' of N such that $|N'| = |N|$ and there is a $a \in N'$ that realises p. By the universality of M, there is an elementary embedding $f : N' \to M$. Now $f^{-1} : f(A) \to A \subset M$ is partial elementary. So, by homogeneity of M, there is a $b \in M$ such that $f^{-1} \cup \{(f(a), b)\}$ is partial elementary. Then b realises p in M. $\square$

Proposition 4.3.7 *Let M be κ-saturated, $A \subset M$ of cardinality less than κ, $\lambda < \kappa$ and M' an elementary extension of M. Then for every sequence $\{\bar{a}_\alpha : \alpha < \kappa\}$ of λ-tuples in M' of length κ there is a sequence $\{\bar{b}_\alpha : \alpha < \kappa\}$ of λ-tuples in M of length κ such that*

$$tp^M(\{\bar{b}_\alpha : \alpha < \kappa\}/A) = tp^{M'}(\{\bar{a}_\alpha : \alpha < \kappa\}/A).$$

Proof We build $\{\bar{b}_\alpha : \alpha < \kappa\}$ by transfinite induction. Suppose $\alpha < \kappa$ and $\bar{b}_\beta \in M^\lambda$, $\beta < \alpha$, have been defined so that

$$tp^M(\{\bar{b}_\beta : \beta < \alpha\}/A) = tp^{M'}(\{\bar{a}_\beta : \beta < \alpha\}/A).$$

Since M is κ-saturated, there exist $\bar{b}'_\beta$, $\beta \leq \alpha$, in M^λ such that

$$tp^M(\{\bar{b}'_\beta : \beta \leq \alpha\}/A) = tp^{M'}(\{\bar{a}_\beta : \beta \leq \alpha\}/A).$$

This implies that

$$tp^M(\{\bar{b}'_\beta : \beta < \alpha\}/A) = tp^{M'}(\{\bar{a}_\beta : \beta < \alpha\}/A) = tp^M(\{\bar{b}_\beta : \beta < \alpha\}/A).$$

Therefore, $\bar{b}'_\beta \to \bar{b}_\beta$, $\beta < \alpha$, is partial elementary over A. Since M is κ-saturated, by Corollary 4.3.3, it is κ-homogeneous. Hence, there exists a $\bar{b}_\beta \in M^\lambda$ such that $\bar{b}'_\beta \to \bar{b}_\beta$, $\beta \leq \alpha$, is partial elementary over A. This implies that

$$tp^M(\{\bar{b}_\beta : \beta \leq \alpha\}/A) = tp^{M'}(\{\bar{a}_\beta : \beta \leq \alpha\}/A).$$

The proof is easily seen now. $\square$

Theorem 4.3.8 *Let* $M \models T$ *be an* $\aleph_0$-*homogeneous model that realises every complete type in* T. *Then* M *is* $\aleph_0$-*saturated.*

Proof Let $\bar{a} \in M^n$ and $p(\bar{x})$ a complete m-type over $\bar{a}$. Consider

$$q[\bar{x}, \bar{y}] = \{\varphi[\bar{x}, \bar{y}] : \varphi[\bar{x}, \bar{a}] \in p\}.$$

By our assumptions, there is a $(\bar{b}, \bar{c}) \in M^{m+n}$ that realises q. This implies that $\bar{c} \to \bar{a}$ is partial elementary. By homogeneity of M, there is a partial elementary extension $(\bar{b}, \bar{c}) \to (\bar{d}, \bar{a})$ of $\bar{c} \to \bar{a}$. This implies that $\bar{d}$ realises p. $\square$

This result is true for all cardinality provided T is complete.

Theorem 4.3.9 *Let* T *be a complete theory. Then every* κ-*homogeneous model* M *of* T *that realises every* $p \in S_n(T)$ *for all* $n \geq 1$ *is* κ-*saturated.*

Proof By Proposition 4.3.6, it is sufficient to prove that M is κ^+-universal. Take $N \models T$ of cardinality $\leq \kappa$. Since T is complete, M and N are elemenatarily equivalent. By hypothesis, for every tuple $\bar{a} \in N$, there is a $\bar{b} \in M$ such that $tp^N(\bar{a}) = tp^M(\bar{b})$. Hence, by Proposition 4.1.8, N is elementarily embeddable in M. $\square$

As a consequence, we now give a characterization of countable complete, theories that has a countable saturated model.

Theorem 4.3.10 *Let* T *be a countable complete theory. Then* T *has a countable saturated model if and only if* $|\cup_n S_n(T)| \leq \aleph_0$.

Proof Let M be a countable, saturated model of T. Since for every $n \geq 1$, $S_n(T) = S_n(M)$, only if part is easy.

For proving if part, start with a countable $N_0 \models T$. Let $\{p_k : k \in \omega\} = \cup_n S_n(T)$. Get an elementary chain $\{N_k : k \in \omega\}$ of countable L-structures such that for every $k \in \omega$, p_k is realised in N_{k+1}. Then $\cup_k N_k \models T$ is countable and realises all $p \in \cup_n S_n(T)$. By Proposition 4.1.6, there is a homogeneous, countable, elementary extension N of $\cup_k N_k$. By the last theorem, N is saturated. $\square$

Below we give another criterion for quantifier elimination. Let M and N be L-structures and $\mathcal{I}(M, N)$ denote the set of all finite partial isomorphisms $M \ni \bar{a} \to \bar{b} \in N$. We say that M, N has *back-and-forth property* if for every $\bar{a} \to \bar{b}$ in $\mathcal{I}(M, N)$ following two conditions are satisfied.

(∗) For every $c \in M$, there is a $d \in N$ such that $\bar{a}c \to \bar{b}d$ is in $\mathcal{I}(M, N)$.
(∗) For every $d \in N$, there is a $c \in N$ such that $\bar{a}c \to \bar{b}d$ is in $\mathcal{I}(M, N)$.

Theorem 4.3.11 *Let T be an L-theory. The following conditions are equivalent.*

(a) T has quantifier elimination.

(b) For every pair of models M, N of T and every $\bar{a} \to \bar{b} \in \mathcal{I}(M,N)$, $tp^M(\bar{a}) = tp^N(\bar{b})$.

(c) For every pair of $\aleph_0$-saturated models M, N of T, $\mathcal{I}(M,N)$ has back-and-forth property.

Proof The equivalence of (a) and (b) is proved in Theorem 2.9.5. Assume (b) and take $\aleph_0$-saturated models M, N of T and a $\bar{a} \to \bar{b} \in \mathcal{I}(M,N)$. Then, by (b), $tp^M(\bar{a}) = tp^N(\bar{b})$. Now take any $c \in M$ and consider

$$p(x) = \{\varphi[\bar{b}, x] : M \models \varphi[\bar{a}, c]\}.$$

Then for any finite set $\varphi_1[\bar{b}, x], \ldots, \varphi_k[\bar{b}, x] \in p(x)$, $M \models \exists x \wedge_{i=1}^{k} \varphi_i[\bar{a}, x]$. Since $tp^M(\bar{a}) = tp^N(\bar{b})$, $N \models \exists x \wedge_{i=1}^{k} \varphi_i[\bar{b}, x]$. Thus $p(x)$ is type in N over $\bar{b}$. Since N is $\aleph_0$-saturated, there is a $d \in N$ that realises $p(x)$. This implies that $\bar{a}c \to \bar{b}d \in \mathcal{I}(M,N)$. Similarly using that M is $\aleph_0$-saturated, (b) implies that for every $d \in N$ there is a $c \in M$ such that $\bar{a}c \to \bar{b}d \in \mathcal{I}(M,N)$.

Assuming (c) we prove (b) now. So, take $M, N \models T$ and $\bar{a} \to \bar{b} \in \mathcal{I}(M,N)$. Let M', N' be $\aleph_0$-saturated elementary extensions of M, N, respectively. It is sufficient to show that $tp^{M'}(\bar{a}) = tp^{N'}(\bar{b})$. By induction on the complexity of L-formulas $\varphi[\bar{x}]$ we show that for every $\bar{a} \to \bar{b} \in \mathcal{I}(M,N)$,

$$M' \models \varphi[\bar{a}] \Leftrightarrow N' \models \varphi[\bar{b}]. \qquad (*)$$

By hypothesis, $(*)$ holds for all atomic φ. Clearly, the set of φ satisfying $(*)$ is closed under $\neg$ and $\vee$. Now let $\psi[\bar{x}, x]$ satisfy $(*)$, and $\varphi[\bar{x}] = \exists x \psi[\bar{x}, x]$. Take a $\bar{a} \to \bar{b} \in \mathcal{I}(M,N)$. Assume that $M' \models \varphi[\bar{a}]$. Since $M' \succeq M$, $M \models \varphi[\bar{a}]$. So, there exists $c \in M$ such that $M \models \psi[\bar{a}, c]$. Hence, $M' \models \psi[\bar{a}, c]$. By back and forth property, there is a $d \in N$ such that $\bar{a}c \to \bar{b}d$ is in $\mathcal{I}(M,N)$. By induction hypothesis, $N' \models \psi[\bar{b}, d]$. Hence, $N' \models \varphi[\bar{b}]$. Similarly, we prove that $N' \models \varphi[\bar{b}]$ implies $M' \models \varphi[\bar{a}]$. □

Theorem 4.3.12 *Let T be a theory such that for every pair of $\aleph_0$-saturated models M, N of T, $\mathcal{I}(M,N)$ is non-empty and has the back-and-forth property. Then T is complete.*

Proof Take any two $M, N \models T$. We are required to show that M and N are elementarily equivalent. By the last result, T has quantifier elimination. Let M' and N' be $\aleph_0$-saturated, elementary extensions of M and N respectively. By our hypothesis, there exists $\bar{a} \to \bar{b} \in \mathcal{I}(M', N')$. Since T has quantifier elimination, $tp^{M'}(\bar{a}) = tp^{N'}(\bar{b})$. In particular, M' and N' are elementarily equivalent. Hence, M and N are elementarily equivalent. □

Theorem 4.3.13 *Let L be a countable language and $\{M_m : m \in \omega\}$ a sequence of L-structures. Suppose $\mathcal{U}$ is a free ultrafilter on ω. Then the ultraproduct $M^{\mathcal{U}} = \times_m M_m / \mathcal{U}$ is $\aleph_1$-saturated.*

Proof Let $A \subset M^{\mathcal{U}}$ be countable and $\Phi[\bar{x}]$ a type over A in $M^{\mathcal{U}}$. Enumerate $A = \{[(a_m^n)] : n \in \omega\}$. Since A and L are countable, $\Phi[\bar{x}]$ is countable. Enumerate $\Phi[\bar{x}] = \{\varphi_k[\bar{x}] : k \in \omega\}$. Set

$$\psi_k[\bar{x}] = \wedge_{i \leq k} \varphi_i[\bar{x}].$$

For each $m \in \omega$, let $\psi_k^m[\bar{x}]$ be the formula obtained from ψ_k by replacing each occurrence of $[(a_i^n)]$ by a_m^n.

Since $\Phi[\bar{x}]$ is a type in $M^{\mathcal{U}}$, each $\psi_k[\bar{x}]$ is realised in $M^{\mathcal{U}}$, say by $([(b_{0m}^k)], [(b_{1m}^k)], [(b_{2m}^k)], \ldots)$. Since $\mathcal{U}$ is free, by Los' fundamental lemma on ultraproduct,

$$V_k = \{m \geq k : M_m \models \psi_k^m[b_{0m}^k, b_{1m}^k, b_{2m}^k, \ldots]\} \in \mathcal{U}.$$

Set $W_k = \cap_{i \leq k} V_i$. Clearly, W_k's are decreasing and $\cap_k W_k = \emptyset$. Further, for each k, $W_k \in \mathcal{U}$.

For each $i \in \omega$, define a sequence $(b_{im}) \in \times_m M_m$ satisfying

$$b_{im} = b_{im}^k$$

for $m \in W_k \setminus W_{k+1}$. For $m \notin W_0$, choose $b_{im} \in M_m$ arbitrarily.

We claim that $([(b_{0m})], [(b_{1m})], [(b_{2m})], \ldots) \in M^{\mathcal{U}}$ realises $\Phi[\bar{x}]$. Take a $k \in \omega$. We prove that

$$M^{\mathcal{U}} \models \psi_k[([(b_{0m})], [(b_{1m})], [(b_{2m})], \ldots)].$$

We show this by proving that

$$W_k = \cup_{i \geq k}(W_i \setminus W_{i+1}) \subset \{m \in \omega : M_m \models \psi_k^m[b_{0m}, b_{1m}, b_{2m}, \ldots]\}.$$

Let $i \geq k$ and $m \in W_i \setminus W_{i+1}$. Then $b_{jm} = b_{jm}^i$ for every $j \in \omega$. But then

$$M_m \models \psi_i^m[b_{0m}^i, b_{1m}^i, b_{2m}^i, \ldots].$$

Since $i \geq k$, it follows that

$$M_m \models \psi_k^m[b_{0m}^i, b_{1m}^i, b_{2m}^i, \ldots].$$

Thus,

$$W_i \setminus W_{i+1} \subset \{m \in \omega : M_m \models \psi_k^m[b_{0m}, b_{1m}, b_{2m}, \ldots]\}$$

for every $i \geq k$. The proof is complete now. $\square$

4.4 Existence of Saturated Structures and Monster Model

In this section, we prove several results on the existence of saturated structures leading to show the existence of monster models for complete theories.

Lemma 4.4.1 *Let L be a countable language, M an L-structure and $\kappa \geq \aleph_0$. Then there is an elementary extension N of M of cardinality $\leq |M|^\kappa$ such that for every $A \subset M$ of cardinality $\leq \kappa$, every $p(x) \in S_1(M/A)$ is realised in N.*

Proof Let $\{p_\alpha : \alpha < |M|^\kappa\}$ be an enumeration of all complete 1-types in M over subsets of M of cardinality $\leq \kappa$. Set $N_0 = M$. There exists an elementary chain $\{N_\alpha : \alpha < |M|^\kappa\}$ of L-structures such that

(∗) $N_\alpha = \cup_{\beta<\alpha} N_\beta$ if $\alpha < |M|^\kappa$ is limit,
(∗) $|N_\alpha| \leq |M|^\kappa$,
(∗) p_α is realised in $N_{\alpha+1}$.

Now take $N = \cup_{\alpha<|M|^\kappa} N_\alpha$. $\qquad\qquad\square$

Theorem 4.4.2 *Let L be a countable language, M an L-structure and $\kappa \geq \aleph_0$. Then there is a κ^+-saturated elementary extension N of M of cardinality $\leq |M|^\kappa$.*

Proof Set $N_0 = M$. Applying the last lemma repeatedly, by transfinite induction, get an elementary chain $\{N_\alpha : \alpha < |M|^\kappa\}$ of L-structures such that

(∗) $N_\alpha = \cup_{\beta<\alpha} N_\beta$ if $\alpha < |M|^\kappa$ is limit,
(∗) $|N_\alpha| \leq |M|^\kappa$,
(∗) each complete 1-type in N_α over a subset of N_α of cardinality $\leq \kappa$ is realised in $N_{\alpha+1}$.

Take $N = \cup_{\alpha<|M|^\kappa} N_\alpha$. $\qquad\qquad\square$

As a consequence of this, under GCH, we show the existence of saturated models of any regular cardinality. Assume that T is a countable theory with an infinite model. Then T has models of every infinite cardinality. In the last theorem, replacing a model M of T be an elementary extension of cardinality 2^κ (which equals κ^+ under GCH), we get the following result.

Proposition 4.4.3 *Assume GCH. Let T be a countable theory with an infinite model and $\kappa \geq \aleph_0$. Then T has a saturated model of cardinality κ^+.*

From this, we can deduce that under the same hypothesis, T has a saturated model of any regular cardinality.

Proposition 4.4.4 *Assume GCH. Let T be a countable theory with an infinite model and $\kappa \geq \aleph_0$ a regular cardinal. Then T has a saturated model of cardinality κ.*

Proof We have already proved the result if in addition κ is a successor cardinal. Assume that κ is not a successor cardinal. Then applying the successor case repeatedly, we get an elementary chain $\{M_\alpha : \aleph_\alpha < \kappa\}$ of models of T such that $M_\alpha = \cup_{\beta<\alpha}M_\beta$ whenever α is a limit ordinal and for each α, M_α is saturated of cardinality $\aleph_\alpha^+$. Now take $M = \cup_{\aleph_\alpha<\kappa}M_\alpha$. $\square$

Next result is a very useful one.

Theorem 4.4.5 *Let $\kappa \geq \aleph_0$ and M an L-structure. Then M has a κ^+-saturated, κ^+-strongly homogeneous elementary extension M_∞.*

Proof Set $M_0 = M$. By Theorem 4.4.2, there exists an elementary chain $\{M_\alpha : \alpha < \kappa^+\}$ of L-structures satisfying the following conditions:

1. For each limit α, $M_\alpha = \cup_{\beta<\alpha}M_\beta$, and
2. for every α, $M_{\alpha+1}$ is $|M_\alpha|^+$-saturated.

Set $M_\infty = \cup_{\alpha<\kappa^+}M_\alpha$.

Now let $A \subset M_\infty$ be of cardinality $\leq \kappa$ and $p \in S_1(M_\infty/A)$. Since κ^+ is regular, there is an $\alpha < \kappa^+$ such that $A \subset M_\alpha$. In particular, $|A| \leq |M_\alpha|$. Since $M_{\alpha+1} \preceq M_\infty$, $p \in S_1(M_{\alpha+1}/A)$. Since $M_{\alpha+1}$ is $|M_\alpha|^+$-saturated, p is realised in $M_{\alpha+1}$. This shows that M_∞ is κ^+-saturated.

Now we proceed to show that M_∞ is κ^+-strongly homogeneous. We start with an $A \subset M_\infty$ of cardinality $\leq \kappa$ and a partial elementary $f : A \to M_\infty$. By regularity of κ^+, there is a $\alpha < \kappa^+$ such that $A \subset M_\alpha$ and $f(A) \subset M_{\alpha+1}$. By tranfinite induction, we define partial elementary maps $f_\beta : M_{\alpha+\beta} \to M_{\alpha+\beta+1}$, $\beta < \kappa^+$, satisfying

1. f_0 extends f,
2. f_β extends f_γ whenever $\gamma < \beta < \kappa^+$,
3. $f_\beta = \cup_{\gamma<\beta}f_\gamma$ if β is limit.
4. $M_{\alpha+\beta} \subset \text{range}(f_{\beta+1})$.

Since $A \subset M_\alpha$, $|A| \leq |M_\alpha|$. Since $M_{\alpha+1}$ is $|M_\alpha|^+$-saturated, by Corollary 4.3.3, there is an elementary extension $f_0 : M_\alpha \to M_{\alpha+1}$ of f.

Suppose f_γ has been defined for all $\gamma < \beta$. If β is limit, take $f_\beta = \cup_{\gamma<\beta}f_\gamma$.

Let $\beta = \gamma + 1$ be a successor ordinal. Then

$$f_\gamma^{-1} : f_\gamma(M_{\alpha+\gamma}) \to M_{\alpha+\gamma+1}$$

is partial elementary, $f_\gamma(M_{\alpha+\gamma}) \subset M_{\alpha+\gamma+1}$ and

$$|f_\gamma(M_{\alpha+\gamma})| = |M_{\alpha+\gamma}| = |f_\gamma(M_{\alpha+\gamma}) \cup M_{\alpha+\gamma}|.$$

We extend f_γ^{-1} to a partial elementary map

$$g : f_\gamma(M_{\alpha+\gamma}) \cup M_{\alpha+\gamma} \to M_{\alpha+\gamma}.$$

Now extend

$$g^{-1} : M_{\alpha+\gamma} \cup g(M_{\alpha+\gamma}) \to M_{\alpha+\gamma+1} \subset M_{\alpha+\gamma+2}$$

to a partial elementary map

$$f_{\gamma+1} : M_{\alpha+\gamma+1} \to M_{\alpha+\gamma}$$

Finally, take $f_\infty = \cup_{\beta<\kappa} f_\beta$. Then f_∞ is an automorphism of M_∞ extending f. □

Remark 4.4.6 This result tells that every consistent theory T with infinite models has a κ^+-saturated, κ^+-strongly homogeneous model $\mathbb{M}$ for arbitrarily large κ. So, depending upon problems at hand, by choosing κ sufficiently large, every model that one is likely to get is of cardinality $\leq \kappa$. Hence, every model one is likely to encounter will be elementarily embedded into $\mathbb{M}$ and all parameters set subsets of $\mathbb{M}$ of cardinality $\leq \kappa$. Indeed, from next chapter onwards, we shall often fix a κ^+-saturated, κ^+-strongly homogeneous model $\mathbb{M}$ for sufficiently large κ and call it a *monster model*.

4.5 Some Consequences of Saturability

Theorem 4.5.1 *Let M and N be saturated L-structures such that $|M| = |N| = \kappa$, say. Then M and N are isomorphic.*

Proof By Corollary 4.3.4, M and N are homogeneous. Fix enumerations $M = \{a_\alpha : \alpha < \kappa\}$ and $N = \{b_\alpha : \alpha < \kappa\}$. Consider $p[x] = tp^M(a_0)$. Since N is saturated, there is a $b \in N$ that realises it. We let $a'_0 = a_0$ and $b'_0 \in N$ to be the first element in the above enumeration of B such that $tp^M(a'_0) = tp^N(b'_0)$. Then $a'_0 \to b'_0$ is partial elementary.

Suppose for $0 < \alpha < \kappa$, $\{a'_\beta \in M : \beta < \alpha\}$ and $\{b'_\beta \in N : \beta < \alpha\}$ have been defined so that for every $\beta < \alpha$, $(a'_\gamma : \gamma < \beta) \to (b'_\gamma : \gamma < \beta)$ is partial elementary. If α is a limit ordinal, then $(a'_\beta : \beta < \alpha) \to (b'_\beta : \beta < \alpha)$ is partial elementary.

Suppose α is an odd successor ordinal, say $\beta + 1$. Let b'_γ be the first element in the enumeration of N different from $b'_\gamma, \gamma < \alpha$. By our assumption, $(b'_\gamma : \gamma < \beta) \to (a'_\gamma : \gamma < \beta)$ is partial elementary. Since M is homogeneous, there is an $a \in M$ such that $(b'_\gamma : \gamma \leq \alpha) \to ((a'_\gamma : \gamma < \alpha), a)$ is partial elementary. We let a'_α denote first such a in the enumeration of M.

If α is an even successor ordinal order, we take a'_α the first element in the enumeration of M different from $a'_\gamma, \gamma < \alpha$. By the same argument, there is a $b \in B$ such that $(a'_\gamma : \gamma' \leq \alpha) \to ((b'_\gamma, \gamma < \alpha), b)$ is partial elementary. We let b'_α the first such element in the above enumeration of B.

Thus, we have defined enumerations $M = \{a'_\alpha : \alpha < \kappa\}$ and $N = \{b'_\alpha : \alpha < \kappa\}$ such for every $\alpha < \kappa$, $(a'_\beta : \beta < \alpha) \to (b'_\beta : \beta < \alpha)$ is partial elementary. It follows that $(a'_\beta : \beta < \kappa) \to (b'_\beta : \beta < \kappa)$ is an isomorphism from M to N. □

Proposition 4.5.2 *Let M be κ-saturated, $A \subset M$ with $|A| < \kappa$ and $tp^M(b/A)$ have only finitely many realisations in M. Then $b \in acl(A)$.*

Proof Suppose the number of realisations of $tp^M(b/A)$ in M is m. Consider

$$\Gamma = \{\varphi[v_i] : \varphi[x] \in tp^M(b/A), i = 0, \ldots, m\} \cup \{\wedge_{0 \le i < j \le m}(v_i \neq v_j)\}.$$

By our assumptions, Γ is not realised in M. Since M is κ-saturated and $|A| < \kappa$, a finite fragment of Γ is not realised in M. Hence, there exist $\varphi_1[x], \ldots, \varphi_k[x] \in tp^M(b/A)$ such that

$$M \models (\wedge_{i=0}^m \wedge_{j=1}^k \varphi_j[v_i]) \rightarrow \vee_{0 \le i < j \le m} v_i = v_j.$$

Now note that the formula $\wedge_{j=1}^k \varphi_j[x]$ witnesses $b \in acl(A)$.   □

As a corollary, we have the following important result.

Theorem 4.5.3 *Let M be a saturated model and $A \subset M$ with $|A| < |M|$. Then the following conditions are equivalent.*

1. $b \in acl(A)$.
2. b has only finitely many conjugates over A, i.e. $b \in ACL(A)$.
3. $tp^M(b/A)$ has only finitely many realisations in M.

Proposition 4.5.4 *Let M be an $\aleph_0$-saturated L-structure and $\varphi[\bar{x}, \bar{a}]$ a minimal L_M-formula in M. Then φ is strongly minimal in M.*

Proof Let N be an elementary extension of M. Take an L_N-formula $\psi[\bar{x}, \bar{b}]$. Since M is $\aleph_0$-saturated, there is a $\bar{c} \in M$ such that $tp^M(\bar{a}, \bar{c}) = tp^N(\bar{a}, \bar{b})$. Hence, for every integer n,

$$M \models \exists_{=n}\bar{x}(\varphi[\bar{x}, \bar{a}] \wedge \psi[\bar{x}, \bar{c}]) \Leftrightarrow N \models \exists_{=n}\bar{x}(\varphi[\bar{x}, \bar{a}] \wedge \psi[\bar{x}, \bar{b}])$$

and

$$M \models \exists_{=n}\bar{x}(\varphi[\bar{x}, \bar{a}] \wedge \neg\psi[\bar{x}, \bar{c}]) \Leftrightarrow N \models \exists_{=n}\bar{x}(\varphi[\bar{x}, \bar{a}] \wedge \neg\psi[\bar{x}, \bar{b}]).$$

Since one of $\varphi(M, \bar{a}) \cap \psi(M, \bar{c})$, $\varphi(M, \bar{a}) \cap \psi(M, \bar{c})^c$ is finite, one of $\varphi(N, \bar{a}) \cap \psi(N, \bar{b})$, $\varphi(N, \bar{a}) \cap \psi(N, \bar{b})^c$ is finite.   □

Theorem 4.5.5 *Let M be a saturated L-structure, $A \subset M$ with $|A| < |M|$ and $X \subset M^n$ be definable. Suppose for all $\sigma \in G_A$, $\sigma(X) = X$. Then X is A-definable.*

Proof Let $\varphi[\bar{x}, \bar{z}]$ be an L-formula and $\bar{m} \in M$ be such that $X = \varphi(M, \bar{m})$. Set

$$\Gamma = \{\varphi[\bar{x}, \bar{m}], \neg\varphi[\bar{y}, \bar{m}]\} \cup \{\psi[\bar{x}] \leftrightarrow \psi[\bar{y}] : \psi \text{ an } L_A - \text{formula}\}.$$

Then Γ is not realised in M: Suppose not. Let $(\bar{a}, \bar{b}) \in M$ realises Γ. In particular, the map f fixing A pointwise and sending $\bar{a}$ to $\bar{b}$ is partial elementary. Since M is saturated, it is strongly homogeneous (Corollary 4.3.4). Hence, f has an extension to a $\sigma \in G_A$. But then $\overline{a} \in X$ and $\overline{b} = \sigma(\overline{a}) \notin X$. This contradicts our hypothesis.

Since M is saturated, there is a finite fragment of Γ that is not realised in M. So, there exist L_A-formulas $\psi_0[\overline{x}], \ldots, \psi_{m-1}[\overline{x}]$ such that

$$M \models \forall \overline{x} \forall \overline{y} (\wedge_{i<m}(\psi_i(\overline{x}) \leftrightarrow \psi_i(\overline{y})) \to (\varphi[\overline{x}, \overline{m}] \leftrightarrow \varphi[\overline{y}, \overline{m}])).$$

For any $s \in 2^m$, set

$$\theta_s[\overline{x}] = \wedge_{s(i)=1}\psi_i(\overline{x}) \wedge \wedge_{s(i)=0}\neg\psi_i(\overline{x}).$$

Let

$$S = \{s \in 2^m : \exists \overline{b} \in X(M \models \theta_s[\overline{b}])\}.$$

We claim that

$$\overline{a} \in X \Leftrightarrow M \models \vee_{s \in S}\theta_s[\overline{a}].$$

Let $\overline{a} \in X$. Define

$$s(i) = 1 \ \text{if} \ M \models \psi_i[\overline{a}]$$
$$= 0 \ \text{if} \ M \models \neg\psi_i[\overline{a}]$$

Then $s \in S$ and $M \models \theta_s[\overline{a}]$

On the other hand, let there be a $s \in S$ such that $M \models \theta_s[\overline{a}]$. Then there exists $\overline{b} \in X$ such that $M \models \theta_s[\overline{b}]$. This implies that $\overline{a} \in X$.

We have shown that the L_A-formula $\vee_{s \in S}\theta_s[\overline{x}]$ defines X. $\qquad \square$

We recast this theorem in the context of a monster model. Let κ be a sufficiently large cardinal and $\mathbb{M}$ a κ^+-saturated, κ^+-strongly homogeneous model of a theory T. According to our convention, every parameter set is of cardinality $\leq \kappa$. We have the following result.

Theorem 4.5.6 *Let $A \subset \mathbb{M}$ and $X \subset \mathbb{M}^n$ be definable. Then X is A-definable if and only if for all $\sigma \in Aut_A(\mathbb{M})$, $\sigma(X) = X$.*

Corollary 4.5.7 *Let M be a saturated L-structure, $A \subset M$ and $|A| < |M|$. Then $DCL(A) = dcl(A)$.*

Proof Earlier we saw that $dcl(A) \subset DCL(A)$. By the last Theorem $DCL(A) \subset dcl(A)$ because M is saturated. $\qquad \square$

We now give an application to elimination of imaginaries and show that ACF has elimination of imaginaries.

Proposition 4.5.8 *Let T be a complete L-theory with infinite models. Suppose for every model M of T, every equivalence L-formula $\theta[\overline{x}, \overline{y}]$ in M and for every $\overline{a} \in M$, there is a tuple $\overline{b} \in M$ such that for every $\sigma \in Aut(M)$, $\sigma(\theta(M, \overline{a})) = \theta(M, \overline{a})$ if and only if $\sigma(\overline{b}) = \overline{b}$. Then T has semi-uniform elimination of imaginaries.*

Proof Take an equivalence formula $\theta[\bar{x}, \bar{y}]$ in T. Suppose $\mathbb{M}$ is a monster model of T and an $\bar{a} \in \mathbb{M}$. Get $\bar{b}$ as in the hypothesis. By Theorem 4.5.5, our hypothesis implies that $\theta(\mathbb{M}, \bar{a})$ is $\bar{b}$-definable. Fix an L-formula $\varphi[\bar{x}, \bar{z}]$ such that

$$\mathbb{M} \models \forall \bar{x}(\theta[\bar{x}, \bar{a}] \leftrightarrow \varphi[\bar{x}, \bar{b}]).$$

Suppose $\bar{c} \in \mathbb{M}$ satisfies $tp^{\mathbb{M}}(\bar{c}) = tp^{\mathbb{M}}(\bar{b})$ and

$$\mathbb{M} \models \forall \bar{x}(\theta[\bar{x}, \bar{a}] \leftrightarrow \varphi[\bar{x}, \bar{c}]).$$

Then $\bar{b} \to \bar{c}$ is partial elementary. Since $\mathbb{M}$ is a monster model, there is a $\sigma \in Aut(\mathbb{M})$ such that $\sigma(\bar{b}) = \bar{c}$. Hence,

$$\sigma(\theta(\mathbb{M}, \bar{a})) = \sigma(\varphi(\mathbb{M}, \bar{b})) = \varphi(\mathbb{M}, \bar{c}) = \theta(\mathbb{M}, \bar{a}).$$

Hence, by our hypothesis $\bar{c} = \bar{b}$.

(I) There is a formula $\psi[\bar{z}] \in tp^{\mathbb{M}}(\bar{b})$ such that

$$\mathbb{M} \models \forall \bar{z} \forall \bar{z}'((\psi[\bar{z}] \wedge \psi[\bar{z}'] \wedge \forall \bar{x}(\theta[\bar{x}, \bar{a}] \leftrightarrow \varphi[\bar{x}, \bar{z}])$$

$$\wedge \forall \bar{x}(\theta[\bar{x}, \bar{a}] \leftrightarrow \varphi[\bar{x}, \bar{z}'])) \to \bar{z} = \bar{z}').$$

To see this, assume to the contrary. For each $\psi[\bar{z}] \in tp^{\mathbb{M}}(\bar{b})$, let $\Psi[\bar{z}, \bar{z}']$ be the formula

$$\psi[\bar{z}] \wedge \psi[\bar{z}'] \wedge \forall \bar{x}(\theta[\bar{x}, \bar{a}] \leftrightarrow \varphi[\bar{x}, \bar{z}]) \wedge \forall \bar{x}(\theta[\bar{x}, \bar{a}] \leftrightarrow \varphi[\bar{x}, \bar{z}']) \wedge \bar{z} \neq \bar{z}'.$$

Now consider
$$p(\bar{z}, \bar{z}') = \{\Psi[\bar{z}, \bar{z}'] : \psi[\bar{z}] \in tp^{\mathbb{M}}(\bar{b})\}.$$

By our assumption, $p(\bar{z}, \bar{z}')$ is finitely satisfiable. By saturability of $\mathbb{M}$, $p(\bar{z}, \bar{z}')$ is realised in $\mathbb{M}$. But this contradicts what we have just proved.

Now consider the formula $\varphi'[\bar{x}, \bar{z}] = \varphi[\bar{x}, \bar{z}] \wedge \psi[\bar{z}]$. So,

$$\mathbb{M} \models \forall \bar{z} \forall \bar{z}'(\forall \bar{x}(\theta[\bar{x}, \bar{a}] \leftrightarrow \varphi'[\bar{x}, \bar{z}]) \wedge \forall \bar{x}(\theta[\bar{x}, \bar{a}] \leftrightarrow \varphi'[\bar{x}, \bar{z}'])) \to \bar{z} = \bar{z}').$$

(II) There exists a finite sequence of L-formulas $\varphi_0[\bar{x}, \bar{z}], \dots, \varphi_n[\bar{x}, \bar{z}]$ such that

$$\mathbb{M} \models \forall \bar{y} \bigvee_{i=0}^{n} \exists_{=1} \bar{z} \forall \bar{x}(\theta[\bar{x}, \bar{y}] \leftrightarrow \varphi_i[\bar{x}, \bar{z}]).$$

Assume to the contrary, we are going to derive a contradiction. For each L-formula $\varphi[\bar{x}, \bar{z}]$, let $\Phi[\bar{y}]$ be the formula

$$\neg \exists_{=1} \bar{z} \forall \bar{x}(\theta[\bar{x}, \bar{y}] \leftrightarrow \varphi[\bar{x}, \bar{z}]).$$

Set

$$q(\bar{y}) = \{\Phi[\bar{y}] : \varphi[\bar{x}, \bar{z}] \text{ an } L-\text{formula}\}.$$

By our assumption, $q(\bar{y})$ is finitely satisfiable in $\mathbb{M}$. Hence, by saturability of $\mathbb{M}$, $q(\bar{y})$ is realised in $\mathbb{M}$, say by $\bar{a}$. This implies that for $\bar{a} \in \mathbb{M}$, there is no L-formula $\varphi[\bar{x}, \bar{z}]$ such that there is a unique $\bar{b} \in \mathbb{M}$ satisfying $\forall \bar{x}(\theta[\bar{x}, \bar{a}] \leftrightarrow \varphi[\bar{x}, \bar{b}])$. We have arrived at a contradiction.

Let $\varphi_0[\bar{x}, \bar{z}], \ldots, \varphi_n[\bar{x}, \bar{z}]$ satisfy our claim. For each $0 \leq i \leq n$, set

$$\psi_i[\bar{y}] = \exists_{=1}\bar{z}\forall\bar{x}(\theta[\bar{x}, \bar{y}] \leftrightarrow \varphi_i[\bar{x}, \bar{z}]) \wedge \wedge_{j<i}\neg\exists_{=1}\bar{z}\forall\bar{x}(\theta[\bar{x}, \bar{y}] \leftrightarrow \varphi_j[\bar{x}, \bar{z}])$$

and

$$\xi_i[\bar{x}, \bar{z}] = \exists\bar{y}(\theta[\bar{x}, \bar{y}] \wedge \psi_i[\bar{y}] \wedge \varphi[\bar{x}, \bar{z}]).$$

Then

$$\mathbb{M} \models \forall\bar{y}\exists_{=1}0 \leq i \leq n\exists_{=1}\bar{z}\forall\bar{x}(\theta[\bar{x}, \bar{z}] \leftrightarrow \xi_i[\bar{x}, \bar{z}]).$$

Since every model M of T is elementarily embedded in a monster model of T, it follows that T has semi-uniform elimination of imaginaries. □

We refer the reader to Sect. B.2 for relevant definitions and results from algebraic geometry used in proving the following theorem.

Theorem 4.5.9 *ACF has uniform elimination of imaginaries.*

Proof By Proposition 1.12.5, it is sufficient to show that *ACF* has semi-uniform elimination of imaginaries. Let $\mathbb{K} \models ACF$ and $\varphi[\bar{x}, \bar{y}]$ be an equivalence formula of $\mathbb{K}$. Take an $\bar{a} \in \mathbb{K}$, $X = \bar{a}/\varphi$, the equivalence class of $\bar{a}$, and Z its closure. Then Z is a constructible set.

Let I be a radical ideal such that $Z = \mathcal{V}(I)$ and k_0 be the smallest subfield such that I is algebraically defined over k_0. Then k_0 is finitely generated, generated by a tuple say $\bar{b}$. Let $p_1, \ldots, p_m$ be polynomials with coefficients among $\bar{b}$ such that

$$Z = \cap_{i=1}^{m}\{\bar{x} : p_i(\bar{x}) = 0\}.$$

Thus, Z is definable over $\bar{b}$.

Let σ be an automorphism of $\mathbb{K}$. By Theorem B.2.14, σ fixes k_0 pointwise if and only if $\sigma(Z) = Z$ if and only if $\sigma(\bar{b}) = \bar{b}$. By Lemma B.2.9, $\sigma(X) = X$ implies $\sigma(Z) = Z$. Conversely, suppose $\sigma(Z) = Z$. and $Y = \sigma(X)$. So, Y and X have the same closure Z. But then by Lemma B.2.12, $X \cap Y \neq \emptyset$. Hence, $X = \sigma(X)$. So, σ fixes X if and only if σ fixes $\bar{b}$ pointwise. By Proposition 4.5.8, *ACF* has semi-uniform elimination of imaginaries. □

We close this section with a technical result that will be used later.

Lemma 4.5.10 *Let M be a κ-saturated structure and $|I|, |J| < \kappa$. Suppose $\{\varphi_i[\bar{x}] : i \in I\}$ and $\{\psi_j[\bar{x}] : j \in J\}$ are L_M-formulas with*

$$\cup_{i \in I} \varphi_i(M) = (\cup_{j \in J} \psi_j(M))^c.$$

Then there is a finite $I_0 \subset I$ such that

$$\cup_{i \in I_0} \varphi_i(M) = \cup_{i \in I} \varphi_i(M).$$

Proof Set

$$A = \cup_{i \in I} \varphi_i(M) \ \& \ B = \cup_{j \in J} \psi_j(M).$$

By our hypothesis, $A = B^c$. Now consider

$$\Gamma[\overline{x}] = \{\neg \varphi_i[\overline{x}] : i \in I\} \cup \{\neg \psi_j[\overline{x}] : j \in J\}.$$

By our assumption, Γ is not realised in M. Note that Γ uses fewer than κ-many parameters. Since M is κ-saturated, $\Gamma[\overline{x}]$ is not a type in M, i.e. a finite subset of it is not realised in M. Let $I_0 \subset I$ and $J_0 \subset J$ be finite sets such that

$$\{\neg \varphi_i[\overline{x}] : i \in I_0\} \cup \{\neg \psi_j[\overline{x}] : j \in J_0\}$$

is not realised in M. Suppose for $\overline{a} \in M$, $M \models \varphi_i[\overline{a}]$ for some $i \in I$. Then $M \models \neg \psi_j[\overline{a}]$ for all $j \in J_0$. Hence, $M \models \varphi_i[\overline{a}]$ for some $i \in I_0$. □

4.6   Type Definable Sets

Let M be an L-structure, $A \subset M$, α an ordinal number and $D \subset M^\alpha$. We call D **invariant over** A if for every automorphism $f \in Aut_A(M)$ that fixes A pointwise, $f(\overline{a}) \in D$ whenever $\overline{a} \in D$. We say that D is **type definable** over A if there is a set $p(\overline{x})$ of L_A-formulas in variables $\overline{x} = \{x_\beta : \beta < \alpha\}$ such that $D = p(M)$, the set of all realisations of p.

Clearly, M^α and $\emptyset$ are type definable over empty set. Let

$$\{B_i = p_i(M) \subset M^\alpha : p_i[\overline{x}] \text{ a set of } L_A\text{-formulas}, i \in I\}$$

be a family of type definable sets over A. Then $\cap_i B_i = q(M)$, where $q = \cup_i p_i$. Next let $B = p(M)$ and $C = q(M)$, where p and q are sets of L_A-formulas. It is easy to check that $B \cup C = r(M)$, where

$$r(\overline{x}) = \{\varphi[\overline{x}] \vee \psi[\overline{x}] : \varphi \in p, \psi \in q\}.$$

Thus, we see that the set of all type definable sets over A is the family of all closed sets of a topology on M^α.

Exercise 4.6.1 Let κ be an infinite cardinal, $\lambda < \kappa$, M a κ-saturated L-structure and $E \subset M^\lambda$ invariant. Set

$$\{p_i(\bar{x}) : i \in I\} = \{tp(\bar{a}/\emptyset) : \bar{a} \in E\}.$$

Show that $E = \cup_{i \in I} p_i(M)$.

Proposition 4.6.2 *Let M be a κ-saturated L-structure, $A \subset M$ of cardinality $< \kappa$, $\alpha < \kappa$, $\bar{x} = \{x_\gamma : \gamma < \alpha\}$ and $\bar{y} = \{y_\gamma : \gamma < \beta\}$ variables. If $D \subset M^\alpha \times M^\beta$ is type definable over A, then $E = proj_{M^\alpha}(D)$ is type definable over A.*

Proof Let $p(\bar{x}, \bar{y})$ be a set of L_A-formulas and $D = p(M)$. Without any loss of generality, we assume that p is closed under finite conjunctions. Consider the following set of L_A-formulas

$$q(\bar{x}) = \{\exists \bar{y} \varphi[\bar{x}, \bar{y}] : \varphi \in p\}.$$

Clearly $proj_{M^\alpha}(D) \subset q(M)$. Conversely, let $\bar{a} \in q(M)$. Set

$$r(\bar{y}) = \{\varphi[\bar{a}, \bar{y}] : \varphi[\bar{x}, \bar{y}] \in p\}.$$

Then $r(\bar{y})$ is a type over a set of cardinality $< \kappa$. Since M is κ-saturated, there is a $\bar{b} \in M$ that realises r. But then $(\bar{a}, \bar{b}) \in D$, implying $\bar{a} \in proj_{M^\alpha}(D)$. $\square$

Proposition 4.6.3 *Let M be a κ-saturated, κ-strongly homogenous L-structure, $A, B \subset M$ of cardinalities less than κ and $\lambda < \kappa$. Suppose $D \subset M^\lambda$ is type definable over B and invariant over A. Then D is type definable over A.*

Proof Enumerate $A = \bar{a} = \{a_\gamma : \gamma < \alpha\}$ and $B = \bar{b} = \{b_\gamma : \gamma < \beta\}$. In what follows $\bar{x} = \{x_\gamma : \gamma < \lambda\}$ and $\bar{y} = \{y_\gamma : \gamma < \beta\}$ are variables.

Let $p(\bar{x}, \bar{y})$ be a set of L-formulas such that $D = p(M, \bar{b})$. We first see that if $tp(\bar{c}/A) = tp(\bar{b}/A)$, then $D = p(M, \bar{c})$: There exists a $f \in Aut_A(M)$ such that $f(\bar{b}) = \bar{c}$. Since D is invariant over A, $\bar{d} \in D$ if and only if $f(\bar{d}) \in D$. Now note that

$$\bar{d} \in D = p(M, \bar{b}) \Leftrightarrow f(\bar{d}) \in p(M, \bar{c}).$$

Consider

$$q(\bar{x}) = \{\exists \bar{y}(\varphi[\bar{y}] \wedge \psi[\bar{x}, \bar{y}]) : \varphi[\bar{y}] \in tp(\bar{b}/\bar{a}) \text{ and } \psi[\bar{x}, \bar{y}] \in p(\bar{x}, \bar{y})\}.$$

Clearly, $D \subset q(M)$. Conversely, let $\bar{c} \in q(M)$. Consider

$$r(\bar{y}) = \{\varphi[\bar{y}] \wedge \psi[\bar{c}, \bar{y}] : \varphi[\bar{y}] \in tp(\bar{b}/\bar{a}) \text{ and } \psi[\bar{x}, \bar{y}] \in p(\bar{x}, \bar{y})\}.$$

Then $r(\bar{y})$ is a type in M over $\bar{a}\bar{c}$ which is of cardinality $< \kappa$. Since M is κ-saturated, there is a $\bar{d}$ that realises $r(\bar{y})$. But then $tp(\bar{d}/\bar{a}) = tp(\bar{b}/\bar{a})$ and $\bar{c} \models p(\bar{x}, \bar{d})$. By the above observation, it follows that $\bar{c} \in D$. $\square$

4.7 $\aleph_0$-Categorical Theories

In this section, we give several characterizations of countable, $\aleph_0$-categorical theories.

Proposition 4.7.1 *If T is a countable, $\aleph_0$-categorical theory with an infinite model, then T is complete.*

Proof If possible, suppose there exists a sentence φ which is undecidable in T. Then by downward Löwenheim–Skolem theorem, there exist infinite, countable $M, N \models T$ such that $M \models \varphi$ and $N \models \neg\varphi$. But then M and N can't be isomorphic, a contradiction. $\square$

Theorem 4.7.2 (Engeler [12], Ryll-Nardzewski [53] and Svenonius [60]) *Let T be a countable complete theory. The following statements are equivalent.*

1. *T is $\aleph_0$-categorical.*
2. *For every $n \geq 1$, every type $p \in S_n(T)$ is isolated.*
3. *For every $n \geq 1$, $S_n(T)$ is finite.*
4. *For every $n \geq 1$, there exist finitely many L-formulas, $\varphi_0[\overline{x}], \ldots, \varphi_k[\overline{x}]$, where $\overline{x} = (x_0, \ldots, x_{n-1})$, such that for every L-formula $\psi[\overline{x}]$,*

$$T \models \forall \overline{x}(\psi[\overline{x}] \leftrightarrow \varphi_i[\overline{x}])$$

   for some $0 \leq i \leq k$.
5. *Every model M of T is atomic.*
6. *Every countable model M of T is atomic.*

Proof (1) implies (2): Let there exist a $p \in S_n(T)$ which is not isolated. Since T is countable, there exists a countable $M \models T$ that realises p and a countable $N \models T$ that omits p. But then M and N are two countable models of T which are not isomorphic.

(2) implies (3): This is so because $S_n(T)$ is compact.

(3) implies (4): Let $p_1, \ldots, p_k$ be all complete n-types in T. Since $S_n(T)$ is Hausdorff, each p_i is isolated, say by θ_i, $1 \leq i \leq k$. It is easily checked that for any L-formula $\psi[\overline{x}]$,

$$T \models \forall \overline{x}(\psi[\overline{x}] \leftrightarrow \vee_{\psi \in p_j} \theta_j[\overline{x}]).$$

The proof is easily seen now.

(4) implies (5): Let $M \models T$ and $\overline{a} \in M^n$. Set

$$\theta[\overline{x}] = \wedge_{M \models \varphi_i[\overline{a}]} \varphi_i[\overline{x}] \wedge \wedge_{M \not\models \varphi_i[\overline{a}]} \neg\varphi_i[\overline{x}].$$

We claim that θ isolates $tp^M(\overline{a})$. Take any L-formula $\psi[\overline{x}]$.

Let $T \models \forall \overline{x}(\theta[\overline{x}] \rightarrow \psi[\overline{x}])$. Since $\theta[\overline{x}] \in tp^M(\overline{a})$, $\psi[\overline{x}] \in tp^M(\overline{a})$.

Conversely, assume that $\psi[\bar{x}] \in tp^M(\bar{a})$ and $\bar{b}$ be any element in M such that $M \models \theta[\bar{b}]$. Then $tp^M(\bar{b}) = tp^M(\bar{a})$. Hence, $\psi[\bar{x}] \in tp^M(\bar{b})$. Thus,

$$\psi \in tp^M(\bar{a}) \Leftrightarrow M \models \forall \bar{x}(\theta[\bar{x}] \to \psi[\bar{x}]).$$

(6) implies (1): Let $M \models T$ be countable. So, M is a countable atomic model of T. Therefore, by Theorem 4.2.4, it is prime. Thus, every countable model of T is prime. By Theorem 4.2.6, any two prime models of a countable complete theory are isomorphic. Thus, T is $\aleph_0$-categorical. $\qquad \square$

Corollary 4.7.3 *Let T be a countable, $\aleph_0$-categorical theory, $M \models T$ and $A \subset M$ finite. Then $acl(A)$ is finite. In particular, finitely generated substructures of models of T are finite.*

Proof Let $A = \{a_0, \ldots, a_{n-1}\}$. Let $\varphi_1[x_0, \ldots, x_n], \ldots, \varphi_k[x_0, \ldots, x_n]$ be a finite set of L-formulas such that any L-formula $\psi[x_0, \ldots, x_n]$ is equivalent in T to some φ_i. Let

$$I = \{i : \varphi_i(a_0, \ldots, a_{n-1}, M) \text{ is finite}\}.$$

Then

$$|acl(A)| \le \sum_{i \in I} |\varphi_i(a_0, \ldots, a_{n-1}, M)| < \infty. \qquad \square$$

Theorem 4.7.4 *Let T be a countable complete theory. Then the following statements are equivalent.*

(a) *T is $\aleph_0$-categorical.*
(b) *Every model M of T is ω-saturated.*
(c) *Every countable $M \models T$ is saturated.*

Proof (a) implies (b): Let $A = \{a_0, \ldots, a_{n-1}\} \subset M$. Since T is countable and $\aleph_0$-categorical, by Theorem 4.7.2, there exist only finitely many L-formulas $\varphi_1[x_0, \ldots, x_n], \ldots, \varphi_k[x_0, \ldots, x_n]$ modulo equivalence in T. Hence, there exist at most k many L_A-formulas $\psi[x_n]$ modulo equivalence in M. This implies that $S_1^M(A)$ is finite. Hence, each $p[x_n] \in S_1^M(A)$ is isolated, say by $\varphi_p[x_n]$. Any element in M that realises $\varphi_p[x_n]$ realises $p[x_n]$.

(c) implies (a): This follows from Theorem 4.5.1. $\qquad \square$

Exercise 4.7.5 If M is a κ-saturated L-structure, show that $|M| \ge \kappa$.

Exercise 4.7.6 Let $\kappa \ge \aleph_0$, M, N L-structures with N κ-saturated and $A \subset M$ with $|A| < \kappa$. Then for every $B \subset M$ of cardinality $< \kappa$, every partial elementary map $f : A \to N$ has a partial elementary extension $g : A \cup B \to N$.

Exercise 4.7.7 Let T be a theory with a constant symbol. Then T has quantifier elimination if and only if whenever $M, N \models T$, N $|M|^+$-saturated, A a substructure of M and $f : A \to N$ an embedding, f admits an elementary extension $g : M \to N$ of f.

Exercise 4.7.8 Let T be the theory of discrete linear orders with no first no last elements. Show that if M and N are $\aleph_0$-saturated models of T, then $\mathcal{I}(M, N)$ is non-empty and has the back and forth property. Conclude that T is a complete theory which is not model complete.

Exercise 4.7.9 Let T be a $\aleph_0$-categorical theory and $M \models T$. Show that for every $n \geq 1$ there is a $m \geq 1$ such that whenever $A \subset M$ is of cardinality n, $acl(A)$ is of cardinality at most m.

Exercise 4.7.10 Call an L-structure M *locally finite* if for every finite $A \subset M$, the substructure of M generated by A is finite. Show that the models of $\aleph_0$-categorical theories are locally finite.

4.8 Stable Theories

For an infinite cardinal κ, we call T κ-*stable* if for all $M \models T$, for all $A \subset M$ of cardinality $\leq \kappa$, $|S_1^M(A)| \leq \kappa$. Note that for $a \neq b$ in A, $tp^M(a/A) \neq tp^M(b/A)$. Hence, if $|A| = \kappa$, $|S_1^M(A)| \geq \kappa$. So, T is κ-stable if and only if for all $M \models T$, for all $A \subset M$ of cardinality κ, $|S_1^M(A)| = \kappa$. If $\kappa = \aleph_0$, κ-stable theories are traditionally called ω-*stable*. A theory T is called *stable* if it is κ-stable for some $\kappa \geq \aleph_0$. The concept of stable theory was introduced by Morley in [43].

Proposition 4.8.1 Let $\kappa \geq \aleph_0$ and T κ-stable. Then for every $n \geq 1$, for every $M \models T$ and for every $A \subset M$ of cardinality κ, $|S_n(M/A)| = \kappa$.

Proof We prove the result by induction on n. Assume the hypothesis for $n - 1$. Take a $M \models T$ and $A \subset M$ of cardinality κ. Get an elementary extension N of M in which each $p \in S_1(M/A)$ is realised. We have

$$S_1(N/A) = \{tp^N(a/A) : a \in N\}.$$

Hence, $|\{tp^N(a/A) : a \in N\}| = \kappa$.

Consider the map $\pi : S_n(N/A) \to S_1(N/A)$ defined by $\pi(p(x_0, \ldots, x_{n-1})) = \{\varphi[x_{n-1}] : \varphi[x_{n-1}] \in p\}$. By Lemma 3.3.4, $|tp^N(a/A)| = |S_{n-1}(N/Aa)|$. By induction hypothesis, $|S_{n-1}(N/Aa)| = \kappa$. It follows that $|S_n(N/A)| = \kappa$. $\qquad\square$

Example 4.8.2 Consider DLO. In Example 3.4.17 we saw that $|S_1(\mathbb{Q}/\mathbb{Q})| = 2^{\aleph_0}$. Hence, DLO is not ω-stable.

Example 4.8.3 Now consider RCF and $RCOF$ and $\mathbb{R}$ with usual interpretations. In Example 3.4.15, we showed that viewing $\mathbb{R}$ as a model of $RCOF$, $|S_1(\mathbb{R})| \geq \mathfrak{c}$. Hence, $RCOF$ is not ω-stable. Since $<$ is definable in the field $\mathbb{R}$, it follows that RCF is not ω-stable.

Example 4.8.4 Let $\mathbb{F} \models ACF, A \subset \mathbb{F}$ and $\mathbb{A}$ the subfield generated by A. Then $|A| \le |\mathbb{A}| \le \max\{|A|, \aleph_0\}$. In Example 3.4.18 we observed that

$$|S_n(\mathbb{F}/A)| = |Spec(\mathbb{A}[\overline{X}])|.$$

By Hilbert basis theorem, every ideal in a polynomial ring over a field is finitely generated. Hence,

$$|Spec(\mathbb{A}[\overline{X}])| = |\mathbb{A}| = \max\{|A|, \aleph_0\}.$$

It follows that ACF is κ-stable for all $\kappa \ge \aleph_0$.

Let $\kappa \ge \aleph_0$. Suppose there is a model M of T and an $A \subset M$ of cardinality κ such that $|S_n(M/A)| > \kappa$. By Theorem 3.7.4, there is a countable $A_0 \subset A$ such that $|S_n(M/A_0)| \ge 2^{\aleph_0}$. This gives us the following important result.

Theorem 4.8.5 *If T is ω-stable, then T is κ-stable for all cardinal $\kappa > \aleph_0$.*

By Theorem 3.7.2, if there is a $M \models T$ and $A \subset M$ such that isolated types are not dense in $S_n(M/A)$, then there is a countable $A_0 \subset A$ such that $|S_n(M/A_0)| \ge 2^{\aleph_0}$. This gives us the following result.

Proposition 4.8.6 *Let T be ω-stable, $M \models T$ and $A \subset M$. Then isolated types are dense in $S_n(M/A)$ for all n.*

Corollary 4.8.7 *If T is a countable complete ω-stable theory, then T has a prime model.*

Proof This follows from the last proposition and Theorem 4.2.8. □

We now show that a countable, complete, ω-stable theory has a saturated model of cardinality κ for each regular cardinal κ.

Proposition 4.8.8 *Let κ be a regular cardinal, T a countable, complete, κ-stable theory and $M \models T$ of cardinality κ. Then M has an elementary saturated extension N of cardinality κ.*

Proof In the proof of Theorem 4.4.2, take $N_\alpha, \alpha < \kappa$, of cardinality κ. □

Corollary 4.8.9 *Let T be a countable, ω-stable complete theory with infinite models. Then for each regular cardinal κ, T has a saturated model of cardinality κ.*

Proof Take any regular cardinal κ. Since T has an infinite model, T has a model M of cardinality κ. Since T is ω-stable, T is κ-stable by Theorem 4.8.5. The result now follows from the last result. □

Let M be an L-structure. A *binary tree of L_M-formulas* is a system $\{\varphi_\epsilon[\overline{x}] : \epsilon \in 2^{<\omega}\}$ of L_M-formulas such that for every $\epsilon \in 2^{<\omega}$ and every $\delta = 0$ or 1,

1. $M \models \varphi_{\epsilon,\delta}[\overline{x}] \rightarrow \varphi_\epsilon[\overline{x}],$

2. $M \models \varphi_{\epsilon,\delta}[\overline{x}] \rightarrow \neg\varphi_{\epsilon,1-\delta}[\overline{x}]$, and
3. $M \models \exists\overline{x}\varphi_\epsilon[\overline{x}]$.

For such a system, the set A_0 of all parameters of φ_ϵ, $\epsilon \in 2^{<\omega}$ is countable and $|S_n(M/A_0)| \geq \mathfrak{c}$.

A theory T is called *totally transcendental* if no model M of T has a binary tree of L_M-formulas. We now have the following result.

Theorem 4.8.10 *If T is ω-stable, T is totally transcendental.*

With essentially the same argument, we have the next two consequence of ω-stability.

Theorem 4.8.11 *If T is ω-stable and $M \models T$ infinite, then there is an L_M-formula $\varphi[x]$ minimal in M.*

Proof Suppose there is no L_M-formula minimal in M. We then show that there is a binary tree of L_M-formulas.

Let $\varphi_\emptyset[x]$ be the formula $x = x$. Suppose for some $\epsilon \in 2^{<\omega}$, $\varphi_\epsilon[x]$ has been defined. Since $\varphi_\epsilon[x]$ is not minimal in M and $\varphi_\epsilon(M)$ is infinite, there is an L_M-formula $\psi[x]$ such that both $\varphi_\epsilon(M) \cap \psi(M)$ and $\varphi_\epsilon(M) \cap \psi(M)^c$ are infinite. Take

$$\varphi_{\epsilon,0}[x] = \varphi_\epsilon[x] \wedge \psi[x] \text{ and } \varphi_{\epsilon,1}[x] = \varphi_\epsilon \wedge \neg\psi[x].$$

Our result follows from the last theorem. □

Proposition 4.8.12 *Let T be ω-stable and $M \models T$ uncountable. Then there is an L_M-formula $\varphi[x]$ such that $\varphi(M)$ is uncountable and for every L_M-formula $\psi[x]$ exactly one of $(\varphi \wedge \psi)(M)$, $(\varphi \wedge \neg\psi)(M)$ is uncountable.*

Proof First note that for every L_M-formula $\varphi[x]$ such that $\varphi(M)$ is uncountable and for every L_M-formula $\psi[x]$, at least one of $(\varphi \wedge \psi)(M)$, $(\varphi \wedge \neg\psi)(M)$ is uncountable.

If possible, suppose a formula φ satisfying the claim of the Lemma does not exist. To complete the proof, we show that there is a binary tree of L_M-formulas.

Take $\varphi_\emptyset[x]$ to be the formula $x = x$. Since M is uncountable, $\varphi_\emptyset(M)$ is uncountable. Suppose for a $\sigma \in 2^{<\omega}$, a L_M formula $\varphi_\sigma[x]$ have been defined so that $\varphi_\sigma(M)$ is uncountable. Then by our assumption, there is an L_M-formula $\psi[x]$ such that both of $(\varphi_\sigma \wedge \psi)(M)$, $(\varphi_\sigma \wedge \neg\psi)(M)$ are uncountable. Set

$$\varphi_{\sigma 0} = \varphi_\sigma \wedge \psi \text{ \& } \varphi_{\sigma 1} = \varphi_\sigma \wedge \neg\psi.$$

The proof is complete now. □

Next we show that models M of a ω-stable theory admit prime model extension of every $A \subset M$.

Theorem 4.8.13 *Let T be ω-stable, $M \models T$ and $A \subset M$. Then there is an elementary substructure N of M which is a prime model extension of A. Further, for every $\overline{a} \in N$, $tp^N(\overline{a}/A)$ is isolated.*

Proof For ordinals α, we define $A_\alpha \subset M$ satisfying

(a) $A_0 = A$.
(b) $\alpha < \beta \Rightarrow A_\alpha \subset A_\beta$.
(c) If α is limit, then $A_\alpha = \cup_{\beta < \alpha} A_\beta$.
(d) If A_α is such that there is an isolated complete 1-type p over A_α realised in $M \setminus A_\alpha$, then $A_{\alpha+1} = A_\alpha \cup \{a_\alpha\}$, where $a_\alpha \in M \setminus A_\alpha$ realises one such p.

Let δ be the first ordinal α such that A_α cannot be further enlarged by (d). Set $N = A_\delta$.

We first show that N is closed under each f^M, f a function symbol. Take $\bar{a} \in N = A_\delta$. Let $a = f^M(\bar{a})$. Since T is ω-stable, by Proposition 4.8.6, there is an isolated type $p \in S_1^M(A_\delta)$ containing the formula $x = f(\bar{a})$. Let $\psi[x]$ be an L_N-formula isolating p. Then a is the only element that realises p. Hence, $a \in A_\delta$. We treat N as a substructure of M canonically.

To show that N is an elementary substructure of M, take an L_N-formula $\varphi[x]$ such that for some $a \in M$, $M \models \varphi[a]$. Since T is ω-stable and $[\varphi] \neq \emptyset$, there is an isolated type p containing φ by Proposition 4.8.6. Let $\psi[x]$ isolate p and $b \in M$ be such that $M \models \psi[b]$. This shows that p is realised in M. Hence, p is realised in N, say by c. Then $N \models \varphi[c]$.

We now show that N is a prime model extension of A. So fix $N' \models T$ and a partial elementary map $f : A \to N'$. We need to get an elementary extension $g : N \to N'$ of f. Set $f_0 = f$. Inductively, for each $\alpha \leq \delta$, we define a partial elementary map $f_\alpha : A_\alpha \to N'$ such that f_α extends f_β whenever $\beta < \alpha$ and for limit α, $f_\alpha = \cup_{\beta < \alpha} f_\beta$.

Let $\alpha < \delta$ and f_α have been defined. We know that $tp^N(a_\alpha/A_\alpha)$ is isolated. Let $\psi[x, \bar{a}]$, $\bar{a} \in A_\alpha$, isolate $tp^N(a_\alpha/A_\alpha)$. In particular, $N \models \exists x \psi[x, \bar{a}]$. Hence, $N' \models \exists x \psi[x, f_\alpha(\bar{a})]$. Choose a $b \in N'$ such that $N' \models \psi[b, f_\alpha(\bar{a})]$. It also follows that $\psi[x, f_\alpha(\bar{a})]$ isolates $tp^{N'}(b/f_\alpha(A_\alpha))$. This implies that $f_{\alpha+1} = f_\alpha \cup \{(a_\alpha, b)\} : A_{\alpha+1} \to N'$ is partial elementary. Now take $g = f_\delta$.

Finally we show that for every $\bar{a} \in N$, $tp^N(\bar{a}/A)$ is isolated. By induction on α, we show that for every $\bar{a} \in A_\alpha$, $tp^N(\bar{a}/A)$ is isolated. We only need to show that if the hypothesis is true for α, it is true for $\alpha + 1$ also. Take $(a_\alpha, \ldots, a_\alpha, \bar{b}) \in A_{\alpha+1}$, with $\bar{b} \in A_\alpha$. By induction hypothesis, $tp^N(\bar{b}/A)$ is isolated and by the construction $tp^N(a_\alpha/A_\alpha)$ is isolated. Hence, by Proposition 3.4.6, $tp^N(a_\alpha, \bar{b}/A)$ is isolated. This proves our claim when the sequence $(a_\alpha, \ldots, a_\alpha)$ is of length 1.

Now take $(a_\alpha, \ldots, a_\alpha)$ of length $n - 1 \geq 1$. Consider the formula

$$\theta_1[x_1, \ldots, x_n] = \wedge_{i=1}^{n-1} (x_i = x_n).$$

Then $\theta_1[x_1, \ldots, x_{n-1}, a_\alpha]$ isolates $p = tp^N(a_\alpha, \ldots, a_\alpha/A_{\alpha+1})$. Let $\theta_2[x_n, \bar{y}]$ be an L_A-formula isolating $tp^N(a_\alpha, \bar{b}/A)$.

Now take a $\varphi[x_1, \ldots, x_{n-1}, x_n, \bar{y}] \in tp^N(a_\alpha, \ldots, a_\alpha, a_\alpha, \bar{b}/A)$. Then

$$\varphi[x_1, \ldots, x_{n-1}, a_\alpha, \bar{b}] \in tp^N(a_\alpha, \ldots, a_\alpha/A_{\alpha+1}).$$

Hence,

$$N \models \forall x_1 \ldots \forall x_{n-1} (\theta_1[x_1, \ldots, x_{n-1}, a_\alpha] \to \varphi[x_1, \ldots, x_{n-1}, a_\alpha, \overline{b}]).$$

i.e. the formula

$$\forall x_1 \ldots \forall x_{n-1} (\theta_1[x_1, \ldots, x_n] \to \varphi[x_1, \ldots \ldots, x_n, \overline{y}]) \in tp^N(a_\alpha, \overline{b}/A).$$

Therefore,

$$N \models \forall x_1 \ldots \forall x_n \forall \overline{y} ((\theta_2[x_n, \overline{y}] \wedge \theta_1[x_1, \ldots, x_n]) \to \varphi[x_1, \ldots, x_n, \overline{y}]).$$

Then the L_A-formula $\theta_2[x_n, \overline{y}] \wedge \theta_1[x_1, \ldots, x_n]$ isolates $tp^N(a_\alpha, \ldots, a_\alpha, a_\alpha, \overline{b}/A).$
□

As a consequence, we have the following useful theorem.

Theorem 4.8.14 *Let T be ω-stable and $M \models T$ uncountable. Then there is a proper elementary extension N of M such that every countable type $q(\overline{x})$ over M realised in N is also realised in M.*

Proof Choose and fix an L_M formula $\varphi[x]$ as in the Proposition 4.8.12. Let

$$p[x] = \{\psi[x] : \psi \text{ an } L_M - \text{formula } \& \varphi[M] \cap \psi[M] \text{ uncountable}\}.$$

If $\psi_1, \ldots, \psi_n \in p[x]$, then each of $\varphi[M] \cap \neg\psi_1[M], \ldots, \varphi[M] \cap \neg\psi_n[M]$ is countable. Hence, $\varphi[M] \cap (\wedge_{i=1}^n \psi_i)[M]$ is uncountable. This shows that $p[x]$ is a 1-type over M. Further, for every L_M formula $\psi[x]$, either ψ or $\neg\psi$ is in p. Hence, $p \in S_1^M(M)$.

If possible, suppose there is a $c \in M$ realising $p[x]$. Since $x \neq c$ is an L_M-formula in $p[x]$, we contradict that c realises $p[x]$. Let M' be an elementary extension of M in which $p[x]$ is realised, say by c. Then $c \notin M$.

By the last theorem, there is an elementary substructure N of M' which is a prime model extension of $M \cup \{c\}$. Further, for every $\overline{b} \in N$, $tp^N(\overline{b}/M \cup \{c\})$ is isolated. In particular, N is a proper elementary extension of M.

Let $q(\overline{x})$ be a countable type over M realised in N, say by $\overline{b}$. Let $\theta[\overline{x}, x]$ be an L_M formula such that $\theta[\overline{x}, c]$ isolates $tp^N(\overline{b}/M \cup \{c\})$. Since $\exists \overline{x} \theta[\overline{x}, x]$ is an L_M formula realised by c, it belongs to p. Further, $q[\overline{x}] \subset tp^N(\overline{b}/M \cup \{c\})$. So, for every $\psi[\overline{x}] \in q$, $\forall \overline{x}(\theta[\overline{x}, x] \to \psi[\overline{x}])$ is realised by c and hence belongs to p.

Thus,

$$\Gamma = \{\exists \bar{x}\theta[\bar{x}, x]\} \cup \{\forall \bar{x}(\theta[\bar{x}, x] \rightarrow \psi[\bar{x}]) : \psi[\bar{x}] \in q\}$$

is a countable subset of $p[x]$. Enumerate $\Gamma = \{\gamma_n[x] : n < \omega\}$. Then for each $n < \omega$,

$$\varphi(M) \setminus \gamma_n(M)$$

is countable. Therefore, there is a $d \in \varphi(M)$ that realises each of $\gamma_n[x]$. Then any $\bar{d} \in M$ such that $M \models \theta[\bar{d}, d]$ realises q.   □

4.9 Morley Rank

We now proceed to systematically introduce key notions of Morley rank and Morley degree introduced by Morley in [43] to prove his famous categoricity theorem. Morley's categoricity theorem will be proved in the next chapter. We shall not present Morley's original proof. We shall present a much simpler proof due to Baldwin and Lachlan that appeared in [5]. Morley rank can be viewed as the generalisation of the notion of dimension to theories more general than minimal theories.

Let M be an L-structure. By induction on ordinals α, for every L_M-formula $\varphi[\bar{x}]$ we define '$MR^M(\varphi) \geq \alpha$' as follows:

1. $MR^M(\varphi) \geq 0$ if $\varphi[M] \neq \emptyset$.
2. If α is limit, $MR^M(\varphi) \geq \alpha$ if for all $\beta < \alpha$, $MR^M(\varphi) \geq \beta$.
3. $MR^M(\varphi) \geq \alpha + 1$ if there exist L_M- formulas $\psi_1[\bar{x}], \psi_2[\bar{x}], \ldots$ such that for each n, $MR^M(\psi_n) \geq \alpha$ and $\psi_1(M), \psi_2(M), \ldots$ are pairwise disjoint subsets of $\varphi(M)$.

If $MR^M(\varphi) \geq \alpha$ for every ordinal α, we write $MR^M(\varphi) = \infty$, and if $\varphi(M) = \emptyset$, we put $MR^M(\varphi) = -\infty$. For an ordinal α, $MR^M(\varphi) = \alpha$ if $MR^M(\varphi) \geq \alpha$ but $MR^M(\varphi) \not\geq \alpha + 1$. We say that an L_M-formula φ has *Morley rank* in M if $MR^M(\varphi) = \alpha$ for some ordinal α. We shall write $MR^M(\varphi) < \infty$ if φ has Morley rank in M or if $\varphi(M) = \emptyset$.

Here are some simple observations.

Remark 4.9.1 $MR^M(\varphi) \geq 1$ if and only if $\varphi(M)$ is infinite. Consequently, $MR^M(\varphi) = 0$ if and only if $\varphi(M)$ is a non-empty finite set and $MR^M(\varphi) = 1$ if φ is minimal in M.

Remark 4.9.2 If $MR^M(\varphi) \geq \alpha$ and $\beta < \alpha$, then there is an L_M-formula $\psi[\bar{x}]$ of Morley rank β such that $\psi(M) \subset \varphi(M)$. This is easily seen by induction on α.

Remark 4.9.3 Let $\varphi[\bar{x}]$ and $\psi[\bar{x}]$ be L_M-formulas and $M \models \forall \bar{x}(\varphi[\bar{x}] \rightarrow \psi[\bar{x}])$. Then $MR^M(\varphi) \leq MR^M(\psi)$. So, if $\varphi(M) = \psi(M)$, $MR^M(\varphi) = MR^M(\psi)$.

Remark 4.9.4 Let $\varphi[\overline{x}]$ be an L_M-formula and $\psi[\overline{x}, \overline{y}] = \varphi[\overline{x}]$. Then $MR^M(\psi) \geq MR^M(\varphi)$. By induction on ordinals α, it is easy to see that $MR^M(\varphi) \geq \alpha \Rightarrow MR^M(\psi) \geq \alpha$. Note that $MR^M(\psi)$ may be strictly larger than $MR^M(\varphi)$. To see this, let M be infinite and $MR^M(\varphi) = 0$. Then $MR^M(\psi) \geq 1$.

Remark 4.9.5 Let $\varphi[\overline{x}]$ and $\psi[\overline{x}]$ be L_M-formulas. Then

$$MR^M(\varphi \vee \psi) = \max\{MR^M(\varphi), MR^M(\psi)\}.$$

By Remark 4.9.3, $MR^M(\varphi \vee \psi) \geq \max\{MR^M(\varphi), MR^M(\psi)\}$.

By induction on ordinals α, we now show that

$$MR^M(\varphi \vee \psi) \geq \alpha \Rightarrow \max\{MR^M(\varphi), MR^M(\psi)\} \geq \alpha. \qquad (*)$$

Since $(\varphi \vee \psi)(M) = \varphi(M) \cup \psi(M)$, this is true for $\alpha = 0$.

Let λ be limit, $(*)$ hold for all $\alpha < \lambda$ and $MR^M(\varphi \vee \psi) \geq \lambda$. Then $MR^M(\varphi \vee \psi) \geq \alpha$ for all $\alpha < \lambda$. By induction hypothesis, for every $\alpha < \lambda$, $\max\{MR^M(\varphi), MR^M(\psi)\} \geq \alpha$ so that either $MR^M(\varphi) \geq \alpha$ or $MR^M(\psi) \geq \alpha$. Hence, at least one of φ, ψ is of Morley rank $\geq \alpha$ for an unbounded in λ set of $\alpha < \lambda$. Thus $\max\{MR^M(\varphi), MR^M(\psi)\} \geq \lambda$.

Now let $(*)$ hold for all ordinals $\leq \alpha$ and $MR^M(\varphi \vee \psi) \geq \alpha + 1$. Get L_M-formulas $\psi_1, \psi_2, \ldots$ of Morley rank $\geq \alpha$ such that $\psi_1(M), \psi_2(M), \ldots$ are pairwise disjoint subsets of $\varphi(M) \cup \psi(M)$. Then $MR^M(\psi_n) = MR^M((\varphi \wedge \psi_n) \vee (\psi \wedge \psi_n)) \geq \alpha$. Therefore, either for infinitely many n, $MR^M(\varphi \cap \psi_n) \geq \alpha$ or for infinitely many n, $MR^M(\psi \cap \psi_n) \geq \alpha$. So, at least one of $MR^M(\varphi)$ or $MR^M(\psi) \geq \alpha + 1$.

Remark 4.9.6 Let $\psi[\overline{x}, \overline{y}]$ be an L_M-formula and $\varphi[\overline{x}] = \exists \overline{y} \psi[\overline{x}, \overline{y}]$. Then $MR^M(\psi) \geq MR^M(\varphi)$. We prove by induction on ordinals α that $MR^M(\varphi) \geq \alpha \Rightarrow MR^M(\psi) \geq \alpha$. This will establish our contention. This is clear for $\alpha = 0$ and for limit ordinals α. So, assume the hypothesis for α. Suppose $MR^M(\varphi) \geq \alpha + 1$. In particular, $MR^M(\varphi) \geq \alpha$. Hence, by induction hypothesis, $MR^M(\psi) \geq \alpha$. Since $MR^M(\varphi) \geq \alpha + 1$, there exist L_M-formulas $\varphi_1[\overline{x}], \varphi_2[\overline{x}], \ldots$, each of rank $\geq \alpha$, such that $\varphi_1(M), \varphi_2(M), \ldots$ are pairwise disjoint, non-empty subsets of $\varphi(M)$. Let $\psi_i[\overline{x}, \overline{y}] = \varphi_i[\overline{x}]$, $i = 1, 2, \ldots$. By Remark 4.9.4, each of $\psi_1, \psi_2, \ldots$ is of rank $\geq \alpha$. Hence, by induction hypothesis, $MR^M(\psi \wedge \psi_i) \geq \alpha$.   But $\psi_1(M) \cap \psi(M), \psi_2(M) \cap \psi(M), \ldots$ are pairwise disjoint subsets of $\psi(M)$. Thus, $MR^M(\psi) \geq \alpha + 1$.

Remark 4.9.7 Let $MR^M(\varphi) = \infty$. Note that there exists an ordinal α such that the Morley rank of every formula having an ordinal Morley rank is of Morley rank $< \alpha$. So, there exist formulas φ_0, φ_1, each of Morley rank ∞, such that $\varphi_0(M), \varphi_1(M)$ partition $\varphi(M)$.

Theorem 4.9.8 *Let M be a $\aleph_0$-homogeneous structure and $\overline{a}, \overline{b} \in M$ be such that $tp^M(\overline{a}) = tp^M(\overline{b})$. Then for every L-formulas $\varphi[\overline{x}, \overline{y}]$,*

$$MR^M(\varphi[\overline{x}, \overline{a}]) = MR^M(\varphi[\overline{x}, \overline{b}]).$$

Proof By induction on ordinals α, we show that for every L-formula $\varphi[\bar{x}, \bar{y}]$,

$$MR^M(\varphi[\bar{x}, \bar{a}]) \geq \alpha \Rightarrow MR^M(\varphi[\bar{x}, \bar{b}]) \geq \alpha. \qquad (*)$$

The result will then follow from symmetry.

Since $tp^M(\bar{a}) = tp^M(\bar{b})$,

$$M \models \exists \bar{x} \varphi[\bar{x}, \bar{a}] \Leftrightarrow M \models \exists \bar{x} \varphi[\bar{x}, \bar{b}].$$

Hence, $(*)$ holds for $\alpha = 0$.

Suppose λ is a limit ordinal and $(*)$ holds for all $\alpha < \lambda$, then

$$\begin{aligned} MR^M(\varphi[\bar{x}, \bar{a}]) \geq \lambda &\Leftrightarrow \forall \alpha < \lambda (MR^M(\varphi[\bar{x}, \bar{a}]) \geq \alpha) \\ &\Leftrightarrow \forall \alpha < \lambda (MR^M(\varphi[\bar{x}, \bar{b}]) \geq \alpha) \\ &\Leftrightarrow MR^M(\varphi[\bar{x}, \bar{b}]) \geq \lambda \end{aligned}$$

Now assume that $(*)$ holds for all $\beta \leq \alpha$ and $\varphi[\bar{x}, \bar{y}]$ is an L-formula. Assume that $MR^M(\varphi[\bar{x}, \bar{a}]) \geq \alpha + 1$. Get L-formulas $\psi_1[\bar{x}, \bar{y}_1], \psi_2[\bar{x}, \bar{y}_2], \ldots$ and $\bar{a}_1, \bar{a}_2, \ldots$ in M such for each n, $MR^M(\psi_n[\bar{x}, \bar{a}_n]) \geq \alpha$ and $\psi_1'(M), \psi_2'(M), \ldots$ are pairwise disjoint subsets of $\varphi(M)$, where $\psi_n'[\bar{x}] = \psi_n[\bar{x}, \bar{a}_n]$.

Since $tp^M(\bar{a}) = tp^M(\bar{b})$, $\bar{a} \to \bar{b}$ is partial elementary. Since M is $\aleph_0$-homogeneous, inductively we define $\bar{b}_1, \bar{b}_2, \ldots$ such that for each n,

$$(\bar{a}, \bar{a}_1, \ldots, \bar{a}_n) \to (\bar{b}, \bar{b}_1, \ldots, \bar{b}_n)$$

is partial elementary. In particular, for each n,

$$tp^M(\bar{a}, \bar{a}_1, \ldots, \bar{a}_n) = tp^M(\bar{b}, \bar{b}_1, \ldots, \bar{b}_n).$$

Since $tp^M(\bar{a}_n) = tp^M(\bar{b}_n)$ and $MR^M(\psi_n[\bar{x}, \bar{a}_n]) \geq \alpha$, by induction hypothesis, $MR^M(\psi_n[\bar{x}, \bar{b}_n]) \geq \alpha$.

Since $tp^M(\bar{a}, \bar{a}_n) = tp^M(\bar{b}, \bar{b}_n)$ and

$$M \models \forall \bar{x}(\psi_n[\bar{x}, \bar{a}_n] \to \varphi[\bar{x}, \bar{a}]),$$

it follows that

$$M \models \forall \bar{x}(\psi_n[\bar{x}, \bar{b}_n] \to \varphi[\bar{x}, \bar{b}]).$$

Let $m \neq n$. Since $tp^M(\bar{a}_n, \bar{a}_m) = tp^M(\bar{b}_n, \bar{b}_m)$ and

$$M \models \neg \exists \bar{x}(\psi_n(\bar{x}, \bar{a}_n) \wedge \psi_m(\bar{x}, \bar{a}_m)),$$

we have

$$M \models \neg \exists \bar{x}(\psi_n(\bar{x}, \bar{b}_n) \wedge \psi_m(\bar{x}, \bar{b}_m)).$$

It follows that $MR^M(\varphi[\bar{x}, \bar{b}]) \geq \alpha + 1$. $\qquad\qquad\qquad\qquad\qquad\square$

Theorem 4.9.9 *Let M be an $\aleph_0$-saturated L-structure and φ an L_M-formula. Then for every $\aleph_0$-saturated elementary extension N of M, $MR^M(\varphi) = MR^N(\varphi)$.*

Proof By induction on ordinals α, we show that for every L_M-formula φ, $MR^M(\varphi) \geq \alpha$ if and only if $MR^N(\varphi) \geq \alpha$.

Since N is an elementary extension of M, the argument at $\alpha = 0$ or limit is clear.

Assume that the assertion holds for all $\alpha \leq \lambda$. Take any L_M-formula φ such that $MR^M(\varphi) \geq \lambda + 1$. Get L_M-formulas $\psi_1, \psi_2, \ldots$ such that $MR^M(\psi_n) \geq \lambda$ for each n and $\psi_1(M), \psi_2(M), \ldots$ are pairwise disjoint subsets of $\varphi(M)$. By induction hypothesis, for each n, $MR^N(\psi_n) \geq \lambda$. Since N is an elementary extension of M, $\psi_1(N), \psi_2(N), \ldots$ are pairwise disjoint subsets of $\varphi(N)$. Thus, $MR^N(\varphi) \geq \lambda + 1$.

Conversely, assume that $MR^N(\varphi) \geq \lambda + 1$. Get L_N-formulas $\psi_1, \psi_2, \ldots$ such that $MR^N(\psi_n) \geq \lambda$ for each n and $\psi_1(N), \psi_2(N), \ldots$ are pairwise disjoint subsets of $\varphi(N)$. Let L-formulas $\theta[\bar{x}, \bar{y}], \theta_1[\bar{x}, \bar{y}_1], \theta_2[\bar{x}, \bar{y}_2], \ldots, \bar{a} \in M$ and $\bar{b}_1, \bar{b}_2, \ldots \in N$ be such that $\varphi = \theta[\bar{x}, \bar{a}]$ and $\psi_n = \theta_n[\bar{x}, \bar{b}_n]$, $n \geq 1$. Using $\aleph_0$-saturatedness and hence $\aleph_0$-homogeneity of M, inductively we now pick $\bar{a}_1, \bar{a}_2, \ldots$ in M such that for all k,

$$tp^M(\bar{a}, \bar{a}_1, \ldots, \bar{a}_k) = tp^N(\bar{a}, \bar{b}_1, \ldots, \bar{b}_k).$$

We take $\bar{a}_1$ that realises $tp^N(\bar{b}_1/\bar{a})$ in M. Suppose $\bar{a}_1, \ldots, \bar{a}_n$ have been defined. Get $\bar{a}'_1, \ldots, \bar{a}'_{n+1}$ that realises $tp^N(\bar{b}_1, \ldots, \bar{b}_{n+1}/\bar{a})$ in M. Then

$$tp^M(\bar{a}, \bar{a}'_1, \ldots, \bar{a}'_n) = tp^N(\bar{a}, \bar{b}_1, \ldots, \bar{b}_n) = tp^M(\bar{a}, \bar{a}_1, \ldots, \bar{a}_n).$$

Therefore, by $\aleph_0$-homogeneity of M, there exists $\bar{a}_{n+1}$ such that

$$tp^M(\bar{a}, \bar{a}_1, \ldots, \bar{a}_{n+1}) = tp^M(\bar{a}, \bar{a}'_1, \ldots, \bar{a}'_{n+1}) = tp^N(\bar{a}, \bar{b}_1, \ldots, \bar{b}_{n+1}).$$

Since N is an elementary extension of M,

$$tp^N(\bar{b}_n) = tp^M(\bar{a}_n) = tp^N(\bar{a}_n).$$

Hence, by the last theorem,

$$MR^N(\theta_n[\bar{x}, \bar{a}_n]) = MR^N(\theta_n[\bar{x}, \bar{b}_n]) \geq \lambda.$$

Therefore, by induction hypothesis, $MR^M(\theta_n[\bar{x}, \bar{a}_n]) \geq \lambda$. Since

$$tp^M(\bar{a}, \bar{a}_1, \ldots, \bar{a}_k) = tp^N(\bar{a}, \bar{b}_1, \ldots, \bar{b}_k),$$

$\theta_1[M, \bar{a}_1], \theta_2[M, \bar{a}_2], \ldots$ are pairwise disjoint subsets of $\varphi(M)$. Thus, $MR^M(\varphi) \geq \lambda + 1$. $\qquad\square$

Proposition 4.9.10 *Let M_0, M_1 be $\aleph_0$-saturated elementary extensions of an L-structure M and φ an L_M-formula. Then $MR^{M_0}(\varphi) = MR^{M_1}(\varphi)$.*

Proof By Proposition 2.8.1 and Theorem 4.4.2, there is a common elementary $\aleph_0$-saturated extension N of M_0 and M_1. Hence, by the last theorem,

$$MR^{M_0}(\varphi) = MR^N(\varphi) = MR^{M_1}(\varphi). \qquad \square$$

All these suggest that we can define rank of a formula independent of the structure from where the parameters come. Fix a first-order theory T. Given a problem at hand, there is a sufficiently large cardinal κ such that every model M of interest will be of cardinality $\leq \kappa$. From now onwards, we assume that all models of interest are elementary substructures of a monster model $\mathbb{M}$ which is κ^+-saturated and κ^+-strongly homogeneous. We fix such a monster model $\mathbb{M}$. All φ will be $L_\mathbb{M}$-formulas. A sentence φ is true will mean that it is true in $\mathbb{M}$. By sets, we shall mean subsets of $\mathbb{M}^n$, $n \geq 1$. Finally, we define

$$MR(\varphi) = MR^\mathbb{M}(\varphi).$$

Let $X \subset \mathbb{M}^n$ be a definable set, defined by say $\varphi[\overline{x}]$. We define

$$MR(X) = MR(\varphi).$$

We have

1. $MR(X)$ is well defined.
2. If $X \subset Y$, then $MR(X) \leq MR(Y)$.
3. $MR(X) = 0$ if and only if X is a finite non-empty set.
4. If X is minimal, then $MR(X) = 1$.
5. $MR(X \cup Y) = \max\{MR(X), MR(Y)\}$.
6. For a limit ordinal α, $MR(X) \geq \alpha$ if and only if for every $\beta < \alpha$, $MR(X) \geq \beta$.
7. $MR(X) \geq \alpha + 1$ if and only if there exist pairwise disjoint definable subsets $Y_1, Y_2, \ldots$ of X each of rank $\geq \alpha$.
8. If $MR(X) \geq \alpha$, then for every $\beta < \alpha$, X has a definable subset Y of rank β.

4.10 Morley Degree

Let α be an ordinal. Call two definable subsets $X, Y \subset \mathbb{M}^n$ α-*equivalent* if $MR(X \triangle Y) < \alpha$. This defines an equivalence relation on definable subsets of $\mathbb{M}^n$. Call a definable set X α-*strongly minimal* if $MR(X) = \alpha$ and for every definable subset Y of X, either Y is of rank $< \alpha$ or $X \setminus Y$ is of rank $< \alpha$. Thus, 0-strongly minimal sets are precisely singletons and 1-strongly minimal sets are precisely minimal sets.

Call a formula φ α-*strongly minimal* if $\varphi(\mathbb{M})$ is α-strongly minimal. Two formulas φ, ψ will be called α-equivalent if $\varphi(\mathbb{M})$ and $\psi(\mathbb{M})$ are α-equivalent. We shall call an L_A-formula φ α-*strongly minimal over* A if $MR(\varphi) = \alpha$ and for every L_A-formula ψ, $MR(\varphi \wedge \psi) < \alpha$ or $MR(\varphi \wedge \neg\psi) < \alpha$.

Theorem 4.10.1 *Every definable subset X of rank $\alpha < \infty$ is a pairwise disjoint union of α-strongly minimal sets $X_1, \ldots, X_d$. The decomposition is unique modulo α-equivalence.*

Proof Set $X_0 = X$. Suppose the assertion is not true. In particular, X_0 is not α-strongly minimal. So, X_0 is a disjoint union of definable sets X_1, Y_1 each of rank α. At least one of these does not admit the above decomposition. Without any loss of generality, assume that Y_1 is not a finite disjoint union of α-strongly minimal sets. Now express Y_1 as a disjoint union of definable sets X_2, Y_2 of rank α such that Y_2 is not a finite disjoint union of α-strongly minimal sets. Proceeding inductively, we define definable subsets $X_1, X_2, \ldots, Y_1, Y_2, \ldots$ of X of rank α such that Y_{n-1} is a disjoint union of X_n and Y_n. But then $X_1, X_2, \ldots$ are pairwise disjoints definable subsets of X of rank α. Hence, $MR(X) \geq \alpha + 1$, a contradiction.

For uniqueness, let X be a disjoint union of α-strongly minimal sets $Y_1, \ldots, Y_e$. We have $Y_1 = \cup_i (Y_1 \cap X_i)$. As Y_1 is α-strongly minimal, there is a unique $1 \leq i \leq d$ such that $MR(Y_1 \cap X_i) = \alpha$. Using the fact that Y_1 and X_i are α-strongly minimal, it is easy to see that $MR(Y_1 \triangle X_i) < \alpha$. Same arguments will also show that 1 is the unique $1 \leq j \leq e$ such that $MR(Y_j \cap X_i) = \alpha$. By rearranging X_is, we can assume that $i = 1$. Proceeding thus, we see the uniqueness. $\square$

For a definable set X having a Morley rank, the positive integer d obtained above is called the *Morley degree* of X, denoted by $MD(X)$. For a formula φ, we define $MD(\varphi) = MD(\varphi(\mathbb{M}))$. If $MR(\varphi) = \alpha$ and $MD(\varphi) = d$, then there exist α-strongly minimal formulas $\varphi_1, \ldots, \varphi_d$ such that $\varphi_1(\mathbb{M}), \ldots, \varphi_d(\mathbb{M})$ partitions $\varphi(\mathbb{M})$ and these $\varphi_1, \ldots, \varphi_d$ are unique modulo α-equivalence. We call $\varphi_1, \ldots, \varphi_d$ *components* of φ.

Now let φ be an L_A-formula of Morley rank $\alpha < \infty$. Then arguing as above, we can get α-strongly minimal L_A-formulas $\varphi_1, \ldots, \varphi_k$ over A such that $\varphi(\mathbb{M})$ is a disjoint union of $\varphi_1(\mathbb{M}), \ldots, \varphi_k(\mathbb{M})$. These $\varphi_1, \ldots, \varphi_k$ are unique (among L_A-formulas) modulo α-equivalence. Note that $k \leq MD(\varphi)$. We shall call $\varphi_1, \ldots, \varphi_k$ *components* of φ *over* A. We call k the Morley degree of φ over A.

Example 4.10.2 A formula φ is definitional if and only if $MR(\varphi)=0$ and $MD(\varphi)=1$.

Example 4.10.3 If φ is an algebraic formula then $MR(\varphi)=0$ and $MD(\varphi)=deg(\varphi)$.

Example 4.10.4 A-definable set $\varphi(\mathbb{M})$ is minimal if and only if $MR(\varphi)=MD(\varphi)=1$.

Example 4.10.5 Let φ be an L_A-formula and $\varphi_1, \ldots, \varphi_k$ its components over A. Then
$$MD(\varphi) = \sum_i MD(\varphi_i).$$

Proposition 4.10.6 *For an L_M-formula $\varphi[\overline{x}]$ the following statements are equivalent.*

(A) $MR(\varphi) = \infty$.
(B) *There is a binary tree $\{\varphi_\epsilon[\overline{x}] : \epsilon \in 2^{<\omega}\}$ of L_M-formulas such that $\varphi_\emptyset[\overline{x}] = \varphi[\overline{x}]$*

Proof If $MR(\varphi) = \infty$, then by induction on the length of $\epsilon \in 2^{<\omega}$, we can easily define a binary tree of L_M-formulas $\{\varphi_\epsilon[\overline{x}] : \epsilon \in 2^{<\omega}\}$ satisfying (a) with $MR(\varphi_\epsilon) = \infty$ for every ϵ.

On the other hand, if (B) holds, then $MR(\varphi_\epsilon) = \infty$ for all ϵ. Suppose not. Then choose an ϵ such that $MR(\varphi_\epsilon)$ is of minimal rank α and of minimal degree among $\varphi_{\epsilon'}$ of rank α. But then $MD(\varphi_{\epsilon,0}) < MD(\varphi_\epsilon)$, a contradiction. In particular, $MR(\varphi) = \infty$. □

Corollary 4.10.7 *Let T be a totally transcendental, countable complete theory with infinite models and $M \models T$. Then for every L_M-formula φ, $MR(\varphi) < \infty$.*

4.11 Rank and Degree of Types

Let $p \in S_n(A)$. We define *the Morley rank of p* by

$$MR(p) = \min\{MR(\varphi) : \varphi \in p\}.$$

If $MR(p) < \infty$, we define *the Morley degree of p* by

$$MD(p) = \min\{MD(\varphi) : \varphi \in p \wedge MR(\varphi) = MR(p)\}.$$

Let $p \in S_n(A)$ have a Morley rank $\alpha < \infty$ and Morley degree d. Choose a $\varphi \in p$ such that $MR(\varphi) = \alpha$ and $MD(\varphi) = d$. Take any L_A-formula ψ. Then both $MR(\varphi \wedge \psi)$ and $MR(\varphi \wedge \neg\psi)$ cannot be of rank α. Because then their degrees will be $< d$. Thus, φ is α-strongly minimal over A. Now assume that $\psi \in p$ is such that $MR(\varphi) = MR(\psi)$ and $MD(\varphi) = MD(\psi)$. Since $MR(\varphi) = \max\{MR(\varphi \wedge \psi), MR(\varphi \wedge \neg\psi)\}$ exactly one of these equals α. Since $MR(\varphi \wedge \psi) = \alpha$, it also follows that $MR(\varphi \wedge \neg\psi) < \alpha$. By the same argument, $MR(\psi \wedge \neg\varphi) < \alpha$. Thus, φ and ψ are α-equivalent.

Theorem 4.11.1 *Let T be a countable, complete theory with infinite models. The following conditions are equivalent.*

1. *T is ω-stable.*
2. *T is totally transcendental theory.*
3. *Every formula has a Morley rank.*

Proof In Theorem 4.8.10, we proved that (1) implies (2). In Corollary 4.10.7, we showed that (2) implies (3).

We now show that (3) implies (1). Let T satisfies (3), $M \models T$ and $A \subset M$ countable. By (3), for every $p \in S_n(M/A)$, $MR(p) < \infty$. Choose $\varphi_p \in M$ with $MR(\varphi_p) = MR(p)$ and $MD(\varphi_p)$ least possible. Then, φ_p determines p:

$$p = \{\psi : \psi \text{ an } L_A - \text{formula } \wedge MR(\varphi_p \wedge \neg\psi) < \alpha\}.$$

So, if $p \neq q \in S_n(A)$, then $\varphi_p \neq \varphi_q$. Since T and A are countable, this shows that $S_n(M/A)$ is countable.   □

Let $\alpha < \infty$. For an α-strongly minimal L_A-formula $\varphi[\overline{x}]$ over A, we define

$$p_\varphi = \{\psi[\overline{x}] : \psi \text{ an } L_A - \text{formula } \wedge MR(\varphi \wedge \neg\psi) < \alpha\}.$$

Note that $p = p_{\varphi_p}$ and for α-strongly minimal φ, $\varphi = \varphi_{p_\varphi}$.
We shall write $MR(\overline{a}/A)$ for $MR(tp(\overline{a}/A))$ and $MD(\overline{a}/A)$ for $MD(tp(\overline{a}/A))$.

Proposition 4.11.2 *Let $\varphi[\overline{x}]$ be a consistent L_A-formula, $|A| < |\mathbb{M}|$. Then*

(a) *For every $\overline{a}$, $MR(\overline{a}/A) \geq 0$.*
(b) $MR(\overline{a}, \overline{b}/A) \geq MR(\overline{a}/A)$
(c) $MR(\varphi) = \max\{MR(p) : \varphi \in p \in S_n(A)\}.$
(d) *If X is a definable set over A, then*

$$MR(X) = \max\{MR(\overline{a}/A) : \overline{a} \in X\}.$$

(e) *If φ has Morley rank, then*

$$MD(\varphi) = \sum \{MD(p) : \varphi \in p \in S_n(A) \wedge MR(p) = MR(\varphi)\}.$$

(f) *If $p \in S_n(A)$ has Morley rank and $B \supset A$, then*

$$MD(p) = \sum \{MD(q) : p \subset q \in S_n(B) \wedge MR(p) = MR(q)\}.$$

(g) *If $p \in S_n(A)$ has Morley rank and $B \supset A$, then p has at least one and at most $MD(p)$ many extensions $q \in S_n(B)$ of the the same rank.*

Proof (a) Since $tp(\overline{a}/A)$ is non-empty and every $\varphi[\overline{x}] \in tp(\overline{a}/A)$ is obviously consistent, (a) is seen trivially.
  (b) Let $\psi[\overline{x}, \overline{y}] \in tp(\overline{a}, \overline{b}/A)$ and $\varphi[\overline{x}] = \exists \overline{y} \psi[\overline{x}, \overline{y}]$. Then $\varphi \in tp(\overline{a}/A)$ and $MR(\psi) \geq MR(\varphi)$. (b) follows.
  (c) First assume that $MR(\varphi) = \infty$. Let p be a complete n-type containing

$$\{\psi : \psi \text{ an } L_A - \text{formula } \wedge MR(\varphi \wedge \neg\psi) < \infty\}.$$

It is easily seen that p contains φ and $MR(p) = \infty$.

Let $MR(\varphi) = \alpha < \infty$ and p a complete n-type over A containing

$$\{\psi[\overline{x}] : \psi \text{ an } L_A - \text{formula } \wedge MR(\varphi \wedge \neg\psi) < \alpha\}.$$

Then $\varphi \in p$ and $MR(p) = \alpha$.

(d) Indeed, (c) and (d) are the same statements because $\mathbb{M}$ is saturated.

(e) Assume that $MR(\varphi) = \alpha < \infty$. Let $\varphi_1, \ldots, \varphi_k$ be the components of φ over A. Then $p_{\varphi_1}, \ldots, p_{\varphi_k}$ are all the complete n-types over A containing φ of Morley rank α. Further, $MD(p_{\varphi_i}) = MD(\varphi_i)$, $1 \leq i \leq k$. Since $MD(\varphi) = \sum_i MD(\varphi_i)$, (e) follows.

(f) Assume that $MR(p) = \alpha < \infty$. Let $\varphi_1, \ldots, \varphi_k$ be the components of φ_p over A. For each φ_i, let $\psi_{i_1}, \ldots, \psi_{i_{j_i}}$ be the components of φ_i over B and $q_{ij} = p_{\psi_{i_j}} \in S_n(B)$, $1 \leq j \leq j_i$, $1 \leq i \leq k$. Then q_{ij}s are all $q \in S_n(B)$ of Morley rank $MR(p)$ that extends p. Further

$$MD(p) = \sum_i MD(\varphi_i) = \sum_i \sum_j MD(\psi_{i_j}) = \sum_i \sum_j MD(q_{ij}).$$

(g) This is a direct consequence of (f).   □

Theorem 4.11.3 *Let $A \subset \mathbb{M}$ be of cardinality less than $|\mathbb{M}|$ and b algebraic over $A \cup \{\overline{a}\}$. Then $MR(\overline{a}, b/A) = MR(\overline{a}/A)$.*

Proof By (b) above, $MR(\overline{a}, b/A) \geq MR(\overline{a}/A)$. By induction on ordinals α, we show that $MR(\overline{a}, b/A) \geq \alpha \Rightarrow MR(\overline{a}/A) \geq \alpha$. For $\alpha = 0$ or limit, the steps are trivial.

Assume that b is algebraic over $A \cup \{\overline{a}\}$ and $MR(\overline{a}, b/A) \geq \alpha + 1$. By induction hypothesis, $MR(\overline{a}/A) \geq \alpha$. If possible, suppose $MR(\overline{a}/A) = \alpha$. Choose a $\varphi[\overline{x}] \in tp(\overline{a}/A)$ of Morley rank α. Let $MD(\varphi) = d$ and $\varphi_1, \ldots, \varphi_d$ be the α-strongly minimal components of φ.

Since b is algebraic over $A \cup \{\overline{a}\}$, there exists an L_A-formula $\psi[\overline{x}, y]$ and an $m \geq 1$ such that

$$\mathbb{M} \models \psi[\overline{a}, b] \wedge \exists_{=m}y\psi[\overline{a}, y].$$

Set

$$\theta[\overline{x}, y] = \varphi[\overline{x}] \wedge \psi[\overline{x}, y] \wedge \exists_{=m}y\psi[\overline{x}, y].$$

Since $\theta \in tp(\overline{a}, b/A)$, $MR(\theta) \geq \alpha + 1$. So, there exist $L_{\mathbb{M}}$-formulas $\theta_1[\overline{x}, y]$, $\theta_2[\overline{x}, y], \ldots$, each of rank $\geq \alpha$, such that $\theta_1(\mathbb{M}), \theta_2(\mathbb{M}), \ldots$ are pairwise disjoint subsets of $\theta(\mathbb{M})$. Set $\xi_k[\overline{x}] = \exists y\theta_k[\overline{x}, y]$, $k \geq 1$.

For every $k \geq 1$, $MR(\xi_k) \geq \alpha$: Since $MR(\theta_k[\overline{x}, y]) \geq \alpha$, there exist $\overline{c}, d$ such that $\mathbb{M} \models \theta_k[\overline{c}, d]$ and $MR(\overline{c}, d/A \cup \{\overline{b}_k\}) \geq \alpha$, where $\overline{b}_k \in \mathbb{M}$ are the parameters occurring in θ_k. So, by induction hypothesis, $MR(\overline{c}/A \cup \{\overline{b}_k\}) \geq \alpha$. Clearly, $\mathbb{M} \models \xi_k[\overline{c}]$. Hence, $MR(\xi_k) \geq \alpha$.

Since $\mathbb{M} \models \xi_k \to \vee_{i=1}^d \varphi_i$, for each $k \geq 1$, there is a i_k such that $MR(\xi_k \wedge \varphi_{i_k}) \geq \alpha$. Note that for infinitely many k, i_ks are the same. So we can assume that for all k, $MR(\xi_k \wedge \varphi_1) \geq \alpha$.

For every $k \geq 1$, $MR(\varphi_1 \wedge \wedge_{i=1}^k \xi_i) \geq \alpha$: Suppose this is not true and k is the first integer for which this inequality fails. Then $k > 1$ and

$$MR(\varphi_1 \wedge \wedge_{i=1}^{k-1} \xi_i), MR(\varphi_1 \wedge \xi_k \wedge \neg \wedge_{i=1}^{k-1} \xi_i) \geq \alpha.$$

This contradicts that φ_1 is α-strongly minimal.

We now have $MR(\wedge_{i=1}^k \xi_i) \geq \alpha$ for each $k \geq 1$. Since $\mathbb{M}$ is saturated, there exists $\overline{c} \in \mathbb{M}$ such that for every k, $\mathbb{M} \models \xi_k[\overline{c}]$. For each k get a d_k such that $\mathbb{M} \models \theta_k[\overline{c}, d_k]$. Since $\theta_1(\mathbb{M}), \theta_2(\mathbb{M}), \dots$ are pairwise disjoint, $d_1, d_2, \dots$ are all distinct. This, in particular, implies that $M \models \psi[\overline{c}, d_k]$ for each k. Since $\mathbb{M} \models \exists_{=m} y \psi[\overline{c}, y]$, we have a contradiction. □

Theorem 4.11.4 *Let T be a strongly minimal theory and $A \subset \mathbb{M}$. Then*

$$MR(a_1, \dots, a_n/A) = dim(a_1, \dots, a_n/A).$$

Proof Recall that by our convention, $|A| < |\mathbb{M}|$. By the last theorem, without loss of generality, we assume that $a_1, \dots, a_n$ are independent over A and show that

$$MR(a_1, \dots, a_n/A) = n.$$

We prove this by induction on n.

Case :$n = 1$. Let a_1 be independent over A. Let $\varphi[x] \in tp(a_1/A)$. As a_1 is independent over A, $\varphi(\mathbb{M})$ is infinite. Hence, $MR(a_1/A) \geq 1$. Since T is strongly minimal, every definable subset of $\mathbb{M}$ is either finite or cofinite in $\mathbb{M}$. So, $MR(\varphi) = 1$. Thus, $MR(a_1/A) = 1$.

inductive step. Assume that the result is true for all $n < m$ and $a_1, \dots, a_m$ are algebraically independent over A.

Let M be an elementary substructure of $\mathbb{M}$ such that $|M| < |\mathbb{M}|$ and $M \supset A \cup \{a_1, \dots, a_{m-1}\}$. As M is algebraically closed in $\mathbb{M}$ (Proposition 1.10.2 (vii)), for every $b \in \mathbb{M} \setminus M$, $a_1, \dots, a_{m-1}, b$ is independent over A. Take $\varphi[x_1, \dots, x_m] \in tp(a_1, \dots, a_m/A)$. As T is strongly minimal,

$$\{b \in \mathbb{M} : \mathbb{M} \models \varphi[a_1, \dots, a_{m-1}, b]\}$$

is cofinite. Thus, there exists a sequence $b_0, b_2, \dots$ such that for each i, $a_1, \dots, a_{m-1}, b_i$ are independent over A and $\mathbb{M} \models \varphi[a_1, \dots, a_{m-1}, b_i]$.

Set $\varphi_i[x_1, \dots, x_m] = \varphi[x_1, \dots, x_m] \wedge x_m = b_i$. Then, by induction hypothesis, $MR(\varphi_i) \geq m - 1$ and $\varphi_0(\mathbb{M}), \varphi_1(\mathbb{M}), \dots$ are pairwise disjoint subsets of $\varphi(\mathbb{M})$. Thus, $MR(a_1, \dots, a_m/A) \geq m$.

We now proceed to show that $MR(a_1, \ldots, a_m/A) \leq m$. Let $\varphi[x_1, \ldots, x_m] \in tp(a_1, \ldots, a_m/A)$. Let $\psi[x_1, \ldots, x_m]$ be an $L_\mathbb{M}$-formula such that $\psi(\mathbb{M}) \subset \varphi(\mathbb{M})$ and $\mathbb{M} \models \neg\psi[a_1, \ldots, a_m]$. We now show that $MR(\psi) < m$. This will complete the proof.

Take any $b_1, \ldots, b_m$ such that $(b_1, \ldots, b_m) \in \psi(\mathbb{M})$. If $b_1, \ldots, b_m$ were algebraically independent over A, $a_i \to b_i$, $1 \leq i \leq m$, would be partial elementary over A and then $tp(a_1, \ldots, a_m/A) = tp(b_1, \ldots, b_m/A)$. But this is not the case. Thus,

$$MR(b_1, \ldots, b_m/A) = dim(b_1, \ldots, b_m/A) < m,$$

whenever $\mathbb{M} \models \psi[b_1, \ldots, b_m]$. So,

$$MR(\psi) = \max\{MR(b_1, \ldots, b_m/A) : (b_1, \ldots, b_m) \in \psi(\mathbb{M})\} < m. \qquad \square$$

4.12 Definable Types

A type $p[\overline{x}] \in S_n(A)$ is called *definable over B* if for every L-formula $\varphi[\overline{x}, \overline{y}]$ there is an L_B-formula $d_p\varphi[\overline{y}]$ such that

$$\forall \overline{a} \in A(\varphi[\overline{x}, \overline{a}] \in p \Leftrightarrow \mathbb{M} \models d_p\varphi[\overline{a}]).$$

Assume that $A = M \preceq \mathbb{M}$ and $B \subset M$. Suppose $\varphi[\overline{x}, \overline{y}]$ an L-formula and $d'_p\varphi[\overline{y}]$ another choice. Then

$$M \models \forall \overline{y}(d_p\varphi[\overline{y}] \leftrightarrow d'_p\varphi[\overline{y}]).$$

Example 4.12.1 Let $p \in S_n(A)$ be isolated. Fix an L-formula $\psi[\overline{x}, \overline{y}]$ and $\overline{a} \in A$ such that $\psi[\overline{x}, \overline{a}]$ isolates p. For every L-formula $\varphi[\overline{x}, \overline{y}]$ take

$$d_p(\overline{y}) = \forall \overline{x}(\psi[\overline{x}, \overline{a}] \to \varphi[\overline{x}, \overline{y}])$$

to witness that p is definable over $\overline{a}$.

In this section, we prove that if T is a complete, ω-stable theory, then every type in $S_n(A)$ is definable over a finite $A_0 \subset A$ and give some consequences of this result on definable sets. We prove some preliminary results first.

A formula $\varphi[\overline{x}, \overline{y}]$ is said to have the *order property* if there exist sequences $\{\overline{a}_m\}, \{\overline{b}_n\}$ in $\mathbb{M}$ such that

$$\forall m \forall n(\mathbb{M} \models \varphi[\overline{a}_m, \overline{b}_n] \Leftrightarrow m < n).$$

Proposition 4.12.2 *An ω-stable theory T has no formula with order property.*

Proof If possible, suppose a formula $\varphi[\overline{x}, \overline{y}]$ has the order property and sequences $\{\overline{a}_m\}, \{\overline{b}_n\}$ in $\mathbb{M}$ are such that

$$\forall m \forall n (\mathbb{M} \models \varphi[\overline{a}_m, \overline{b}_n] \Leftrightarrow m < n).$$

For each rational r, add new and distinct constant symbols $\overline{c}_r, \overline{d}_r$ and axioms so that the interpretations of these in every model are distinct.

Clearly the theory $T \cup \{\varphi[\overline{c}_r, \overline{d}_s] : r < s\} \cup \{\neg\varphi[\overline{c}_r, \overline{d}_s] : r \geq s\}$ is finitely satisfiable. Hence, by compactness theorem, it has a model, say M. For brevity, we denote the interpretations of $\overline{c}_r, \overline{d}_s$ in M by $\overline{c}_r, \overline{d}_s$ respectively. Thus,

$$\forall r \forall s (M \models \varphi[\overline{c}_r, \overline{d}_s] \Leftrightarrow r < s).$$

For each rational s, we have

$$(-\infty, s) = \{r \in \mathbb{Q} : M \models \varphi[\overline{c}_r, \overline{d}_s]\}.$$

Since T is ω-stable, by Theorem 4.11.1, $MR(\varphi[\overline{x}, \overline{d}_s]) < \infty$. Now choose an L_M-formula $\psi[\overline{x}]$ of minimal Morley rank and minimal Morley degree such that the set

$$\{r \in \mathbb{Q} : M \models \psi[\overline{c}_r]\}$$

is an infinite interval. Let r belong to the interior of this set. Consider the formulas

$$\psi_0[\overline{x}] = \varphi[\overline{x}, \overline{d}_r] \wedge \psi[\overline{x}] \ \& \ \psi_1[\overline{x}] = \neg\varphi[\overline{x}, \overline{d}_r] \wedge \psi[\overline{x}].$$

Then $\{q \in \mathbb{Q} : M \models \psi_i[\overline{c}_q]\}$, $i = 0, 1$, are infinite intervals in $\mathbb{Q}$. By the minimality of the rank of ψ, it follows that $MR(\psi_0) = MR(\psi) = MR(\psi_1)$. But then $MD(\psi_0) < MD(\psi)$ and we have arrived at a contradiction. $\qquad \square$

Lemma 4.12.3 *Let $M \models T$ and $\varphi[\overline{x}]$ be an L_M-formula and $\psi[\overline{x}]$ an L_M-formula. Suppose $MR(\varphi) = MR(\varphi \wedge \psi) = \alpha$. Then there is an $\overline{a} \in M$ such that $\mathbb{M} \models \varphi[\overline{a}] \wedge \psi[\overline{a}]$.*

Proof Since there is an irreducible component φ' of φ over M such that $MR(\varphi' \wedge \psi) = MR(\varphi \wedge \psi)$, without any loss of generality, we assume that $MD(\varphi) = 1$.

We prove the result by induction on α. Let $\alpha = 0$. Since $MR(\varphi \wedge \psi) = 0$, there exists an $\overline{a} \in \mathbb{M}$ such that $\mathbb{M} \models \varphi[\overline{a}] \wedge \psi[\overline{a}]$. Since M is an elementary submodel of $\mathbb{M}$ and $MR(\varphi) = 0$, $\emptyset \neq \varphi(M) = \varphi(\mathbb{M})$. Hence, $\overline{a} \in M$.

Assume that the result is true for all $\beta < \alpha$ and φ, ψ are as in the hypothesis of the result. Since $MD(\varphi) = 1$, $MR(\varphi \wedge \neg\psi) = \beta < \alpha$. Now we get L_M-formulas $\varphi_0[\overline{x}], \varphi_1[\overline{x}], \ldots$ of Morley rank β such that $\varphi_0(M), \varphi_1(M), \ldots$ are pairwise disjoint subsets of $\varphi(M)$. Since $MR(\varphi \wedge \neg\psi) = \beta$, $MR(\varphi_k \wedge \neg\psi) < \beta$ for all but finitely many k. Fix such a k. Then $MR(\varphi_k \wedge \psi) = \beta$. So, by induction hypothesis, there is an $\overline{a} \in M$ such that $\mathbb{M} \models \varphi_k[\overline{a}] \wedge \psi[\overline{a}]$. This implies that $\mathbb{M} \models \varphi[\overline{a}] \wedge \psi[\overline{a}]$. $\qquad \square$

The following is the main technical result of this section.

Theorem 4.12.4 *Let T be ω-stable, $M \models T$ $\aleph_0$-saturated of cardinality $< |\mathbb{M}|$ and $\varphi[\bar{x}, \bar{y}]$, $\psi[\bar{x}]$ L_M-formulas with $MR(\psi) = \alpha$. Then the set*

$$X = \{\bar{a} \in M : MR(\varphi[\bar{x}, \bar{a}] \wedge \psi[\bar{x}]) = \alpha\}$$

is M-definable. Moreover, if φ and ψ are L_A-formulas, $A \subset M$, then X is A-definable.

Proof Let $\psi_1, \ldots, \psi_d$ be α-strongly minimal components of ψ over M (or over A for the last part of the result). Then

$$X = \cup_{i=1}^{d} \{\bar{a} \in M : MR(\varphi[\bar{x}, \bar{a}] \wedge \psi_i[\bar{x}]) = \alpha\}.$$

This shows that we need to prove the result under the assumption that the Morley degree of ψ over M (over A for the last part of the result) is 1.

Claim: For each $\bar{a} \in M$ such that $MR(\varphi[\bar{x}, \bar{a}] \wedge \psi[\bar{x}]) = \alpha$, there is a finite set $X_{\bar{a}} \subset \varphi[\mathbb{M}, \bar{a}] \cap \psi[M]$ such that for all $\bar{b}$,

$$X_{\bar{a}} \subset \varphi[\mathbb{M}, \bar{b}] \Rightarrow MR(\varphi[\bar{x}, \bar{b}] \wedge \psi[\bar{x}]) = \alpha.$$

Assuming that our claim is false, we show that φ has the order property. Since T is ω-stable, this will contradict Proposition 4.12.2.

Let $\bar{a} \in M$ witness that our claim is false. In particular, $MR(\varphi[\bar{x}, \bar{a}] \wedge \psi[\bar{x}]) = \alpha$. Hence, by Lemma 4.5.10, there is an $\bar{c}_0 \in M$ such that $\mathbb{M} \models \varphi[\bar{c}_0, \bar{a}] \wedge \psi[\bar{c}_0]$. Since $\bar{c}_0 \in \varphi[\mathbb{M}, \bar{a}] \cap \psi[M]$, by our assumption, there is a $\bar{b}_0$ such that $\bar{c}_0 \in \varphi[\mathbb{M}, \bar{b}_0]$ and $MR(\varphi[\bar{x}, \bar{b}_0] \wedge \psi[\bar{x}]) < \alpha$.

Now assume that $\bar{c}_0, \bar{b}_0, \ldots, \bar{c}_m, \bar{b}_m$ have been defined such that for all $0 \leq i \leq m$,

$$\bar{c}_i \in \varphi[\mathbb{M}, \bar{a}] \cap \psi(M) \ \& \ MR(\varphi[\bar{x}, \bar{b}_i] \wedge \psi[\bar{x}]) < \alpha.$$

Since

$$MR(\varphi[\bar{x}, \bar{a}] \wedge \psi[\bar{x}] \wedge \wedge_{i=0}^{m} \neg \varphi[\bar{x}, \bar{b}_i]) = \alpha,$$

by Lemma 4.5.10, there exists a

$$\bar{c}_{m+1} \in \varphi[\mathbb{M}, \bar{a}] \cap \psi[M] \setminus \cup_{i=0}^{m} \varphi[\mathbb{M}, \bar{b}_m].$$

Since $\{\bar{c}_0, \ldots, \bar{c}_{m+1}\} \subset \varphi[\mathbb{M}, \bar{a}] \cap \psi[M]$, by our assumption, there is a $\bar{b}_{m+1}$ such that

$$\{\bar{c}_0, \ldots, \bar{c}_{m+1}\} \subset \varphi[\mathbb{M}, \bar{b}_{m+1}] \ \& \ MR(\varphi[\bar{x}, \bar{b}_{m+1}] \wedge \psi[\bar{x}]) < \alpha.$$

For $\bar{a}_i = \bar{c}_{i+1}$, $i \in \omega$, $\mathbb{M} \models \varphi[\bar{a}_i, \bar{b}_j] \Leftrightarrow i < j$. Thus, our claim is proved.

Now consider

$$\mathcal{X} = \{X \subset \psi(\mathbb{M}) : |X| < \aleph_0 \wedge (X \subset \varphi(\mathbb{M}, \overline{a}) \Rightarrow MR(\varphi[\overline{x}, \overline{a}] \wedge \psi[\overline{x}]) = \alpha)\}.$$

For each $X \in \mathcal{X}$, set

$$\varphi_X[\overline{y}] = \wedge_{\overline{b} \in X} \varphi[\overline{b}, \overline{y}].$$

By our claim.

$$MR(\varphi[\overline{x}, \overline{a}] \wedge \psi[\overline{x}]) = \alpha \Leftrightarrow \exists X \in \mathcal{X}(\mathbb{M} \models \varphi_X[\overline{a}]).$$

Arguing similarly with $\neg\varphi[\overline{x}, \overline{y}]$, we see that '$MR(\neg\varphi[\overline{x}, \overline{a}] \wedge \psi[\overline{x}]) = \alpha$' is equivalent to finite or infinite disjunction of L_M-formulas. Since Morley degree of ψ is 1,

$$MR(\varphi[\overline{x}, \overline{a}] \wedge \psi[\overline{x}]) = \alpha \Leftrightarrow MR(\neg\varphi[\overline{x}, \overline{a}] \wedge \psi[\overline{x}]) \neq \alpha$$

Hence, by Lemma 4.5.10, '$MR(\varphi[\overline{x}, \overline{a}] \wedge \psi[\overline{x}]) = \alpha$' is equivalent to disjunction of finitely many φ_X. This proves the first part of our result.

Now assume that ψ is an L_A-formula, $A \subset M$. Take any $\sigma \in G_A$. Since M is $\aleph_0$-homogeneous, by Theorem 4.9.8,

$$MR(\varphi[\overline{x}, \overline{a}] \wedge \psi[\overline{x}]) = \alpha \Leftrightarrow MR(\varphi[\overline{x}, \sigma(\overline{a})] \wedge \psi[\overline{x}]) = \alpha.$$

So, any automorphism of M that fixes A pointwise, fixes

$$\{\overline{a} \in M : MR(\varphi[\overline{x}, \overline{a}] \wedge \psi[\overline{x}]) = \alpha\}$$

setwise. Hence, by Theorem 4.5.5, this set is A-definable.   □

As a consequence of this theorem, we now have

Theorem 4.12.5 *If T is a complete, ω-stable theory $M \models T$ and $A \subset M$, then every type in $S_n(M/A)$ is definable over some finite subset A_0 of A.*

Proof Let $p[\overline{x}] \in S_n(A)$ be of Morley rank α. Fix $\psi[\overline{x}] \in p$ of rank α and of minimal degree. Let $A_0 \subset A$ be a finite set such that ψ is an L_{A_0}-formula. For any L-formula $\varphi[\overline{x}, \overline{y}]$ and any $\overline{a} \in A$,

$$\varphi[\overline{x}, \overline{a}] \in p \Leftrightarrow MR(\varphi[\overline{x}, \overline{a}] \wedge \psi[\overline{x}]) = \alpha.$$

The result now follows from the last theorem.   □

Corollary 4.12.6 *Let T be an ω-stable complete theory, $M \models T$, $A \subset M$ and $D \subset M^n$ A-definable. Then every definable $X \subset D^m$ is $A \cup D$-definable.*

Proof Let $\varphi[\bar{x}, \bar{y}]$ be an L-formula and $\bar{b} \in M$ be such that $\varphi[\bar{x}, \bar{b}]$ defines X. Then

$$X = \{\bar{a} \in D^m : \varphi[\bar{a}, \bar{y}] \in tp(\bar{b}/D)\}.$$

Since $tp(\bar{b}/D)$ is definable over D by the last theorem, X is definable over $A \cup D$. $\qquad\square$

4.13 Forking Independence

Throughout this section, we assume that T is a countable, complete, ω-stable theory.

The main purpose of this section is to generalise the notion of independence to ω-stable theories. The key notion of forking independence as well as all results of this section is due to Shelah [54].

Let $A \subset B$, $p \in S_n(A)$, $q \in S_n(B)$ and $p \subset q$. We say that q is a *forking extension* of p (or that q *forks over* p) if $MR(q) < MR(p)$. Otherwise, q is called a *non-forking extension* of p or we say that q *does not fork over* p.

In Proposition 4.11.2 (g), we proved the following.

Theorem 4.13.1 *Let $p \in S_n(A)$ and $MR(p) < \infty$. Then p has at least one and at most $MD(p)$ many non-forking extensions over $S_n(B)$ for every $B \supset A$.*

We say that $\bar{a}$ is *independent from B over A* if $MR(\bar{a}/A \cup B) = MR(\bar{a}/A)$, i.e. $tp(\bar{a}/A \cup B)$ does not fork over $tp(\bar{a}/A)$, and write

$$\bar{a} \underset{A}{\downarrow} B.$$

Remark 4.13.2 If T is strongly minimal, by Theorem 4.11.4,

$$\bar{a} \underset{A}{\downarrow} B \Leftrightarrow dim(\bar{a}/A \cup B) = dim(\bar{a}/A).$$

Thus, this is a generalisation of the notion of independence we introduced for strongly minimal theories. For stable theories, we shall see that forking independence enjoys several properties of algebraic independence.

Proposition 4.13.3 (Monotonicity) $\bar{a} \underset{A}{\downarrow} B \Rightarrow \forall C \subset B(\bar{a} \underset{A}{\downarrow} C)$.

Proof This follows from

$$MR(\bar{a}/A) \geq MR(\bar{a}/A \cup C) \geq MR(\bar{a}/A \cup B).$$

$\qquad\square$

Proposition 4.13.4 (Finite Basis) $\bar{a} \downarrow_A B \Leftrightarrow \forall$ *finite* $B_0 \subset B(\bar{a} \downarrow_A B_0)$.

Proof Only if part follows from monotonicity. Now assume that $\bar{a} \not\downarrow_A B$. Then there exists a $\varphi[\bar{x}] \in tp(\bar{a}/A \cup B)$ such that $MR(\varphi) < MR(\bar{a}/A)$. Let $B_0 \subset B$ be a finite set such that φ is an $L_{A \cup B_0}$-formula, then $\bar{a} \not\downarrow_A B_0$.   □

Proposition 4.13.5 (Transitivity) $\bar{a} \downarrow_A \bar{b}, \bar{c} \Leftrightarrow \bar{a} \downarrow_A \bar{b} \wedge \bar{a} \downarrow_{A, \bar{b}} \bar{c}$.

Proof We have $MR(\bar{a}/A) \geq MR(\bar{a}/A, \bar{b}) \geq MR(\bar{a}/A, \bar{b}, \bar{c})$. The result is easily seen now.   □

Proposition 4.13.6 (Symmetry) $\bar{a} \downarrow_A \bar{b} \Leftrightarrow \bar{b} \downarrow_A \bar{a}$.

Proof Assume that $\bar{a} \downarrow_A \bar{b}$. Let $MR(\bar{a}/A) = \alpha = MR(\bar{a}/A, \bar{b})$, $MR(\bar{b}/A) = \beta$.

Case 1. A is a $\aleph_0$-saturated model.

If possible, suppose $MR(\bar{b}/A, \bar{a}) < \beta$. We shall arrive at a contradiction. Choose $\varphi[\bar{x}] \in tp(\bar{a}/A)$ such that $MR(\varphi) = \alpha$, $\psi[\bar{y}] \in tp(\bar{b}/A)$ such that $MR(\psi) = \beta$ and $\gamma[\bar{x}, \bar{y}] \in tp(\bar{a}, \bar{b}/A)$ with $MR(\gamma[\bar{a}, \bar{y}]) < \beta$.

By Theorem 4.12.4, there is an L_A-formula $\xi[\bar{x}]$ defining $\{\bar{c} : MR(\psi[\bar{y}] \wedge \gamma[\bar{c}, \bar{y}]) \neq \beta\}$. Since $MR(\psi) = \beta$, ξ defines $\{\bar{c} : MR(\psi[\bar{y}] \wedge \gamma[\bar{c}, \bar{y}]) < \beta\}$. Note that $\xi \in tp(\bar{a}/A)$.

Since $MR(\bar{a}/A, \bar{b}) = \alpha$ and the formula $\varphi[\bar{x}] \wedge \xi[\bar{x}] \wedge \gamma[\bar{x}, \bar{b}] \in tp(\bar{a}/A, \bar{b})$, $MR(\varphi[\bar{x}] \wedge \xi[\bar{x}] \wedge \gamma[\bar{x}, \bar{b}]) = \alpha$. Therefore, by Lemma 4.12.3, there is an $\bar{a}' \in A$ such that

$$\mathbb{M} \models \varphi[\bar{a}'] \wedge \xi[\bar{a}'] \wedge \gamma[\bar{a}', \bar{b}].$$

In particular, $MR(\psi[\bar{y}] \wedge \gamma[\bar{a}', \bar{y}]) < \beta$. But $\psi[\bar{y}] \wedge \gamma[\bar{a}', \bar{y}] \in tp(\bar{b}/A)$ and $MR(\bar{b}/A) = \beta$. We have arrived at a contradiction and our result is proved in this case.

General Case. Let M be a saturated model containing A such that $|M| < |\mathbb{M}|$.

Choose $\bar{b}' \in \mathbb{M}$ realising a non-forking extension of $tp(\bar{b}/A)$ to M.

Claim. There is a $\bar{c} \in \mathbb{M}$ such that $tp(\bar{a}, \bar{b}/A) = tp(\bar{c}, \bar{b}'/A)$.

To see this, set
$$q[\bar{x}] = \{\rho[\bar{x}, \bar{b}'] : \rho[\bar{x}, \bar{y}] \in tp(\bar{a}, \bar{b}/A)\}.$$

Then q is a type over $A, \bar{b}'$. Hence, by saturability of $\mathbb{M}$, there is a $\bar{c} \in \mathbb{M}$ realising q. Let $\bar{a}' \in \mathbb{M}$ realise a non-forking extension of $tp(\bar{c}/A\bar{b}')$ to $M\bar{b}'$.

Claim. $MR(\bar{c}/A\bar{b}') = MR(\bar{a}/A\bar{b})$.

To see this, take a $\lambda[\bar{x}, \bar{y}] \in tp(\bar{c}, \bar{b}'/A)$. Then $\rho = \lambda[\bar{x}, \bar{b}'] \in tp(\bar{c}/A\bar{b}')$ and $\lambda[\bar{x}, \bar{b}] \in tp(\bar{a}/A\bar{b})$.

Since $tp(\bar{b}/A) = tp(\bar{b}'/A)$, $MR(\lambda[\bar{x}, \bar{b}]) = MR(\lambda[\bar{x}, \bar{b}'])$. then

$$MR(\bar{a}'/M) \geq MR(\bar{a}'/M\bar{b}') = MR(\bar{c}/A\bar{b}') = MR(\bar{a}/A\bar{b}) = \alpha.$$

and

$$MR(\bar{a}'/M) \leq MR(\bar{a}'/A) = MR(\bar{c}/A) = MR(\bar{a}/A) = \alpha.$$

Thus, $\bar{a}' \underset{M}{\downarrow} \bar{b}'$. By case 1, $\bar{b}' \underset{M}{\downarrow} \bar{a}'$.

Hence, as $tp(\bar{a}, \bar{b}/A) = tp(\bar{a}', \bar{b}'/A)$, $MR(\bar{b}/A\bar{a}) = MR(\bar{b}'/A\bar{a}')$. So,

$$\beta = MR(\bar{b}/A) \geq MR(\bar{b}/A\bar{a}) = MR(\bar{b}'/A\bar{a}') \geq MR(\bar{b}'/M\bar{a}') = \beta.$$

Thus, $\bar{b} \downarrow \bar{a}$. $\qquad\square$

Proposition 4.13.7 $\bar{a} \underset{A}{\downarrow} acl(A)$.

Proof Take any finite tuple $\bar{b} \in acl(A)$. By Theorem 4.11.3, we have $0 \leq MR(\bar{b}/A\bar{a}) \leq MR(\bar{b}/A) = 0$, i.e. $\bar{b} \underset{A}{\downarrow} \bar{a}$. Hence, by symmetry, $\bar{b} \underset{A}{\downarrow} \bar{a}$. The result now follows from the finite basis property. $\qquad\square$

Proposition 4.13.8 $\bar{a}, \bar{b} \underset{A}{\downarrow} C \Leftrightarrow (\bar{a} \underset{A}{\downarrow} C \wedge \bar{b} \underset{A,\bar{a}}{\downarrow} C)$.

Proof By finite basis property of forking, without any loss of generality, we assume that C is a finit tuple $\bar{c}$. Now.

$$\bar{a}, \bar{b} \underset{A}{\downarrow} \bar{c} \Leftrightarrow \bar{c} \underset{A}{\downarrow} \bar{a}, \bar{b} \qquad \text{(symmetry)}$$
$$\Leftrightarrow \bar{c} \underset{A}{\downarrow} \bar{a} \wedge \bar{c} \underset{A,\bar{a}}{\downarrow} \bar{b} \text{ (transitivity)}$$
$$\Leftrightarrow \bar{a} \underset{A}{\downarrow} \bar{c} \wedge \bar{b} \underset{A,\bar{a}}{\downarrow} \bar{c} \text{ (symmetry)}$$

$\qquad\square$

Chapter 5
Morley Categoricity Theorem

Abstract In this chapter, we present the proof of Baldwin and Lachlan of the Morley categoricity theorem. The proof uses among other things indiscernibles and Vaughtian pair of models. Morley's theorem is a very important milestone in model theory. It heralded the modern era of model theory. The concept of indiscernibles was introduced by Ehrenfeucht and Mostowski [11].

5.1 Existence of Indiscernibles

Let M be an L-structure, $A \subset M$, λ an ordinal and $X = \{\overline{x}_i : i \in I\} \subset M^\lambda$. We call X a *set of indiscernibles over A* if for every L_A-formula $\varphi[\overline{v}_1, \ldots, \overline{v}_n]$, $\overline{v}_1, \ldots, \overline{v}_n$ λ-tuples of distinct variables, and n-tuples $(\overline{x}_{i_1}, \ldots, \overline{x}_{i_n})$, $(\overline{x}_{j_1}, \ldots, \overline{x}_{j_n})$ in X,

$$M \models \varphi[\overline{x}_{i_1}, \ldots, \overline{x}_{i_n}] \leftrightarrow \varphi[\overline{x}_{j_1}, \ldots, \overline{x}_{j_n}].$$

If $A = \emptyset$ and X is a set of A-indiscernibles, then we simply call X a set of indiscernibles. If for some $i \neq j \in I$, $\overline{x}_i = \overline{x}_j$, then all $\overline{x}_i \in X$ are the same. Therefore, a set of A-indiscernibles elements are usually assumed to be distinct.

Next suppose that $(I, <)$ is a linearly ordered set, $(\overline{x}_i : i \in I)$ a sequence in M^λ and $A \subset M$. We call $(\overline{x}_i : i \in I)$ a set of *order indiscernibles over A* if for every L_A-formula $\varphi[\overline{v}_1, \ldots, \overline{v}_n]$ and $i_1 < \ldots < i_n$, $j_1 < \ldots < j_n$ in I

$$M \models \varphi[\overline{x}_{i_1}, \ldots, \overline{x}_{i_n}] \leftrightarrow \varphi[\overline{x}_{j_1}, \ldots, \overline{x}_{j_n}].$$

If $A = \emptyset$, we call $\{\overline{x}_i : i \in I\}$ a sequence of order indiscernibles.

Again, if $(\overline{x}_i : i \in I)$ is a set of order indiscernibles and if for some $i < j \in I$, $\overline{x}_i = \overline{x}_j$, then all $\overline{x}_i$ are the same. Hence, we always assume that elements of a set of order indiscernibles are all distinct.

We shall be proving results mostly for $\lambda = 1$. Readers should observe that most of the results we prove hold for general λ with exactly the same proof.

© Springer Nature Singapore Pte Ltd. 2017
H. Sarbadhikari and S.M. Srivastava, *A Course on Basic Model Theory*,
DOI 10.1007/978-981-10-5098-5_5

Here is a surprising result.

Theorem 5.1.1 *Let T be a theory with an infinite model. Then, for every linearly ordered set $(I, <)$, T has a model M with a set of order indiscernibles $(x_i : i \in I)$, x_i distinct.*

Proof Let M be an infinite model of T. Fix a linear order $<$ on M. Take new constant symbols $\{c_i : i \in I\}$ and let T' be the extension of T whose new axioms are the following sentences:

1. $c_i \neq c_j, i \neq j$.
2. $\varphi[c_{i_1}, \ldots, c_{i_k}] \to \varphi[c_{j_1}, \ldots, c_{j_k}]$, $\varphi[v_1, \ldots, v_n]$ an L-formula, $i_1 < \ldots < i_k$, $j_1 < \ldots < j_k$.

We show that T' is finitely satisfiable. Then, by compactness theorem, T' has model N in which $(c_i^N : i \in I)$ is a sequence of distinct order indiscernibles.

Let $T_0 \subset T'$ be finite. Let I_0 be the set of all $i \in I$ such that c_i appears in T_0 and $\varphi_1, \ldots, \varphi_m$ be all the L-formulas that appear in T_0 under clause (2). Let $v_1, \ldots, v_n$ be all the variables that have a free occurrence in $\varphi_1, \ldots, \varphi_m$.

Let $[M]^n$ denote the set of all finite subsets of M of cardinality n. We define $F : [M]^n \to \mathcal{P}(\{1, \ldots, m\})$ as follows. Let $A = \{a_1, \ldots, a_n\} \in [M]^n$ with $a_1 < \ldots < a_n$. Then,

$$F(A) = \{j \leq m : M \models \varphi_j[a_1, \ldots, a_n]\}.$$

By Ramsey theorem (Theorem 8.5.2), M has an infinite homogeneous subset X, i.e. $F \restriction [X]^n$ takes a constant value, say $\eta \subset \{1, \ldots, m\}$. For each $i \in I_0$, choose $b_i \in X$ such that whenever $i < i'$, $b_i < b_{i'}$. Interpret $c_i^M = b_i, i \in I_0$.

For every $i_1 < \ldots < i_n$ in I_0 and every $1 \leq j \leq m$,

$$M \models \varphi_j[b_{i_1}, \ldots, b_{i_n}] \Leftrightarrow j \in \eta.$$

Thus, we have got a model of T_0. $\qquad\qquad\qquad\qquad\qquad\qquad\qquad\qquad\qquad\qquad\Box$

Let M be a model of T, $(I, <)$ a linearly ordered set and $X = \{x_i : i \in I\}$, $A \subset M$. We define *the Ehrenfeucht–Mostowski type* of X over A, denoted by $EM(X/A)$, to be the set of L_A-formulas $\varphi[v_0, \ldots, v_{n-1}]$ satisfying

$$\forall i_0 < \ldots < i_{n-1}(M \models \varphi[x_{i_0}, \ldots, x_{i_{n-1}}]).$$

We shall write $EM(X)$ instead of $EM(X/\emptyset)$.

Note that if $X = (x_i : i \in I)$ is an infinite set of order indiscernibles over $A \subset M$, then a L_A-formula $\varphi[v_1, \ldots, v_n] \in EM(X/A)$ if and only if $M \models \varphi[x_{i_1}, \ldots, x_{i_n}]$ for some $i_1 < \ldots < i_n$.

The following is a very general result on the existence of order indiscernibles.

Theorem 5.1.2 *Let T be an L-theory and $(I, <)$ an infinite linearly ordered set. Let $M \models T$, $X = \{x_i : i \in I\} \subset M$ with $i \neq i' \Rightarrow x_i \neq x_{i'}$. Then, for every infinite* linearly ordered set $(J, <)$, there is an elementary extension N of M having an infinite set $Y = \{y_j : j \in J\}$ of order indiscernibles indexed by J such that $EM(Y) = EM(X)$.

Proof Add new constant symbols $\{c_j : j \in J\}$. Let T' be the theory having following axioms:

$$Diag_{el}(M),$$

$$\{c_j \neq c_{j'} : j \neq j' \in J\},$$

$$\{\varphi[c_{j_1}, \ldots, c_{j_n}] : \varphi[v_1, \ldots, v_n] \in EM(X), j_1 < \ldots < j_n\}$$

and

$$\{\varphi[c_{j_1}, \ldots, c_{j_n}] \leftrightarrow \varphi[c_{j'_1}, \ldots, c_{j'_n}] : \varphi[v_1, \ldots, v_n] \text{ an } L - \text{formula},$$

$$j_1 < \ldots < j_n, j'_1 < \ldots < j'_n\}.$$

We show that T' has a model, say N. Take $y_j = c_j^N, j \in J$. Then, $Y = \{y_j : j \in J\}$ is a set of order indiscernibles in N with $EM(Y) = EM(X)$.

By compactness theorem, it is sufficient to show that T' is finitely satisfiable. Let $C \subset J$ be finite and F a finite set of L-formulas. Without any loss of generality, we assume that no variable other than $v_0, \ldots, v_{n-1}$ is free in any $\varphi \in F$. So, each $\varphi \in F$ can be represented as $\varphi[v_0, \ldots, v_{n-1}]$. Suppose T_0 consists of

$$\{c_j \neq c_{j'} : j \neq j' \in C\},$$

$$\{\varphi[c_{j_0}, \ldots, c_{j_{n-1}}] : \varphi \in F \cap EM(X) \wedge j_0 < \ldots < j_{n-1} \in C\}$$

and

$$\{\varphi[c_{j_0}, \ldots, c_{j_{n-1}}] \leftrightarrow \varphi[c_{j'_0}, \ldots, c_{j'_{n-1}}] : \varphi \in F$$

$$\wedge j_0 < \ldots < j_{n-1}, j'_0 < \ldots < j'_{n-1} \in C\}.$$

To complete the proof we now show that $M \models T_0$.

Let

$$[X]^n = \{(x_{i_k} : k < n) : i_0 < \ldots < i_{n-1} \in I\}.$$

We define an equivalence relation $\sim$ on $[X]^n$ by

$$\overline{x} \sim \overline{x'} \Leftrightarrow \forall \varphi \in F(M \models \varphi[\overline{x}] \leftrightarrow \varphi[\overline{x'}]).$$

This is a finite equivalence relation on $[X]^n$. Hence, by Ramsey's theorem (Theorem 8.5.2), there exists an infinite $X' \subset X$ homogeneous with respect to $\sim$. Now interpret c_j in M, $j \in C$, by $x_{i_j} \in X'$ such that $j \rightarrow i_j$ is order-preserving. Thus, M models T_0. $\qquad\square$

We give another proof of the existence of order indiscernibles. The technique used here is important and will be imitated later.

Theorem 5.1.3 *Let $N \preceq M$ be L-structures, $\aleph_0 \leq |N| < |M|$, $\lambda \leq |N|$ and M be $|N|$-saturated. Then, there is a sequence $\overline{c}_0, \overline{c}_1, \overline{c}_2, \overline{c}_3, \ldots$ of distinct elements in M^λ which are order indiscernibles over N.*

Proof Take any $\overline{c}_0 \in M^\lambda \setminus N^\lambda$. Let $\mathcal{U}$ be an ultrafilter on N^λ containing $\{\varphi(N) : \varphi[\overline{x}] \in tp^M(\overline{c}_0/N)\}$.

Let $\{\xi_i[\overline{x}] : i \in I\}$ be an enumeration of all $N\overline{c}_0$-formulas. For each $i \in I$, set

$$\theta_i[\overline{x}] = \begin{cases} \xi_i[\overline{x}] & \text{if } \xi_i(N) \in \mathcal{U}, \\ \neg\xi_i[\overline{x}] & \text{if } \neg\xi_i(N) \in \mathcal{U}. \end{cases}$$

Then, $q_0(\overline{x}) = \{\theta_i[\overline{x}] : i \in I\}$ is a type in M over $N\overline{c}_0$. Since M is $|N|$-saturated, q_0 is realised in M. Choose any $\overline{c}_1 \models q_0$ in M.

Note that $\overline{c}_1 \neq \overline{c}_0$. If not, then the formula $\overline{x} = \overline{c}_0$ is in q_0. But this implies $\overline{c}_0 \in N^\lambda$.

Suppose $\overline{c}_0, \overline{c}_1, \ldots, \overline{c}_n$ have been defined. Let $\{\eta_j[\overline{x}] : j \in J\}$ be an enumeration of all $N\overline{c}_0\overline{c}_1 \ldots \overline{c}_n$-formulas. For each $j \in J$, set

$$\theta_j[\overline{x}] = \begin{cases} \eta_j[\overline{x}] & \text{if } \eta_j(N) \in \mathcal{U}, \\ \neg\eta_j[\overline{x}] & \text{if } \neg\eta_j(N) \in \mathcal{U}. \end{cases}$$

Then, $q_n(\overline{x}) = \{\theta_j[\overline{x}] : j \in J\}$ is a type in M over $N\overline{c}_0\overline{c}_1 \ldots \overline{c}_n$. By the same reason, q_n is realised in M. Take any $\overline{c}_{n+1} \models q_n$ in M.

By induction on n, we show that for every $0 \leq i_1 < \ldots < i_n, 0 \leq j_1 < \ldots < j_n$,

$$tp^M(\overline{c}_{i_1}, \ldots, \overline{c}_{i_n}/N) = tp^M(\overline{c}_{j_1}, \ldots, \overline{c}_{j_n}/N).$$

<u>Initial Case:</u> Let $m \geq 1$. Then, both $\overline{c}_0$ and $\overline{c}_m$ satisfy the same set of L_N-formulas, viz. those that define sets in $\mathcal{U}$. Hence, $tp^M(\overline{c}_0/N) = tp^M(\overline{c}_m/N)$.

Inductive Step. Assume the hypothesis for all pairs of increasing sequences of length n. Take any $i_1 < \ldots < i_n < i_{n+1}$ and $j_1 < \ldots < j_n < j_{n+1}$. By induction hypothesis,

$$tp^M(\overline{c}_{i_1}, \ldots, \overline{c}_{i_n}/N) = tp^M(\overline{c}_{j_1}, \ldots, \overline{c}_{j_n}/N).$$

Hence, there exists a $f \in Aut_N(M)$ such that $f(\overline{c}_{i_p}) = \overline{c}_{j_p}$ for all $p \leq n$. So, for every L_N-formula $\varphi[\overline{x}, \overline{x}_1, \ldots, \overline{x}_n]$ and every $\overline{a} \in M$

$$M \models \varphi(\overline{a}, \overline{c}_{i_1}, \ldots, \overline{c}_{i_n}) \Leftrightarrow M \models \varphi(f(\overline{a}), \overline{c}_{j_1}, \ldots, \overline{c}_{j_n}).$$

Therefore,

$$tp(\overline{c}_{i_{n+1}}, \overline{c}_{i_1}, \ldots, \overline{c}_{i_n}/N) = tp(f(\overline{c}_{i_{n+1}}), \overline{c}_{j_1} \ldots, \overline{c}_{j_n}/N).$$

It follows that both $\overline{c}_{j_{n+1}}$ and $f(\overline{c}_{i_{n+1}})$ satisfy the same set of $N\overline{c}_{j_1} \ldots, \overline{c}_{j_n}$-formulas, viz. those that define sets in $\mathcal{U}$. Hence,

$$tp(\overline{c}_{i_{n+1}}, \overline{c}_{i_1}, \ldots, \overline{c}_{i_n}/N) = tp(f(\overline{c}_{i_{n+1}}), \overline{c}_{j_1} \ldots, \overline{c}_{j_n}/N) = tp(\overline{c}_{j_{n+1}}, \overline{c}_{j_1}, \ldots, \overline{c}_{j_n}/N).$$

$\square$

Proposition 5.1.4 *Let $\kappa > \aleph_0$, M a κ-saturated L-structure, $A \subset M$ of cardinality less than κ and $\lambda < \kappa$. Suppose $\overline{a}_0, \overline{a}_1, \overline{a}_2, \ldots$ is a sequence of order indiscernibles over A in M^λ. Then, there exist $\overline{a}_{-1}, \overline{a}_{-2}, \overline{a}_{-3}, \ldots$ such that*

$$\ldots, \overline{a}_{-3}, \overline{a}_{-2}, \overline{a}_{-1}, \overline{a}_0, \overline{a}_1, \overline{a}_2, \overline{a}_3, \ldots$$

is a sequence of order indiscernibles over A.

Proof We first show that there exists a $\overline{a}_{-1} \in M^\lambda$ such that $\overline{a}_{-1}, \overline{a}_0, \overline{a}_1, \overline{a}_2, \overline{a}_3, \ldots$ is a sequence of order indiscernibles over A.

Using this repeatedly, by induction, we get $\overline{a}_{-1}, \overline{a}_{-2}, \overline{a}_{-3}, \ldots \in M^\lambda$ such that for every k,

$$\overline{a}_{-k}, \ldots, \overline{a}_{-3}, \overline{a}_{-2}, \overline{a}_{-1}, \overline{a}_0, \overline{a}_1, \overline{a}_2, \overline{a}_3, \ldots$$

is a sequence of order indiscernibles over A. This will complete the proof.

Take any $k \geq 1$ and any L_A-formula $\varphi_k[\overline{x}, \overline{x}_1, \ldots, \overline{x}_k]$. For $\overline{i} = i_1 < \ldots < i_k$ and $\overline{j} = j_1 < \ldots < j_k$, let $\psi(\varphi_k, \overline{i}, \overline{j})$ denote the L_A-formula

$$\varphi_k[\overline{x}, \overline{a}_{i_1}, \ldots, \overline{a}_{i_k}] \leftrightarrow \varphi_k[\overline{x}, \overline{a}_{j_1}, \ldots, \overline{a}_{j_k}].$$

For $\overline{i} = i_1 < \ldots < i_k$ and $\overline{j} = j_0 < j_1 < \ldots < j_k$, let $\xi(\varphi_k, \overline{i}, \overline{j})$ denote the L_A-formula

$$\varphi_k[\overline{x}, \overline{a}_{i_1}, \ldots, \overline{a}_{i_k}] \leftrightarrow \varphi_k[\overline{a}_{j_0}, \overline{a}_{j_1}, \ldots, \overline{a}_{j_k}].$$

Let $p(\overline{x})$ consist of all formulas of the form $\psi(\varphi_k, \overline{i}, \overline{j})$ and $\xi(\varphi_k, \overline{i}, \overline{j})$ for all possible $k, \varphi_k, \overline{i}$ and $\overline{j}$. Since $\overline{a}_0, \overline{a}_1, \overline{a}_2, \ldots$ are order indiscernibles over A, it is easy to check that $p(\overline{x})$ is a finitely satisfiable set of formulas over a set of parameters of cardinality $< \kappa$. Since M is κ-saturated, $p(\overline{x})$ is realised in M. Any $\overline{a}_{-1} \in M$ realising $p(\overline{x})$ witnesses our claim. $\square$

Using the same idea one can easily prove the following theorem. Its proof is left to the reader as a simple exercise.

Proposition 5.1.5 *Let $\kappa > \aleph_0$, M a κ-saturated L-structure, $A \subset M$ of cardinality less than κ and $\lambda < \kappa$. Suppose $\overline{a}_0, \overline{a}_1, \overline{a}_2, \ldots$ is a sequence of order indiscernibles over A in M^λ. Then, for each $\omega_0 \leq \alpha < \kappa$ there exists a $\overline{a}_\alpha \in M^\lambda$ such that*

$$\overline{a}_0, \overline{a}_1, \overline{a}_2, \overline{a}_3, \ldots \overline{a}_\alpha, \ldots, \quad \alpha < \kappa$$

are order indiscernibles over A.

5.2   Applications of Indiscernibility

From now on, if $X = \{x_i : i \in I\}$ is a sequence of order indiscernibles, then we linearly order X by $x_i < x_j \Leftrightarrow i < j$ and replace I by X itself.

Theorem 5.2.1 *Let T be a complete, strongly minimal theory, $M \models T$ and $A = \{a_i : i \in I\}$ an independent set in M. Then, for every positive integer k and k-tuples $\overline{a}, \overline{b}$ in A, $tp^M(\overline{a}) = tp^M(\overline{b})$. Thus, A is a set of indiscernibles.*

Proof This follows from Proposition 2.12.3 by taking $\psi[x]$ to be $x = x$, $N = M$ and $X = \emptyset$ to conclude that the bijection $\overline{a} \to \overline{b}$ is partial elementary. $\square$

Assume that T has built-in Skolem functions, $M \models T$ and X an infinite set of order indiscernibles in M. Then, every element of the Skolem hull $\mathcal{H}(X)$ of X can be represented as $t[x_{i_0}, \ldots, x_{i_{n-1}}]$, where $t[\overline{v}]$ is an L-term and $x_{i_0} < \ldots < x_{i_{n-1}}$ in X. We have already observed that $\mathcal{H}(X)$ is an elementary substructure of M (Theorem 1.8.2). Therefore, $\mathcal{H}(X)$ is a model of T. Such models are called *Ehrenfeucht–Mostowski models*.

Theorem 5.2.2 *Let T be a theory with built-in Skolem functions, $M, N \models T$, X, Y infinite sets of order indiscernibles in M and N, respectively, such that $EM(X) = EM(Y)$. Suppose $\sigma : X \to Y$ is an order-preserving map. Then, σ can be extended to an elementary embedding $\tau : \mathcal{H}(X) \to \mathcal{H}(Y)$. Moreover, if σ is onto, τ will be an isomorphism.*

Proof Let $t[\overline{v}]$ be an L-term and $x_0 < \ldots < x_{n-1}$ an increasing sequence in X. Define

$$\tau(t[\overline{x}]) = t[\sigma(\overline{x})].$$

Then,

(1) τ is well defined and one-to-one: Let $\overline{x}$ and $\overline{x'}$ be increasing sequences in X. Since σ is order-preserving, X and Y are sets of indicernibles and $EM(X) = EM(Y)$, we have

$$M \models t[\overline{x}] = s[\overline{x'}]$$

if and only if

$$N \models t[\sigma(\overline{x})] = s[\sigma(\overline{x'})].$$

(2) Finally, let $\varphi[v_1, \ldots, v_n]$ be an L-formula $t_1[\overline{v}], \ldots, t_n[\overline{v}]$ L-terms and $\overline{x} \in X$ an increasing sequence in X. Since $EM(X) = EM(Y)$, we have

$$M \models \varphi[t_1[\overline{x}], \ldots, t_m[\overline{x}]] \Leftrightarrow N \models \varphi[t_1[\sigma(\overline{x})], \ldots, t_m[\sigma(\overline{x})]].$$

(3) Assume that σ is onto. Then, $t[\overline{y}] = \tau(t[\sigma^{-1}(\overline{y})])$, $\overline{y}$ an increasing sequence in Y. So, τ is onto.

Our proof is complete. $\square$

Theorem 5.2.3 *Let T be a countable theory with an infinite model. Then, for every infinite cardinal κ, T has a model M of cardinality κ such that for every countable $A \subset M$, M realises only countably many types in $S_1^M(A)$.*

Proof Let T^* be the Skolemisation of T and $N \models T^*$ having a sequence $X = \{x_\alpha : \alpha < \kappa\}$ of order indiscernibles of type κ. Since T, and hence T^*, has an infinite model, such an N exists. Take $M = \mathcal{H}(X)$. Then, $|M| = \kappa$. Let $A \subset M$ be countable. Then, each $a \in A$ can be expressed as $t_a[\overline{x}^a]$, where $\overline{x}^a$ is a finite increasing sequence in X. Let Y be the set of all $x \in X$ that appear in $\overline{x}^a$, $a \in A$. Then, Y is countable.

For $y_1 < \ldots < y_n$ and $z_1 < \ldots < z_n$ in X, we define

$$\overline{y} \sim_Y \overline{z} \Leftrightarrow \forall i \forall y \in Y((y_i = y \leftrightarrow z_i = y) \wedge (y_i < y \leftrightarrow z_i < y)).$$

The importance of this equivalence relation is the following. If $\overline{y} \sim_Y \overline{z}$, then for every $a_1 = t_1[\overline{x}], \ldots, a_m = t_m[\overline{x}]$ in A with $\overline{x}$ an increasing sequence in Y, for every L-formula $\varphi[u, v_1, \ldots, v_n]$ and for every term $t[w_1, \ldots, w_n]$,

$$M \models \varphi[t[\overline{y}], a_1, \ldots, a_m] \leftrightarrow \varphi[t[\overline{z}], a_1, \ldots, a_m].$$

Thus, the cardinality of types over A realised in M is $\leq |\cup_n X^n/\sim_Y|$.

But $\overline{y} \sim_Y \overline{z}$ if and only if for all i,

(1) $y_i \in Y$ and $y_i = z_i$

or

(2) $y_i \notin Y$ and $z_i \notin Y$ and there is no element of Y between y_i and z_i.

Thus, each $\sim_Y$-equivalence class gives a finite partition of Y into intervals. Since Y is well-ordered and countable, there are only countably many finite partitions of Y into intervals. $\square$

As an application of indiscernibles, we give the following important sufficient condition for ω-stability of a theory.

Theorem 5.2.4 *Let T be a countable theory with an infinite model. Suppose T is κ-categorical, where κ is an uncountable cardinal. Then, T is ω-stable.*

Proof Suppose T is not ω-stable. Then, there is a $N' \models T$ and countable $A \subset N'$ such that $S_1^{N'}(A)$ is uncountable. If $|N'| > \kappa$, we replace it by an elementary substructure containing A of cardinality κ which exists by downward Löwenheim–Skolem theorem. On the other hand, if $|N'| < \kappa$, we replace it by an elementary extension of cardinality κ which exists by upward Löwenheim–Skolem theorem. Thus, without any loss of generality, we assume that $|N'| = \kappa$. Now take an elementary extension N of N' of cardinality κ that realises $\aleph_1$ many types in $S_1^N(A)$.

On the other hand, by the last theorem, T has a model M of cardinality κ such that M realises only countably many types in $S_1^M(A)$ for every countable $A \subset M$. But then M and N are not isomorphic. This contradicts that T is κ-categorical. $\square$

We now show that if T is a stable theory, then every infinite set of order indiscernibles in a model of T is in fact a set indiscernibles. We need the following lemma.

Lemma 5.2.5 *For every infinite cardinal κ, there is an order-dense linearly ordered set $(I, <)$ of cardinality greater than κ that has a dense subset J of cardinality at most κ.*

Proof Let λ be the least cardinal such that $\kappa < 2^\lambda$. Take $I = \mathbb{Q}^\lambda$ with lexicographic ordering. So, for $f, g \in I$, $f < g$ if $f(\alpha) < g(\alpha)$, where $\alpha < \lambda$ is the least ordinal β such that $f(\beta) \neq g(\beta)$. Take

$$J = \{f \in I : \exists \alpha < \lambda \forall \gamma \geq \alpha (f(\gamma) = 0)\}.$$

It is easy to see that $|I| > \kappa$, I is order dense, $|J| \leq \kappa$ and J is dense in I. $\square$

Theorem 5.2.6 *Let T be a countable κ-stable theory, $\kappa \geq \aleph_0$. Suppose $M \models T$ and $X \subset M$ is an infinite set of order indiscernibles. Then, X is a set of indiscernibles in M.*

Proof Let $x_1 < \ldots < x_n$ be in X, $\varphi[v_1, \ldots, v_n]$ a formula without parameters and $M \models \varphi[x_1, \ldots, x_n]$. Let S_n denote the permutation group of $\{1, \ldots, n\}$. Consider

$$\Gamma = \{\sigma \in S_n : M \models \varphi[x_{\sigma(1)}, \ldots, x_{\sigma(n)}]\}.$$

We need to show that $\Gamma = S_n$. Assume to the contrary. We shall contradict that T is κ-stable. Get $\sigma \in \Gamma$, $\tau \in S_n \setminus \Gamma$ and a transposition $\epsilon = (i, i+1) \in S_n$, $i < n$, such that $\tau = \sigma \circ \epsilon$.

Let I and J be as in the last example. By Theorem 5.1.2, we have a model $N \models T$ having a set of order indiscernibles Y of order type I such that $EM(X) = EM(Y)$. Let Z be a dense subset of Y with $|Z| \leq \kappa$. Take $\psi[v_1, \ldots, v_n] = \varphi[v_{\sigma(1)}, \ldots, v_{\sigma(n)}]$. Since $EM(X) = EM(Y)$, there exist $y_1 < \ldots < y_n$ in Y such that

$$N \models \psi[y_1, \ldots, y_i, y_{i+1}, \ldots, y_n]$$

and

$$N \models \neg\psi[y_1, \ldots, y_{i+1}, y_i, \ldots, y_n].$$

Let $x < y$ be in Y. Since Z is dense in Y, there exist $z_1 < \ldots < z_{n-1}$ in Z such that

$$z_{i-1} < x < z_i < y < z_{i+1}.$$

But then

$$N \models \psi[z_1, \ldots, z_{i-1}, x, z_i, \ldots, z_{n-1}]$$

and

$$N \models \neg\psi[z_1, \ldots, z_{i-1}, y, z_i, \ldots, z_{n-1}].$$

Hence, $tp^N(x/Z) \neq tp^N(y/Z)$. Since $|Y| > \kappa$ and $|Z| \leq \kappa$, this contradicts that T is κ-stable. $\square$

For the next result, we need an example.

Example 5.2.7 *Let κ be an infinite cardinal, $I = \kappa \times \mathbb{Q}$ with lexicographic order. For all $A \subset \kappa$, define an order isomorphism $\sigma_A : I \to I$ by*

$$\sigma_A(\alpha, r) = (\alpha, r) \qquad if \ \alpha \in A$$
$$= (\alpha, r+1) \ if \ \alpha \in \kappa \setminus A.$$

Then, for $A \neq B$, $\sigma_A \neq \sigma_B$. Thus, there are 2^κ many automorphisms of I.

Theorem 5.2.8 *Let T be a countable theory with an infinite model. Then, for every $\kappa \geq \aleph_0$, T has a model of cardinality κ with 2^κ many automorphisms.*

Proof Let T^* be the Skolemisation of T. By Theorem 5.1.2, T^* has a model M with a set of order indiscernibles X of order type $I = \kappa \times \mathbb{Q}$ described above. Take $N = \mathcal{H}(X)$.

By Theorem 5.2.2, $N \models T$, $|N| = \kappa$ and N has 2^κ many automorphisms.   □

Theorem 5.2.9 *Let T be a theory with built-in Skolem functions, $M \models T$, $p \in S_n^M(\emptyset)$ such that M omits p. Assume that M contains an infinite set X of order indiscernibles. Then, for every infinite cardinal κ, T has a model of cardinality κ that omits p.*

Proof Under our hypothesis, by Theorem 5.1.2, T has a model N' having an infinite set Y of order indiscernibles of order type κ such that $EM(X) = EM(Y)$. Now take $N = \mathcal{H}(Y)$. Then, $N \models T$ and $|N| = \kappa$.

If possible, suppose $(t_1(\overline{y}), \ldots, t_n(\overline{y}))$, $\overline{y}$ an increasing tuple of elements in Y, realises p. Take an increasing tuple $\overline{x}$ in X of the same length as that of $\overline{y}$. Since $EM(X) = EM(Y)$, for every L-formula $\varphi[v_1, \ldots, v_n]$,

$$M \models \varphi[t_1(\overline{x}), \ldots, t_n(\overline{x})] \Leftrightarrow N \models \varphi[t_1(\overline{y}), \ldots, t_n(\overline{y})].$$

So, $(t_1(\overline{x}), \ldots, t_n(\overline{x}))$ in M realises p. This is a contradiction.   □

5.3 Vaughtian Pair of Models

Throughout this section we assume that T is a countable complete theory with infinite models.

Let $\kappa > \lambda \geq \aleph_0$. We say that T has a (κ, λ)-*model* if there is a model $M \models T$ and an L-formula $\varphi[x]$ such that $|M| = \kappa$ and $|\varphi(M)| = \lambda$.

Proposition 5.3.1 *Let $\kappa > \aleph_0$ and T be κ-categorical. Then, for no $\kappa > \lambda \geq \aleph_0$, T has a (κ, λ)-model.*

Proof If possible, let for some $\kappa > \lambda \geq \aleph_0$, there exist a $M \models T$ and an L-formula φ such that $|M| = \kappa$ and $|\varphi(M)| = \lambda$. Add κ many new constants $\{c_\alpha : \alpha < \kappa\}$ to L. Consider the theory

$$T' = T \cup \{c_\alpha \neq c_\beta : \alpha < \beta < \kappa\} \cup \{\varphi[c_\alpha] : \alpha < \kappa\}.$$

Then, M is a model of each finite part of T'. Hence, by compactness and downward Löwenheim–Skolem theorem, there is a $N \models T'$ of cardinality κ. Further, $|\varphi(N)| = \kappa$. Since $|\varphi(M)| = \lambda < \kappa$, M and N are not isomorphic. This contradicts that T is κ-categorical.   □

We say that (N, M) is a *Vaughtian pair of models* of T if M is a proper elementary substructure of N and there is an L_M-formula $\varphi[x]$ such that $\varphi(N) = \varphi(M)$ and infinite.

Proposition 5.3.2 *If T has a (κ, λ)-model, then T has a Vaughtian pair of models.*

Proof Let $N \models T$ and φ an L-formula such that $|N| = \kappa$ and $|\varphi(N)| = \lambda$. By downward Löwenheim–Skolem theorem, there is an elementary substructure M of N containing $\varphi(N)$ of cardinality λ. Clearly, $\varphi[x]$ witnesses that (N, M) is a Vaughtian pair of models of T.   □

Using downward Löwenheim–Skolem theorem, we get the following result.

Proposition 5.3.3 *If T has no Vaughtian pair of models, $M \models T$ and $X \subset M$ infinite and definable. Then, $|X| = |M|$.*

Proof Let an L-formula $\varphi[x, \overline{x}]$ and $\overline{a} \in M$ be such that $X = \varphi(M, \overline{a})$. If possible, suppose $|X| < |M|$. By downward Löwenheim–Skolem theorem, there is an elementary substructure N of M containing $X \cup \{\overline{a}\}$ such that $|N| = |X|$. But then φ witnesses that (M, N) is a Vaughtian pair of models of T which contradicts our hypothesis.   □

Theorem 5.3.4 *Let T have no Vaughtian pair of models, $M \models T$ and $X \subset M$ infinite and X-definable. Then, M has no proper elementary substructure containing X. Moreover, if T is ω-stable, M is a prime model extension of X.*

Proof If there is a proper elementary substructure N of M containing X, then (M, N) is a Vaughtian pair of models of T, witnessed by an L_X-formula φ that defines X.

Assume that T is ω-stable. By Theorem 4.8.13, there is a prime model extension N over X which, without any loss of generality, we can assume to be an elementary substructure of M. But then $N = M$. Thus, M is a prime model extension of X.   □

Set $L^* = L \cup \{U\}$, where U is a unary predicate symbol. A Vaughtian pair of models (N, M) will be canonically regarded as an L^*-structure with universe N and $U^N = M$.

To each L-formula $\varphi[x_0, \ldots, x_{n-1}]$ we associate an L^*-formula φ^U as follows:

$$\varphi^U[x_0, \ldots, x_{n-1}] = \varphi[x_0, \ldots, x_{n-1}]$$

if φ is atomic;

$$(\neg\varphi)^U = \neg\varphi^U, \quad (\varphi \vee \psi)^U = \varphi^U \vee \psi^U;$$

and

$$(\exists x\varphi)^U = \exists x(U[x] \wedge \varphi^U).$$

It is easy to see that if (N, M) is a Vaughtian pair of models regarded canonically as an L^*-structure, $\varphi[\overline{x}]$ an L-formula and $\overline{a} \in M$, then

$$M \models \varphi[\overline{a}] \Leftrightarrow (N, M) \models \varphi^U[\overline{a}].$$

Proposition 5.3.5 *If T has a Vaughtian pair of models, then T has a Vaughtian pair (N_0, M_0) with N_0 countable.*

Proof Let (N, M) be a Vaughtian pair of models of T and $\varphi[x]$ an L_M-formula such that $\varphi[M]$ is infinite and $\varphi(N) = \varphi(M)$. Let $\overline{a} \in M$ be the parameters occurring in φ. We regard (N, M) as an L^*-structure. By Löwenheim–Skolem theorem, (N, M) has a countable elementary substructure (N_0, M_0) such that $\overline{a} \in N_0$. Since $(N, M) \models U[a_i]$ for all i, $(N_0, M_0) \models U[a_i]$ for all i. Hence, $\overline{a} \in M_0$.

Since $(N, M) \models \exists x\neg U[x]$, $(N_0, M_0) \models \exists x\neg U[x]$. Hence, M_0 is a proper subset of N_0. For every L-formula $\psi[\overline{v}]$, we have

$$(N, M) \models \forall\overline{v}((\psi[\overline{v}] \wedge \wedge_i U[v_i]) \rightarrow \psi^U[\overline{v}]).$$

Hence,

$$(N_0, M_0) \models \forall\overline{v}((\psi[\overline{v}] \wedge \wedge_i U[v_i]) \rightarrow \psi^U[\overline{v}]).$$

This shows that M_0 is an elementary L-substructure of N_0. Since $\varphi(N)$ is infinite, for each k,

$$(N, M) \models \exists x_1 \ldots \exists x_k(\wedge_{i=1}^k \varphi[x_i] \wedge \wedge_{i<j} x_i \neq x_j).$$

Hence,

$$(N_0, M_0) \models \exists x_1 \ldots \exists x_k(\wedge_{i=1}^k \varphi[x_i] \wedge \wedge_{i<j} x_i \neq x_j).$$

Thus, $\varphi(N_0)$ is infinite.

$$(N, M) \models \forall x(\varphi[x] \rightarrow U[x]).$$

Hence,

$$(N_0, M_0) \models \forall x(\varphi[x] \rightarrow U[x]).$$

This shows that $\varphi(N_0) = \varphi(M_0)$.   □

Lemma 5.3.6 *Let T be an L-theory with no Vaughtian pair of models. Suppose M is a model of T and $\varphi[\overline{x}, \overline{y}]$ an L_M formula. Then, there is a positive integer n such that for all $\overline{b} \in M$, $|\varphi(M, \overline{b})| > n \Rightarrow |\varphi(M, \overline{b})| \geq \aleph_0$.*

Proof If M is finite, any $n > |M|$ will do our job. Now assume that M is infinite. Suppose such an integer n does not exist. For every n, fix a $\overline{b}_n \in M$ such that $n < |\varphi(M, \overline{b}_n)| < \aleph_0$. Fix a proper elementary extension N of M which exists because M is infinite. Since N is an elementary extension of M, for each n, $\varphi(N, \overline{b}_n) = \varphi(M, \overline{b}_n)$.
    Let

$$L^* = L_N \cup \{U\} \cup \{c_{00}, \ldots, c_{0m-1}, c_{10}, \ldots, c_{1m-1}, \ldots\} \cup \{d_0, \ldots, d_{p-1}\},$$

where m is the arity of $\overline{x}$ and p is the arity of $\overline{y}$. Let T^* consist of the following L^* sentences:

1. $Diag_{el}(N, M)$.
2. $\overline{c}_k \neq \overline{c}_l, k < l$.
3. $\varphi[\overline{c}_k, \overline{d}], k \in \omega$.
4. $\wedge_j U[d_j]$.
5. $\forall \overline{x}(\varphi[\overline{x}, \overline{d}] \to \wedge_i U[x_i])$,

where $\overline{c}_k = (c_{k0}, \ldots, c_{km-1})$, $k \in \omega$, and $\overline{d} = (d_0, \ldots, d_{p-1})$.
    Given any finite set of elements in T^*, we can choose n large enough to see that these finitely many formulas are realised in (N, M). So, by compactness theorem, there is an elementary extension (N', M') of (N, M) and a $\overline{b} \in N'$ realising T^*. But then (N', M') is a Vaughtian pair of models of T. This contradiction proves the result.   □

Theorem 5.3.7 *If T has no Vaughtian pair of models, $M \models T$ and $\varphi[\overline{x}]$ an L_M formula minimal in M. Then, φ is strongly minimal over M.*

Proof If possible, suppose φ is not strongly minimal. Then, there is an elementary extension N of M, an L-formula $\psi[\overline{x}, \overline{y}]$ and $\overline{b} \in N$ such that both $\varphi(N) \cap \psi(N, \overline{b})$ and $\varphi(N) \cap \neg\psi(N, \overline{b})$ are infinite, i.e. for every positive integer n,

$$N \models \exists \overline{y}(\exists_{>n} \overline{x}(\varphi[\overline{x}] \wedge \psi[\overline{x}, \overline{y}]) \wedge \exists_{>n} \overline{x}(\varphi[\overline{x}] \wedge \neg\psi[\overline{x}, \overline{y}])).$$

Since φ is minimal in M, by the last Lemma, there ia a positive integer n such that

$$M \models \forall \overline{y}(\exists_{\leq n} \overline{x}(\varphi[\overline{x}] \wedge \psi[\overline{x}, \overline{y}]) \vee \exists_{\leq n} \overline{x}(\varphi[\overline{x}] \wedge \neg\psi[\overline{x}, \overline{y}])).$$

Since N is an elementary extension of M, we have a contradiction.   □

Corollary 5.3.8 *If T is a ω-stable theory with no Vaughtian pair of models and $M \models T$, then there is a strongly minimal L_M-formula.*

Proof This follows from Theorem 4.8.11 and the last theorem. □

Before we prove our next theorem, we make a couple of observations.

Lemma 5.3.9 *Let* (N, M) *be a countable Vaughtian pair of models of* T, $\overline{a} \in N$ *and* p *an* n-*type in* N *over* $\overline{a}$. *Then, there is a countable elementary extension* (N', M') *of* (N, M) *and a* $\overline{b} \in N'$ *realising* p.

Proof It is easily seen that $q = p \cup Diag_{el}(N, M)$ is finitely satisfiable in (N, M). Therefore, there is a countable elementary extension (N', M') of (N, M) realising q. This shows that there is a $\overline{b} \in N'$ realising p. □

Lemma 5.3.10 *Let* (N, M) *be a countable Vaughtian pair of models of* T. *Then, there exists a countable elementary extension* (N', M') *of* (N, M) *such that if* $\overline{b} \in M$ *and* p *a complete type in* N *over* $\overline{b}$ *realised in* N, *then* p *is realised in* M'.

Proof Since N (and M) are countable, there are only countably many p above. So, if we prove that for each complete type p over some $\overline{b} \in M$ realised in N there is a countable elementary extension (N', M') of (N, M) such that p is realised in M', then by iterating the construction we shall get a desired extension.

Now consider

$$q[\overline{v}] = Diag_{el}(N, M) \cup \{\varphi^U[\overline{v}, \overline{b}] : \varphi[\overline{v}, \overline{b}] \in p\}.$$

For $\varphi_1, \ldots, \varphi_k \in p$

$$N \models \exists \overline{v} \wedge_{i=1}^{k} \varphi_i[\overline{v}, \overline{b}].$$

Since M is an elementary substructure of N,

$$M \models \exists \overline{v} \wedge_{i=1}^{k} \varphi_i[\overline{v}, \overline{b}].$$

Therefore,

$$(N, M) \models \exists \overline{v}(\wedge_{j=1}^{m} U(v_j) \wedge \wedge_{i=1}^{k} \varphi_i^U[\overline{v}, \overline{b}]),$$

where $\overline{v} = (v_1, \ldots, v_m)$. Thus, $q[\overline{v}]$ is finitely satisfiable in (N, M). Hence, there is a countable elementary extension (N', M') of (N, M) in which $q[\overline{v}]$ is realised. Any $\overline{c} \in N'$ realising $q[\overline{v}]$ belongs to M' and realises p. □

Theorem 5.3.11 *Let* (N, M) *be a countable Vaughtian pair of models of* T. *Then, there is a countable Vaughtian pair of models* (N_∞, M_∞) *of* T *which is an elementary extension of* (N, M) *such that both* N_∞ *and* M_∞ *are homogeneous and realise the same complete types in* T. *In particular,* M_∞ *and* N_∞ *are isomorphic by Proposition 4.1.7.*

Proof Set $N_0 = N$ and $M_0 = M$. We shall now define an elementary chain $\{(N_k, M_k)\}$ of countable L^*-structures satisfying the following:

(a) If $p \in S_n(T)$ is realised in N_{3k}, then it is realised in M_{3k+1}.

(b) If $\overline{a}, \overline{b}, c \in M_{3k+1}$ and $tp^{M_{3k+1}}(\overline{a}) = tp^{M_{3k+1}}(\overline{b})$, then there is a $d \in M_{3k+2}$ such that $tp^{M_{3k+2}}(\overline{a}, c) = tp^{M_{3k+2}}(\overline{b}, d)$.

(c) If $\overline{a}, \overline{b}, c \in N_{3k+2}$ and $tp^{N_{3k+2}}(\overline{a}) = tp^{N_{3k+2}}(\overline{b})$, then there is a $d \in N_{3k+3}$ such that $tp^{N_{3k+3}}(\overline{a}, c) = tp^{N_{3k+3}}(\overline{b}, d)$.

(a) clearly follows from 5.3.10. To see (b), consider

$$p[x] = \{\varphi[\overline{b}, x] : M_{3k+1} \models \varphi[\overline{a}, c]\}.$$

For $\varphi_1, \ldots, \varphi_m \in p$, $M_{3k+1} \models \exists x \wedge_{i=1}^{m} \varphi_i[\overline{a}, x]$. Hence, $M_{3k+1} \models \exists x \wedge_{i=1}^{m} \varphi_i[\overline{b}, x]$. (b) now follows from 5.3.10. Similarly, (c) follows from 5.3.9.

Finally, we take $N_\infty = \cup_k N_k$ and $M_\infty = \cup_k M_k$. □

Theorem 5.3.12 *If T has a Vaughtian pair of models, then T has a $(\aleph_1, \aleph_0)$ model. In particular, if T has a (κ, λ) model for some $\kappa > \lambda \geq \aleph_0$, T has a $(\aleph_1, \aleph_0)$ model.*

Proof Since T has a Vaughtian pair of models, by Proposition 5.3.5 and the last theorem, T has a countable Vaughtian pair of models (N, M) such that M and N are homogeneous and realise same types in $S_n(T)$, $n \geq 1$. Therefore, M and N are isomorphic by Proposition 4.1.7. Let $\varphi[x]$ be an L_M-formula such that $\varphi(M) = \varphi(N)$ is infinite.

Set $N_0 = N$. By induction we are going to define a strictly ascending elementary chain of models $\{N_\alpha : \alpha < \omega_1\}$ such that for all $\alpha < \omega_1$, N_α is isomorphic to M and $\varphi(N_{\alpha+1}) = \varphi(N_\alpha)(= \varphi(M))$. Then, $\cup_{\alpha < \omega_1} N_\alpha$ will be a $(\aleph_1, \aleph_0)$ model for T witnessed by φ.

Let α be limit and suppose, for all $\beta < \alpha$, N_β has been defined satisfying the above conditions. Take $N_\alpha = \cup_{\beta < \alpha} N_\beta$. Since each N_β is homogeneous, N_α is homogeneous. Also N_α and M realise the same types. Since both N_α and M are countable, by Proposition 4.1.7 they are isomorphic. It is also clear that $\varphi(N_\alpha) = \varphi(M)$.

Suppose N_α have been defined. Fix an isomorphism $g : M \to N_\alpha$. Take $N_{\alpha+1} = (N \setminus M) \cup N_\alpha$. We can canonically make $N_{\alpha+1}$ into an L-structure isomorphic to N with an isomorphism from $N \to N_{\alpha+1}$ mapping M onto N_α. The result is easily seen now. □

5.4 Morley Categoricity Theorem

We are now in a position to give a proof of Morley Categoricity Theorem—a corner stone of modern model theory.

Theorem 5.4.1 *If an ω-stable theory T has a $(\aleph_1, \aleph_0)$ model, it has a $(\kappa, \aleph_0)$ model for all $\kappa > \aleph_1$.*

Proof Let $M \models T$ be of cardinality $\aleph_1$ and $\varphi[x]$ an L-formula such that $|\varphi(M)| = \aleph_0$. Get a proper elementary extension N of M as in Theorem 4.8.14. Now consider

$$q[x] = \{\varphi[x]\} \cup \{x \neq a : a \in \varphi(M)\}.$$

Since $|\varphi(M)| = \aleph_0$, $q[x]$ is a countable 1-type over M. Further, it is omitted in M. Hence, it is omitted in N. Thus, $\varphi(N) = \varphi(M)$.

If $|N| > \kappa$, we replace N by an elementary substructure N' of N containing $\varphi(M)$ of cardinality κ. Then, N' is a $(\kappa, \aleph_0)$ model of T witnessed by $\varphi[x]$.

If $|N| < \kappa$, we can iterate the process and get a strictly ascending elementary chain $\{N_\alpha : \alpha \leq \delta\}$ such that $N_0 = M$, $N_\alpha = \cup_{\beta < \alpha} N_\beta$ if α is limit, $\varphi(N_\alpha) = \varphi(M)$ for all $\alpha \leq \delta$ till $|N_\delta| = \kappa$. Then, N_δ is a $(\kappa, \aleph_0)$ model of T witnessed by φ. □

Theorem 5.4.2 *If T is κ-categorical for some uncountable κ, then T has no Vaughtian pair of models.*

Proof Let T be κ-categorical for some uncountable κ. By Theorem 5.2.4, T is ω-stable. If possible, assume that T has a Vaughtian pair of models. Then, by Theorem 5.3.12, T has a $(\aleph_1, \aleph_0)$ model. Since T is ω-stable, by the last theorem, it has a $(\kappa, \aleph_0)$ model, say M, witnessed by an L-formula $\varphi[x]$.

Add κ-many distinct new constants $\{c_\alpha : \alpha < \kappa\}$ to L_M and consider

$$\Gamma = Diag_{el}(M) \cup \{c_\alpha \neq c_\beta : \alpha < \beta < \kappa\} \cup \{\varphi[c_\alpha] : \alpha < \kappa\}.$$

Clearly, Γ is finitely satisfiable in M. Hence, by the compactness theorem, Γ has a model, say N'. We then have $M \preceq N'$ and $|N'| \geq \kappa$. Next take an elementary substructure N of N' containing $X = \{c_\alpha^{N'} : \alpha < \kappa\}$ of cardinality κ. We also have $|\varphi(N)| = \kappa$. But then M and N cannot be isomorphic. □

Finally, we prove Baldwin–Lachlan theorem that immediately proves Morley categoricity theorem.

Theorem 5.4.3 (Baldwin–Lachlan) *Let κ be any uncountable cardinal. Then, T is κ-categorical if and only if T is ω-stable and has no Vaughtian pair of models.*

Proof Only if part follows from Theorem 5.2.4 and the last theorem. Conversely, assume that T is ω-stable and has no Vaughtian pair of models. Since T is ω-stable, by Corollary 4.8.7 it has a prime model, say M_0. Let $M, N \models T$, each of cardinality κ. Without any loss of generality, we assume that M_0 is a common elementary substructure of both M and N.

Fix an L-formula $\varphi[x, \overline{x}]$ and $\overline{a} \in M_0^n$ such that $\varphi[x, \overline{a}]$ is strongly minimal over M_0. Such a formula exists by Corollary 5.3.8. Set

$$\psi[x, \overline{a}] = \varphi[x, \overline{a}] \vee \vee_{i < n} x = a_i.$$

Then, ψ is also strongly minimal. Since T has no Vaughtian pair of models, $|\psi(M)| = |\psi(N)| = \kappa$. Since κ is uncountable and T countable, $dim(\psi(M)) = dim(\psi(N)) = \kappa$.

By Proposition 3.4.19, there is a partial elementary bijection $f : \psi(M) \to \psi(N)$. But M is a prime model extension of $\psi(M)$ by Theorem 5.3.4. Hence, there is an elementary map $g : M \to N$ extending f. Since T has no Vaughtian pair of models, N has no proper elementary substructure containing $\psi(N)$. Hence, g is onto N. Thus, we have proved that g is an isomorphism.   □

Corollary 5.4.4 (Morley Categoricity Theorem) *Let $\kappa, \lambda > \aleph_0$ and T a countable complete theory with infinite models. Then, T is κ-categorical if and only if it is λ-categorical.*

Chapter 6
Strong Types

Abstract In this chapter, we make a systematic study of Lascar strong types and Kim–Pillay strong types. We also introduce Galois topological group of a complete theory. We close this chapter by showing connections of these with descriptive set theory.

6.1 General Facts on Bounded, Invariant, Equivalence Relations

Throughout this chapter, we assume that T is a countable complete theory. We also fix a monster model $\mathbb{M}$ of T which is κ-saturated and κ-strongly homogeneous, where κ is a (fixed) large strongly inaccessible cardinal. A subset A of $\mathbb{M}$ will be called small if $|A| < \kappa$. Also, a sequence $\bar{a}$ in $\mathbb{M}$ is small if its length is $< \kappa$.

Let $\lambda < \kappa$ and E an equivalence relation on $\mathbb{M}^\lambda$. We shall use $\mathbb{M}^\lambda/E$ to denote the set of all E-equivalence classes and $\pi : \mathbb{M}^\lambda \to \mathbb{M}^\lambda/E$ the quotient map. For $\bar{a} \in \mathbb{M}^\lambda$, often we shall write $[\bar{a}]$ instead of $\pi(\bar{a})$. Also, in what follows $\bar{x}, \bar{y}, \bar{z}$, etc. will denote sequences of distinct variables of length λ and $|T| = \max\{\lambda, \aleph_0\}$.

We call E *invariant* if whenever $\bar{a} E \bar{b}$ and $\sigma \in Aut(\mathbb{M})$, $\sigma(\bar{a}) E \sigma(\bar{b})$. We call E *bounded* if $|\mathbb{M}^\lambda/E| < \kappa$.

In this chapter, we introduce several bounded, invariant, equivalence relations on $\mathbb{M}^\lambda$, $\lambda < \kappa$. For each of these equivalence relations E, for every $\bar{a} \in \mathbb{M}^\lambda$, the set $\{\bar{b} \in \mathbb{M}^\lambda : tp(\bar{b}/\emptyset) = tp(\bar{a}/\emptyset)\}$ is E invariant. So, each E-equivalence class C is contained in a set of the form $\{\bar{b} \in \mathbb{M}^\lambda : tp(\bar{b}/\emptyset) = tp(\bar{a}/\emptyset)\}, \bar{a} \in \mathbb{M}^\lambda$. For this reason, E-equivalence classes are called *strong types*.

Three very important strong types are the so-called Shelah strong types, Kim–Pillay strong types and Lascar strong types. These strong types and associated Galois groups (which will be introduced later in the chapter) play an important role in stable, simple and NIP theories. These topics also have connections with descriptive set theory which we shall point out at the end of this chapter.

Let E be a bounded, invariant, equivalence relation on $\mathbb{M}^\lambda$. Assume that $|\mathbb{M}^\lambda/E| = \mu < \kappa$. Let $S = \{\bar{a}_\alpha \in \mathbb{M}^\lambda : \alpha < \mu\}$ be a cross section of $\mathbb{M}^\lambda/E$, i.e. S intersects each E-equivalence class in exactly one point. Then, for every $D \subset \mathbb{M}^\lambda/E$, $\pi^{-1}(D)$ is

© Springer Nature Singapore Pte Ltd. 2017
H. Sarbadhikari and S.M. Srivastava, *A Course on Basic Model Theory*,
DOI 10.1007/978-981-10-5098-5_6

S-invariant: Take $\overline{a} \in \pi^{-1}(D)$ and $f \in Aut_S(\mathbb{M})$. Suppose $\overline{a}_\alpha \, E \, \overline{a}$. Since E is invariant, $f(\overline{a}_\alpha) \, E f(\overline{a})$. But $f(\overline{a}_\alpha) = \overline{a}_\alpha$. Hence, $f(\overline{a}) \, E \, \overline{a}$ implying that $f(\overline{a}) \in \pi^{-1}(D)$.

Call $D \subset \mathbb{M}^\lambda / E$ *closed* if $\pi^{-1}(D)$ is type-definable over a small set. Since $\pi^{-1}(D)$ is S-invariant, by Proposition 4.6.3, D is closed if and only if $\pi^{-1}(D)$ is type-definable over S. Now it is entirely routine to check that the set of all closed sets in $\mathbb{M}^\lambda / E$ is precisely the set of all closed sets of a topology on $\mathbb{M}^\lambda / E$. We shall call this topology the *logic topology*.

Remark 6.1.1 Let E be an invariant equivalence relation on $\mathbb{M}^\lambda$. By Exercise 4.6.1, there is a set $\{p_i(\overline{x}, \overline{y}) : i \in I\}$ of complete types over emptyset such that $E = \cup_i p_i(\mathbb{M})$. Suppose $\mathbb{M}'$ is another monster model of T and $E' = \cup_i p_i(\mathbb{M}')$. Then, E' is invariant and an equivalence relation on $\mathbb{M}'^\lambda$.

For instance, we see that E' is symmetric. Suppose $\overline{a}' \, E' \, \overline{b}'$. So, there exists a $i_0 \in I$ such that $(\overline{a}', \overline{b}') \models p_{i_0}$. If possible, suppose $\neg(\overline{b}' E' \overline{a}')$. For each $i \in I$ there exists a $\varphi_i \in p_i$ such that $\mathbb{M}' \models \neg\varphi_i[\overline{b}', \overline{a}']$. Then, $(\overline{a}', \overline{b}') \models q(\overline{x}, \overline{y})$, where $q(\overline{x}, \overline{y})$ is the set of L-formulas

$$p_{i_0}(\overline{x}, \overline{y}) \cup \{\neg\varphi_i[\overline{y}, \overline{x}] : i \in I\}.$$

Hence, it is a type over empty set. Since $\mathbb{M}$ is κ-saturated, it is realised by some $(\overline{a}, \overline{b}) \in \mathbb{M}$. But then $\overline{a} \, E \, \overline{b}$ and $\neg(\overline{b} E \overline{a})$. This is a contradiction since E is symmetric. Thus, we have proved that E' is symmetric.

We now prove a surprising result that for every bounded, invariant, equivalence relation E on $\mathbb{M}^\lambda$, $|\mathbb{M}^\lambda / E|$ has a common upper bound less than κ.

Theorem 6.1.2 *Let E be a bounded, invariant, equivalence relation on $\mathbb{M}^\lambda$, $\lambda < \kappa$. Then,*

$$|\mathbb{M}^\lambda / E| \le 2^{(|T|^{(2^{|T|})})}.$$

Proof Since E is invariant, by Proposition 4.6.1, there is a family $\{p_j(\overline{x}, \overline{y}) : j \in J\}$ of complete types over empty set such that $E = \cup_{j \in J} p_j(\mathbb{M})$. Note that

$$|\times_{j \in J} p_j| \le |T|^{(2^{|T|})} = \nu,$$

say. Then, $2^{(|T|^{(2^{|T|})})} = \beth_1(\nu)$.

Set $\mu = \beth_1(\nu)^+$. If possible, suppose there is a sequence $\{\overline{a}_\alpha : \alpha < \mu\}$ of pairwise E-inequivalent elements in $\mathbb{M}^\lambda$ of length μ. Then, for every $\alpha < \beta < \mu$, there is a

$$\Phi_{(\alpha, \beta)} \in \times_{j \in J} p_j$$

such that for every $j \in J$,

$$\mathbb{M} \models \neg\Phi_{(\alpha, \beta)}(j)[\overline{a}_\alpha, \overline{a}_\beta].$$

By Erdös–Rado theorem (Theorem 8.5.4), this correspondence has an infinite homogeneous set, i.e. there is an infinite subset $I \subset \mu$ such that for every $\alpha < \beta$ in I, $\Phi_{(\alpha,\beta)}$ is the same, say $\Phi \in \times_{j \in J} p_j$.

Now consider the following set of formulas without parameters in variables $\{\overline{x}_\alpha : \alpha < \kappa\}$ with each $\overline{x}_\alpha = \{x_{\alpha\beta} : \beta < \lambda\}$ (with $\{x_{\alpha\beta} : \beta < \lambda, \alpha < \kappa\}$ distinct):

$$q((\overline{x}_\alpha)_{\alpha<\kappa}) = \{\overline{x}_\alpha \neq \overline{x}_\beta : \alpha < \beta < \kappa\} \cup \{\neg\Phi(j)[\overline{x}_\alpha, \overline{x}_\beta] : j \in J, \alpha < \beta < \kappa\}.$$

Then, $\{\overline{a}_\alpha : \alpha \in I\}$ witnesses that q is finitely satisfiable in $\mathbb{M}$. Hence, there is an elementary extension $\mathbb{M}'$ of $\mathbb{M}$ in which q is realised, say by $(\overline{b}'_\alpha)_{\alpha<\kappa}$.

By Proposition 4.3.7, there is a sequence $(\overline{c}_\alpha)_{\alpha<\kappa}$ of λ-tuples in $\mathbb{M}$ such that

$$tp^{\mathbb{M}}((\overline{c}_\alpha)_{\alpha<\kappa}/\emptyset) = tp^{\mathbb{M}'}((\overline{b}'_\alpha)_{\alpha<\kappa}/\emptyset).$$

This implies that for all $\alpha < \beta < \kappa$, $[\overline{c}_\alpha] \neq [\overline{c}_\beta]$. This contradicts that E is bounded. $\qquad\square$

Theorem 6.1.3 *Let $\lambda < \kappa$ and E be a bounded, invariant, type-definable, equivalence relation on $\mathbb{M}^\lambda$. Then,*

$$|\mathbb{M}^\lambda/E| \leq 2^{|T|}.$$

Proof Since E is invariant and type-definable, by Proposition 4.6.3, E is type-definable over empty set. Hence, in the last proof, we can replace a family of types by a single type $p(\overline{x}, \overline{y})$ over empty set which is of cardinality at most $|T|$. So, in this case corresponding Φ function takes values in p, a set of cardinality at most $|T|$. We take $\mu = \beth_1(|T|)^+$ and repeat the same argument to arrive at a contradiction. $\qquad\square$

Remark 6.1.4 Let E be a bounded, invariant equivalence relation. Assume that $E = \cup_{i \in I} p_i(\mathbb{M})$ where $\{p_i(\overline{x}, \overline{y})\}$ is a family of types over empty set. Take another monster model $\mathbb{M}'$ and define $E' = \cup_{i \in I} p_i(\mathbb{M}')$. In Remark 6.1.1 we saw that E' is an equivalence relation on $\mathbb{M}'$. From the above argument, it follows that E' is also bounded.

Let $A \subset \mathbb{M}$ be small and $\lambda < \kappa$. For $\overline{a}, \overline{b} \in \mathbb{M}^\lambda$, define

$$\overline{a} \equiv_A \overline{b} \Leftrightarrow tp^{\mathbb{M}}(\overline{a}/A) = tp^{\mathbb{M}}(\overline{b}/A).$$

Proposition 6.1.5 *$\equiv_A$ is a bounded, A-invariant equivalence relation on $\mathbb{M}^\lambda$.*

Proof Clearly, $\equiv_A$ is an A-invariant equivalence relation on $\mathbb{M}^\lambda$. Let the set of all L_A-formulas $\varphi[\overline{x}, \overline{y}]$ be of cardinality $\mu < \kappa$. Enumerate all such formulas by $\{\varphi_\alpha : \alpha < \mu\}$. Set $\varphi_\alpha^0 = \varphi_\alpha$ and $\varphi_\alpha^1 = \neg\varphi_\alpha$.

Now note that for every $\equiv_A$-equivalence class C, there is a unique function $\epsilon : \mu \to \{0, 1\}$ such that

$$C = \cap_{\alpha<\mu} \varphi_\alpha^{\epsilon(\alpha)}(\mathbb{M}).$$

Further, this correspondence is one-to-one. Hence,

$$|\mathrm{M}^{\lambda}/\equiv_A| \leq 2^{\mu}.$$

The result follows because κ is strongly inaccessible. $\square$

Corollary 6.1.6 $\equiv_{\emptyset}$ *is a bounded, invariant, equivalence relation on* M^{λ}.

The proof of the following result is an imitation of the proof of Theorem 5.1.3.

Theorem 6.1.7 *Let M be a small elementary substructure of* M. *Assume that* $\overline{a} \equiv_M$ $\overline{b}$, $\overline{a}, \overline{b} \in \mathrm{M}^{\lambda}$. *Then, there is a sequence* $\overline{c}_1, \overline{c}_2, \overline{c}_3, \ldots$ *in* M^{λ} *such that both the sequences* $\overline{a}, \overline{c}_1, \overline{c}_2, \overline{c}_3, \ldots$ *and* $\overline{b}, \overline{c}_1, \overline{c}_2, \overline{c}_3, \ldots$ *are order indiscernibles over M.*

Proof Let $\mathcal{U}$ be an ultrafilter on M^{λ} containing $\{\varphi(M) : \varphi[\overline{x}] \in tp^{\mathrm{M}}(\overline{a}/M)\}$.

Let $\{\xi_i[\overline{x}] : i \in I\}$ be an enumeration of all $M\overline{a}\overline{b}$-formulas. For each $i \in I$, set

$$\theta_i[\overline{x}] = \begin{cases} \xi_i[\overline{x}] & \text{if } \xi_i(M) \in \mathcal{U}, \\ \neg\xi_i[\overline{x}] & \text{if } \neg\xi_i(M) \in \mathcal{U}. \end{cases}$$

Then, $q_0(\overline{x}) = \{\theta_i[\overline{x}] : i \in I\}$ is a type in M over $M\overline{a}\overline{b}$. Since M is κ-saturated, q_0 is realised in M. Choose any $\overline{c}_1 \models q_0$ in M.

Suppose $\overline{c}_1, \cdots, \overline{c}_n$ have been defined. Let $\{\eta_j[\overline{x}] : j \in J\}$ be an enumeration of all $M\overline{a}\overline{b}\overline{c}_1 \cdots \overline{c}_n$-formulas. For each $j \in J$, set

$$\theta_j[\overline{x}] = \begin{cases} \eta_j[\overline{x}] & \text{if } \eta_j(M) \in \mathcal{U}, \\ \neg\eta_j[\overline{x}] & \text{if } \neg\eta_j(M) \in \mathcal{U}. \end{cases}$$

Then, $q_n(\overline{x}) = \{\theta_j[\overline{x}] : j \in J\}$ is a type in M over $M\overline{a}\overline{b}\overline{c}_1 \cdots \overline{c}_n$. By the same reason, q_n is realised in M. Take any $\overline{c}_{n+1} \models q_n$ in M.

Set $\overline{c}_0 = \overline{a}$ or $\overline{b}$. To conclude the result, we show that $\overline{c}_0, \overline{c}_1, \overline{c}_2, \ldots$ is a sequence of order indiscernibles over M.

Fix $\overline{c}_0 = \overline{a}$. By induction on n, we show that for every $0 \leq i_1 < \cdots < i_n, 0 \leq j_1 < \cdots < j_n,$

$$tp^{\mathrm{M}}(\overline{c}_{i_1}, \cdots, \overline{c}_{i_n}/M) = tp^{\mathrm{M}}(\overline{c}_{j_1}, \cdots, \overline{c}_{j_n}/M).$$

<u>Initial Case:</u> Let $m \geq 1$. Then, both $\overline{a}$ and $\overline{c}_m$ satisfy the same set of L_M-formulas, viz. those that define sets in $\mathcal{U}$. Hence, $tp^{\mathrm{M}}(\overline{a}/M) = tp^{\mathrm{M}}(\overline{c}_m/M)$.

Inductive Step. Assume the hypothesis for all pairs of increasing sequences of length n. Take any $i_1 < \cdots < i_n < i_{n+1}$ and $j_1 < \cdots < j_n < j_{n+1}$. By induction hypothesis,

$$tp^{\mathrm{M}}(\overline{c}_{i_1}, \cdots, \overline{c}_{i_n}/M) = tp^{\mathrm{M}}(\overline{c}_{j_1}, \cdots, \overline{c}_{j_n}/M).$$

Hence, there exists a $f \in Aut_M(\mathrm{M})$ such that $f(\overline{c}_{i_p}) = \overline{c}_{j_p}$ for all $p \leq n$. So, for every L_M-formula $\varphi[\overline{x}, \overline{x}_1, \cdots, \overline{x}_n]$,

$$\varphi(M, \overline{c}_{i_1}, \cdots, \overline{c}_{i_n}) = \varphi(M, \overline{c}_{j_1}, \cdots, \overline{c}_{j_n}).$$

We also have

$$tp(\overline{c}_{i_{n+1}}, \overline{c}_{i_1}, \cdots, \overline{c}_{i_n}/M) = tp(f(\overline{c}_{i_{n+1}}), \overline{c}_{j_1} \cdots, \overline{c}_{j_n}/M).$$

It follows that both $\overline{c}_{j_{n+1}}$ and $f(\overline{c}_{i_{n+1}})$ satisfy the same set of $M\overline{c}_{j_1} \cdots, \overline{c}_{j_n}$-formulas, viz. those that define sets in $\mathcal{U}$. Hence,

$$tp(\overline{c}_{i_{n+1}}, \overline{c}_{i_1}, \cdots, \overline{c}_{i_n}/M) = tp(f(\overline{c}_{i_{n+1}}), \overline{c}_{j_1} \cdots, \overline{c}_{j_n}/M) = tp(\overline{c}_{j_{n+1}}, \overline{c}_{j_1}, \cdots, \overline{c}_{j_n}/M).$$

$\square$

6.2   Shelah Strong Types and Kim–Pillay Strong Types

A finite equivalence relation is an equivalence relation with finitely many equivalence classes.

(I) Let E_{Sh} be the intersection of all finite equivalence relations on $\mathbb{M}^\lambda$ which are definable over empty set. Since the cardinality of the set of all L-formulas without parameters is $|T|$ and since the number of equivalence classes of a finite equivalence relation is less than $\aleph_0$, we can easily see that the number of E_{Sh}-equivalence classes is at most $\aleph_0^{|T|}$. Since each $\emptyset$-definable equivalence relation is invariant, E_{Sh} is invariant too. Also, note that $\equiv_\emptyset$ is the intersection of all equivalence relations of the $\{\varphi(\mathbb{M}), \neg\varphi(\mathbb{M})\}$, where $\varphi[\overline{x}]$ varies over all L-formulas without parameters, $E_{Sh} \subseteq \equiv_\emptyset$. It follows that each E_{Sh}-equivalence class is contained in a $\equiv_\emptyset$-class. This equivalence relation was introduced by Shelah and plays a fundamental role in stability theory. In his honour, E_{Sh}-equivalence classes are called *Shelah strong types*. Readers should see books of Shelah [54] and of Pillay [47] for this topic.

(II) Let E_{KP} be the intersection of all type-definable, bounded, invariant equivalence relations on $\mathbb{M}^\lambda$. By Proposition 4.6.3, every type-definable, invariant set is $\emptyset$-type-definable. Hence, E_{KP} is the intersection of all bounded, $\emptyset$-type-definable equivalence relations. Clearly, E_{KP} is invariant and $E_{KP} \subset E_{Sh}$. By Theorem 6.1.3, for each bounded, $\emptyset$-type-definable equivalence relation E on $\mathbb{M}^\lambda$, $|\mathbb{M}^\lambda/E| \leq 2^{|T|}$. Since the set of all types over empty set is of cardinality $\leq 2^{|T|}$, it follows that

$$|\mathbb{M}^\lambda/E_{KP}| \leq (2^{|T|})^{(2^{|T|})} < \kappa,$$

because κ is strongly inaccessible. Thus, E_{KP} is a bounded, invariant, equivalence relation. Since the intersection of any family of type-definable equivalence relations is type-definable, we have the following proposition.

Proposition 6.2.1 E_{KP} *is the finest, bounded, type-definable equivalence relation on $\mathbb{M}^\lambda$ and $|\mathbb{M}^\lambda/E_{KP}| \leq 2^{|T|}$.*

This equivalence relation was introduced by Kim and Pillay in [25]. In their honour, E_{KP}-equivalence classes are called *Kim–Pillay strong types*. We shall make more historical comments on this topic later in this chapter.

6.3 Lascar Strong Types

Let E_L be the intersection of all bounded, invariant, equivalence relations on $\mathbb{M}^\lambda$. By Proposition 4.6.1, for every bounded, invariant, equivalence relation E on $\mathbb{M}^\lambda$, there is a set $\{p_i(\overline{x}, \overline{y}) : i \in I\}$ of types over empty set such that $E = \cup_{i \in I} p_i(\mathbb{M})$. Hence, the set of all bounded, invariant, equivalence relations is of cardinality at most $2^{2^{|T|}} = \mu$, say. In Theorem 6.1.2, we showed that for every bounded, invariant, equivalence relation E on $\mathbb{M}^\lambda$, $|\mathbb{M}^\lambda/E| \leq 2^{(2^{|T|})} = \nu$, say. It follows that $|\mathbb{M}^\lambda/E_L| \leq \nu^\mu < \kappa$. Thus, E_L is the finestest bounded, invariant equivalence relation on $\mathbb{M}^\lambda$ and $\mathbb{M}^\lambda$, $|\mathbb{M}^\lambda/E_L| \leq 2^{(2^{|T|})}$. In particular, $E_L \subset E_{KP}$.

We now proceed to give several other descriptions of E_L. Let $\overline{a}, \overline{b} \in \mathbb{M}^\lambda$. We define

$$\overline{a} E_0 \overline{b} \Leftrightarrow \exists M \preceq \mathbb{M}(|M| < \kappa \ \& \ \overline{a} \equiv_M \overline{b}),$$

and

$$\overline{a} E_1 \overline{b} \Leftrightarrow \exists \overline{a}_0, \overline{a_1}, \overline{a_2}, \ldots \in \mathbb{M}^\lambda (\overline{a}_0 = \overline{a}, \overline{a}_1 = \overline{b} \ \& \ \overline{a}_0, \overline{a_1}, \overline{a_2}, \ldots \ \text{are order indiscernibles}).$$

We make a series of easily provable observations now.

1. E_0 is reflexive, symmetric and invariant.
2. The transitive closure of E_0, denoted by $trcl(E_0)$, is an invariant equivalence relation.
3. Fix any small $M \preceq \mathbb{M}$. Then, $\equiv_M \subset trcl(E_0)$. Hence, $trcl(E_0)$ is bounded. In particular, $E_L \subset trcl(E_0)$.
4. E_1 is reflexive and invariant.
5. Let $\overline{a} E_1 \overline{b}$. Take a sequence $\overline{a}_0, \overline{a_1}, \overline{a_2}, \ldots \in \mathbb{M}^\lambda$ such that $\overline{a}_0 = \overline{a}$, $\overline{a}_1 = \overline{b}$ and $\overline{a}_0, \overline{a_1}, \overline{a_2}, \ldots$ are order indiscernibles. By Proposition 5.1.4, there exist $\overline{a}_{-1}, \overline{a}_{-2}, \overline{a}_{-3}, \ldots \in \mathbb{M}^\lambda$ such that

$$\ldots, \overline{a}_{-3}, \overline{a}_{-2}, \overline{a}_{-1}, \overline{a}_0, \overline{a}_1, \overline{a}_2, \overline{a}_3, \ldots$$

   are order indiscernibles. This implies that $\overline{b}, \overline{a}, \overline{a}_{-1}, \overline{a}_{-2}, \ldots$ are order indiscernibles. It follows that E_1 is symmetric. Hence, $trcl(E_1)$ is an invariant equivalence relation.
6. $E_1 \subset E_L$: Suppose not. Get $\overline{a}, \overline{b}$ such that $\overline{a} E_1 \overline{b}$ and $\neg(\overline{a} E_L \overline{b})$. By Proposition 5.1.5, for each $\alpha < \kappa$ there exists a $\overline{a}_\alpha \in \mathbb{M}^\lambda$ such that $\overline{a}_0 = \overline{a}$, $\overline{a}_1 = \overline{b}$ and

$$\overline{a}_0, \overline{a}_1, \overline{a}_2, \overline{a}_3, \cdots \overline{a}_\alpha, \ldots, \quad \alpha < \kappa$$

are order indiscernibles. In particular, for every $\alpha < \beta < \kappa$, $tp^{\mathbb{M}}(\bar{a}, \bar{b}) = tp^{\mathbb{M}}$ $(\bar{a}_\alpha, \bar{a}_\beta)$. Hence, there exists $f \in Aut(\mathbb{M})$ such that $f(\bar{a}) = \bar{a}_\alpha$ and $f(\bar{b}) = \bar{a}_\beta$. Since E_L is invariant, it follows that $\neg(\bar{a}_\alpha E_L \bar{a}_\beta)$. This contradicts that E_L is bounded. We now see that $trcl(E_1) \subset E_L$.

7. $E_0 \subset trcl(E_1)$: Let $\bar{a} E_0 \bar{b}$. Get $M \preceq \mathbb{M}$ small such that $\bar{a} \equiv_M \bar{b}$. By Theorem 6.1.7, there is a sequence $\bar{c}_1, \bar{c}_2, \bar{c}_3, \ldots$ in $\mathbb{M}^\lambda$ such that both the sequences $\bar{a}, \bar{c}_1, \bar{c}_2, \bar{c}_3, \ldots$ and $\bar{b}, \bar{c}_1, \bar{c}_2, \bar{c}_3, \ldots$ are order indiscernibles. In particular, $\bar{a} E_1 \bar{c}_1$ and $\bar{b} E_1 \bar{c}_1$. This implies that $\bar{a}\ trcl(E_1)\ \bar{b}$.

We have proved the following theorem.

Theorem 6.3.1 $E_L = trcl(E_0) = trcl(E_1)$.

As a consequence of these descriptions of Lascar strong types, we can realise them as orbits under the action of a subgroup of the automorphism group of the monster model.

Let $Autf_L(\mathbb{M})$ denote the smallest subgroup of $Aut(\mathbb{M})$ containing each of $Aut_M(\mathbb{M})$, $M \preceq \mathbb{M}$, M small. Each element of $Autf_L(\mathbb{M})$ is of the form $\sigma_n \circ \cdots \circ \sigma_0$, where $\sigma_i \in Aut_{M_i}(\mathbb{M})$, $M_i \preceq \mathbb{M}$ small, $0 \leq i \leq n$. Automorphisms of $\mathbb{M}$ belonging to $Autf_L(\mathbb{M})$ are sometimes called *strong automorphisms*. The group $Autf_L(\mathbb{M})$ was introduced by Lascar in [33].

Proposition 6.3.2 *For any $\sigma \in Aut(\mathbb{M})$, the following statements are equivalent:*

1. $\sigma \in Autf_L(\mathbb{M})$.
2. σ *fixes all Lascar strong types setwise.*
3. *For all small $M \preceq \mathbb{M}$, σ fixes the Lascar strong types containing $\bar{m}$, where $\bar{m}$ is an enumeration of M.*
4. *There exists a small $M \preceq \mathbb{M}$, σ fixes the Lascar strong types containing $\bar{m}$, where $\bar{m}$ is an enumeration of M.*

Proof Let $\sigma = \sigma_n \circ \cdots \circ \sigma_0$, where $\sigma_i \in Aut_{M_i}(\mathbb{M})$, $M_i \preceq \mathbb{M}$ small, $0 \leq i \leq n$. Take any $\bar{a} \in \mathbb{M}^\lambda$. Set $\bar{a}_0 = \bar{a}$. Now suppose $\bar{a}_{i+1} = \sigma_i(\bar{a}_i)$, $0 \leq i \leq n$. Then, for each $0 \leq i \leq n$, $\bar{a}_{i+1} \equiv_{M_i} \bar{a}_i$. Hence, $\sigma(\bar{a}) = \bar{a}_{n+1}\ trcl(E_0)\ \bar{a}$. This shows that σ fixes the Lascar strong type containing $\bar{a}$. We have shown that (1) implies (2).

To complete the proof, we only need to show that (4) implies (1). Let $M \preceq \mathbb{M}$, $|M| < \kappa$ and suppose $\sigma \in Aut(\mathbb{M})$ fixes the Lascar strong types containing $\bar{m}$, where $\bar{m}$ is an enumeration of M. Since $E_L = trcl(E_0)$ and $\sigma(\bar{m}) E_L \bar{m}$, $\sigma(\bar{m})\ trcl(E_0)\ \bar{m}$. Set $\bar{m}_0 = \sigma(\bar{m})$. We get small $M_i \preceq \mathbb{M}$, $0 \leq i \leq n$, and $\bar{m}_1, \cdots, \bar{m}_n, \bar{m}_{n+1} = \bar{m}$ such that for all $0 \leq i \leq n, \bar{m}_i \equiv_{M_i} \bar{m}_{i+1}$. Then, for each $0 \leq i \leq n$, there exists a $\sigma_i \in Aut_{M_i}(\mathbb{M})$ such that $\bar{m}_{i+1} = \sigma_i(\bar{m}_i)$. Hence, $\bar{m} = (\sigma_n \circ \cdots \circ \sigma_0 \circ \sigma)(\bar{m})$ implying $\sigma_n \circ \cdots \circ \sigma_0 \circ \sigma \in Aut_M(\mathbb{M}) \subset Autf_L(\mathbb{M})$. It follows that $\sigma \in Autf_L(\mathbb{M})$. $\square$

Proposition 6.3.3 $Autf_L(\mathbb{M})$ *is a normal subgroup of $Aut(\mathbb{M})$.*

Proof Take any $M \preceq \mathbb{M}$, $|M| < \kappa$, $\sigma \in Aut_M(\mathbb{M})$ and $\tau \in Aut(\mathbb{M})$. It is sufficient to show that $\tau^{-1} \circ \sigma \circ \tau \in Autf_L(\mathbb{M})$. Fix an enumeration $\bar{m}$ of M. By (2) of the last proposition, $\sigma(\tau(\bar{m})) E_L \tau(\bar{m})$. Since E_L is invariant, $\tau^{-1}(\sigma(\tau(\bar{m}))) E_L \bar{m}$. Hence, by (4) of the last proposition, $\tau^{-1} \circ \sigma \circ \tau \in Autf_L(\mathbb{M})$. $\square$

Theorem 6.3.4 E_L *equals the orbit equivalence relation on* $\mathbb{M}^\lambda$ *under the action of* $Autf_L(\mathbb{M})$.

Proof Take any small $M \preceq \mathbb{M}$, $\sigma \in Aut_M(\mathbb{M})$ and $\bar{a} \in \mathbb{M}^\lambda$. Suppose $\bar{b} = \sigma(\bar{a})$. Then, $\bar{b} \equiv_M \bar{a}$. But $\equiv_M \subset E_0 \subset trcl(E_0) = E_L$. Now it is clear that for every $\sigma \in Autf_L(\mathbb{M})$, $\bar{a} E_L \sigma(\bar{a})$.

Conversely, let $\bar{a} E_L \bar{b}$. Then, $\bar{a} trcl(E_0) \bar{b}$. Then, there exist small $M_0, \cdots, M_{n-1} \preceq \mathbb{M}$, and $\bar{a}_1, \cdots, \bar{a}_{n-1} \in \mathbb{M}^\lambda$ such that $\bar{a} \equiv_{M_0} \bar{a}_1$, $\bar{a}_1 \equiv_{M_1} \bar{a}_2$, $\cdots$, $\bar{a}_{n-2} \equiv_{M_{n-2}} \bar{a}_{n-1}$, $\bar{a}_{n-1} \equiv_{M_{n-1}} \bar{b}$. It follows that there exist $\sigma_i \in Aut_{M_i}(\mathbb{M})$, $0 \le i \le n$, such that $\sigma_0(\bar{a}) = \bar{a}_1$, $\sigma_i(\bar{a}_i) = \bar{a}_{i+1}$, $1 \le i < n - 1$, and $\sigma_{n-1}(\bar{a}_{n-1}) = \bar{b}$. Hence, $\bar{b} = \sigma(\bar{a})$, where $\sigma = \sigma_{n-1} \circ \cdots \circ \sigma_0 \in Autf_L(\mathbb{M})$. $\qquad\square$

We give a few simple consequences of this theorem.

Proposition 6.3.5 *Let* $\sigma \in Aut(\mathbb{M})$. *The following statements are equivalent:*

1. $\sigma \in Autf_L(\mathbb{M})$.
2. *For every invariant set* $Y \subset \mathbb{M}^\lambda$, $\lambda < \kappa$, *for every bounded, invariant equivalence relation* E *on* Y *and for every* $\bar{a} \in Y$, $\bar{a} E \sigma(\bar{a})$.
3. *For some small elementary substructure* $\bar{m} = M$ *of* $\mathbb{M}$, $\bar{m} E_L \sigma(\bar{m})$.

Proof Since Y is invariant, extending E on $\mathbb{M}^\lambda$ by declaring any two elements not in Y equivalent, we get a bounded, invariant equivalence relation on $\mathbb{M}^\lambda$ containing E. Since Y is also invariant, it follows that $E_L|Y \subset E$. By the last Theorem 6.3.4, $\bar{a} E_L \sigma(\bar{a})$ if $\sigma \in Autf_L(\mathbb{M})$. Hence, (1) implies (2).

(2) implies (3) by taking $Y = p(M)$, where $p = tp(\bar{m})$. (1) follows from (3) by Proposition 6.3.2. $\qquad\square$

We now give another description of the logic topology on $\mathbb{M}^\lambda/E$, where E is a bounded, invariant equivalent relation on $\mathbb{M}^\lambda$. Fix any small $M \preceq \mathbb{M}$. Consider $S_\lambda(\mathbb{M}/M)$ with Stone topology. This makes $S_\lambda(\mathbb{M}/M)$ a compact, Hausdorff, zero-dimensional topological space. Let $\bar{a}, \bar{b} \in \mathbb{M}^\lambda$ be such that $tp^\mathbb{M}(\bar{a}/M) = tp^\mathbb{M}(\bar{b}/M)$, i.e. $\bar{a} E_0 \bar{b}$. Since $E_0 \subset E_L$, $\bar{a} E_L \bar{b}$. Since $E_L \subset E$, it follows that $\bar{a} E \bar{b}$. Thus, we have a surjection $f_M : S_\lambda(\mathbb{M}/M) \to \mathbb{M}^\lambda/E$ defined by $f_M(tp^\mathbb{M}(\bar{a}/M)) = [\bar{a}]$, $\bar{a} \in \mathbb{M}^\lambda$.

Proposition 6.3.6 $D \subset \mathbb{M}^\lambda/E$ *is closed in the logic topology if and only if* $f_M^{-1}(D)$ *is closed in the Stone topology.*

Proof Define $g : \mathbb{M}^\lambda \to S_\lambda(\mathbb{M}/M)$ by $g(\bar{a}) = tp^\mathbb{M}(\bar{a}/M)$, $\bar{a} \in \mathbb{M}^\lambda$. Then, $\pi = f_M \circ g$, where $\pi : \mathbb{M}^\lambda \to \mathbb{M}^\lambda/E$ is the quotient map.

Assume that $f_M^{-1}(D)$ is closed in the Stone topology. Get a set $p(\bar{x})$ of L_M-formulas such that

$$f_M^{-1}(D) = \cap_{\varphi \in p}[\varphi].$$

Then, for any $\bar{a} \in \mathbb{M}^\lambda$,

$$\bar{a} \in \pi^{-1}(D) \Leftrightarrow tp^{\mathbb{M}}(\bar{a}/M) \in f_M^{-1}(D)$$
$$\Leftrightarrow \forall \varphi \in p(tp^{\mathbb{M}}(\bar{a}/M) \in [\varphi])$$
$$\Leftrightarrow \bar{a} \models p.$$

This shows that $\pi^{-1}(D) = p(\mathbb{M})$. Hence, D is closed in the logic topology.

Next assume that $D \subset \mathbb{M}^\lambda$ is closed in the logic topology, i.e. $\pi^{-1}(D)$ is type-definable. Since $E_L \subset E$, by Theorem 6.3.4, $\pi^{-1}(D)$ is invariant under the action of $Autf_L(\mathbb{M})$. In particular, it is invariant over M. Hence, by Proposition 4.6.3, there is a set of L_M-formulas $p(\bar{x})$ such that $\pi^{-1}(D) = p(\mathbb{M})$. From here, it is entirely routine to check that $f_M^{-1}(D) = \cap_{\varphi \in p}[\varphi]$. So, $f_M^{-1}(D) \subset S_\lambda(\mathbb{M}/M)$ is closed in the Stone topology. $\qquad \square$

The equivalence relation E_L for n-tuples was introduced by Lascar in [33] as the orbit equivalence under the action of $Autf_L(\mathbb{M})$. In his honour, E_L-equivalence classes are called *Lascar strong types*. A detailed study of it was made in [25] by Kim and Pillay where, in particular, it was shown that E_L is the finest bounded invariant equivalence relation. Significantly, Kim and Pillay showed that if T is so-called simple, then $E_L = E_{KP}$. It is pertinent to point out that stable theories are simple. Thus, the paper of Kim and Pillay is of fundamental importance for Lascar strong types that brought this topic into model theory. Readers should see the books of Casanovas [7] and of Wagner [67] as well as the papers of Kim and Pillay [25], Lascar and Pillay [34], Newelski [44], Casanovas, Lascar, Pillay and Ziegler [8] and Hart, Kim and Pillay [17] for more on this topic.

6.4 The Galois Group $Gal_L(T)$

Note that we have fixed a large strongly inaccessible cardinal κ and that by a monster model of T we mean a κ-saturated, κ-strongly homogeneous model of T.

For any monster model $\mathbb{M}$ of T, we set $Gal_L(\mathbb{M}) = Aut(\mathbb{M})/Autf_L(\mathbb{M})$. In this section, we show that for any two monster models $\mathbb{M}$ and $\mathbb{N}$ of T, $Gal_L(\mathbb{M})$ and $Gal_L(\mathbb{N})$ are isomorphic. In the next section, we shall give a topology to $Gal_L(\mathbb{M})$ which will make it into a compact topological group. Further, for any two monster models $\mathbb{M}$ and $\mathbb{N}$ of T, we shall show that there is an isomorphism from $Gal_L(\mathbb{M})$ to $Gal_L(\mathbb{N})$ which is also a homeomorphism. By $Gal_L(T)$, we shall mean $Gal_L(\mathbb{M})$ for some monster model $\mathbb{M}$ of T.

Fix a monster model $\mathbb{M}$ of T. As before, let $\pi : Aut(\mathbb{M}) \to Gal_L(\mathbb{M})$ be the quotient map. For $\sigma \in Aut(\mathbb{M})$, we shall often write $[\sigma]$ for $\pi(\sigma)$.

Let $\mathbb{M}' \succeq \mathbb{M}$. Choose any small $M, N \preceq \mathbb{M}$ and $N' \preceq \mathbb{M}'$ having enumerations $\bar{n} = N$ and $\bar{n}' = N'$ such that $tp(\bar{n}'/M) = tp(\bar{n}/M)$. Take any $\sigma \in Aut(\mathbb{M}')$. Then, there exists a sequence $\bar{a}$ in $\mathbb{M}$ such that

$$\bar{a} \models tp(\sigma(\bar{n}')/M).$$

In particular,

$$tp(\bar{a}) = tp(\sigma(\bar{n}')) = tp(\bar{n}') = tp(\bar{n}).$$

Hence, there is a $\tau \in Aut(\mathbb{M})$ such that $\tau(\bar{n}) = \bar{a}$. Note that

$$tp(\tau(\bar{n})/M) = tp(\sigma(\bar{n}')/M).$$

We now make a series of very crucial observations which will help to define an isomorphism between $Gal_L(\mathbb{M})$ and $Gal_L(\mathbb{N})$.

1. Let τ' be another automorphism of $\mathbb{M}$ such that $\tau'(\bar{n}) = \bar{a}$. Then, τ and τ' agree on the small elementary substructure N of $\mathbb{M}$. Hence, $[\tau] = [\tau']$.
2. Next assume that $N_0 \preceq \mathbb{M}$ and $N_0' \preceq \mathbb{M}'$ are another pair of substructures with enumerations $\bar{n}_0$ and $\bar{n}_0'$, respectively, such that $tp(\bar{n}_0'/M) = tp(\bar{n}_0/M)$. Also, assume that $\tau_0 \in Aut(\mathbb{M})$ is such that $tp(\tau_0(\bar{n}_0)/M) = tp(\sigma(\bar{n}_0')/M)$. We show that $[\tau] = [\tau_0]$.

Since N' and N_0' are small and elementarily equivalent, by Proposition 2.8.1, there is a small $N_1' \preceq \mathbb{M}'$ which is a common elementary extension of both N' and N_0'.

Fix an enumeration $\bar{n}_1'$ of N_1'. By Proposition 4.3.7, there is a sequence $\bar{n}_1 \in \mathbb{M}$ such that $tp(\bar{n}_1/M) = tp(\bar{n}_1'/M)$. Let $\delta \in Aut(\mathbb{M})$ be such that $tp(\delta(\bar{n}_1)/M) = tp(\sigma(\bar{n}_1')/M)$. To complete the proof, we now show that $[\tau] = [\delta] = [\tau_0]$.

Proof of $[\delta] = [\tau]$: Without any loss of generality, we assume that $\bar{n}_1' = \bar{n}'\bar{a}'$. Let $\bar{n}_1 = \bar{b}\bar{a}$, where the length of $\bar{b}$ equals the length of $\bar{n}'$ and the length of $\bar{a}$ equals the length of $\bar{a}'$. We have the following:

$$tp(\bar{n}/M) = tp(\bar{n}'/M) = tp(\bar{b}/M), \tag{1}$$

and

$$tp(\tau(\bar{n})/M) = tp(\sigma(\bar{n}')/M) = tp(\delta(\bar{b})/M). \tag{2}$$

By (1), there is a $g \in Aut_M(\mathbb{M}) \subset Autf_L(\mathbb{M})$ such that $g(\bar{n}) = \bar{b}$. Hence, by (2), $tp(\tau(\bar{n})/M) = tp(\delta(g(\bar{n}))/M)$. Therefore, by observation (1), $[\tau] = [\delta \circ g] = [\delta]$.

Similarly, we prove that $[\tau_0] = [\delta]$.

3. Next take a different $M_0 \preceq \mathbb{M}$. By Proposition 2.8.1, there is a small $M_1 \preceq \mathbb{M}$ which is a common elementary extension of both M and M_0. Get $\tau \in \mathbb{M}$ such that $tp(\tau(\bar{n})/M_1) = tp(\sigma(\bar{n}')/M_1)$. In particular, $tp(\tau(\bar{n})/M) = tp(\sigma(\bar{n}')/M)$ and $tp(\tau(\bar{n})/M_0) = tp(\sigma(\bar{n}')/M_0)$. This shows that $[\tau]$ does not depend on M either.

Thus, we have a well-defined map

$$\alpha_{\mathbb{M}}^{\mathbb{M}'} : Aut(\mathbb{M}') \to Gal_L(\mathbb{M})$$

defined as follows: Choose any pair of small $M, N \preceq \mathbb{M}$ and $N' \preceq \mathbb{M}'$ with enumerations $\bar{n} = N$ and $\bar{n}' = N'$ such that $tp(\bar{n}'/M) = tp(\bar{n}/M)$. For any $\sigma \in Aut(\mathbb{M}')$. Take any $\tau \in Aut(\mathbb{M})$ such that $tp(\tau(\bar{n})/M) = tp(\sigma(\bar{n}')/M)$. Define $\alpha_{\mathbb{M}}^{\mathbb{M}'}(\sigma) = [\tau]$.

Let $\sigma = id$ be the identity automorphism of $Aut(\mathbb{M}')$. In this case, we can choose τ to be the identity automorphism of $\mathbb{M}$. Thus, $\alpha_{\mathbb{M}}^{\mathbb{M}'}(id) = e$, the identity element of $Gal_L(\mathbb{M})$.

Next let $\sigma, \sigma' \in Aut(\mathbb{M}')$, $M, N = \bar{n} \preceq \mathbb{M}$, $N' = \bar{n}' \preceq \mathbb{M}'$ with $tp(\bar{n}'/M) = tp(\bar{n}/M)$. Get $\tau \in Aut(\mathbb{M})$ such that $tp(\tau(\bar{n})/M) = tp(\sigma(\bar{n}')/M)$. Now get $\tau' \in Aut(\mathbb{M})$ such that $tp(\tau'(\tau(\bar{n}))/M) = tp(\sigma'(\sigma(\bar{n}'))/M)$. This shows that

$$\alpha_{\mathbb{M}}^{\mathbb{M}'}(\sigma' \circ \sigma) = \alpha_{\mathbb{M}}^{\mathbb{M}'}(\sigma') \cdot \alpha_{\mathbb{M}}^{\mathbb{M}'}(\sigma).$$

Thus, we have shown that

4. $\alpha_{\mathbb{M}}^{\mathbb{M}'} : Aut(\mathbb{M}') \to Gal_L(\mathbb{M})$ is a homomorphism.

Further, assume that $\mathbb{M}'$ is λ^+-saturated and λ^+-strongly homogeneous, where $\lambda = |\mathbb{M}|$. We have the following:

(a) $\alpha_{\mathbb{M}}^{\mathbb{M}'}$ is onto: Fix small elementary substructures M and $\bar{n} = N$ of $\mathbb{M}$. Take $\bar{n}' = \bar{n}$. Let $\tau \in Aut(\mathbb{M})$. Since $\mathbb{M}'$ is λ^+-strongly homogeneous, there is an extension $\sigma \in Aut(\mathbb{M}')$ of τ. Then, $\alpha_{\mathbb{M}}^{\mathbb{M}'}(\sigma) = [\tau]$.

(b) $ker(\alpha_{\mathbb{M}}^{\mathbb{M}'}) \subset Autf_L(\mathbb{M}')$: Let $\sigma \in Aut(\mathbb{M}')$ be such that $\alpha_{\mathbb{M}}^{\mathbb{M}'}(\sigma) = e$. Take small elementary substructures $M, \bar{n} = N$ of $\mathbb{M}$. By our hypothesis, there is a $\tau \in Autf_L(\mathbb{M})$ such that $tp(\sigma(\bar{n})/M) = tp(\tau(\bar{n})/M) = tp(\tau'(\bar{n})/M)$, where $\tau' \in Aut(\mathbb{M}')$ is an extension of τ. Note that if $\tau \in Autf_L(\mathbb{M})$, then every extension τ' of τ belongs to $Autf_L(\mathbb{M}')$. There is a $\delta \in Aut_M(\mathbb{M}') \subset Autf_L(\mathbb{M}')$ such that $\sigma(\bar{n}) = \delta(\tau'(\bar{n}))$. This implies that $[\sigma] = [\delta \circ \tau'] = e$, i.e. $\sigma \in Autf_L(\mathbb{M}')$.

(c) $Autf_L(\mathbb{M}') \subset ker(\alpha_{\mathbb{M}}^{\mathbb{M}'})$: Let $N' \preceq \mathbb{M}'$ be small and $\sigma \in Aut_{N'}(\mathbb{M}')$. Suffices to show that $\alpha_{\mathbb{M}}^{\mathbb{M}'}(\sigma) = e$. Fix any enumeration $\bar{n}'$ of N'. For any small $M \preceq \mathbb{M}$, by Proposition 4.3.7, there is a sequence $\bar{n}$ in $\mathbb{M}$ such that $tp(\bar{n}/M) = tp(\bar{n}'/M) = tp(\sigma(\bar{n}')/M)$. Therefore, corresponding to σ we can choose $\tau = id$. Hence, $\alpha_{\mathbb{M}}^{\mathbb{M}'}(\sigma) = e$.

We have now proved that

5. If $\mathbb{M}'$ is a $|\mathbb{M}|^+$-saturated, $|\mathbb{M}|^+$-strongly homogeneous elementary extension of $\mathbb{M}$, then $\alpha_{\mathbb{M}}^{\mathbb{M}'}$ induces an isomorphism, denoted by $\beta_{\mathbb{M}}^{\mathbb{M}'}$, from $Gal_L(\mathbb{M}')$ to $Gal_L(\mathbb{M})$.

Theorem 6.4.1 *Let $\mathbb{M}$ and $\mathbb{N}$ be two monster models of T. Then, $Gal_L(\mathbb{M})$ and $Gal_L(\mathbb{N})$ are isomorphic.*

Proof Recall that we have fixed a large strongly inaccessible cardinal κ and $\mathbb{M}$ and $\mathbb{N}$ are κ-saturated, κ^+-strongly homogeneous models of T. In particular, they are elementarily equivalent. Let $\lambda = \max\{|\mathbb{M}|^+, |\mathbb{N}|^+\}$. Using Theorem 4.4.5, get a λ-saturated, λ-strongly homogeneous model $\mathbb{M}'$ of T which is a common elementary extension of both $\mathbb{M}$ and $\mathbb{N}$. By observation 5 above, $Gal_L(\mathbb{M})$ as well as $Gal_L(\mathbb{N})$ are isomorphic to $Gal_L(\mathbb{M}')$. $\qquad\square$

6.5   Topology on $Gal_L(T)$

We now proceed to define a topology on $Gal_L(\mathbb{M})$ which will make it into a topological group. Let $M \preceq \mathbb{M}$ be small. We fix an enumeration $\overline{m}$ of M. Set

$$S(\overline{m}) = \{tp^{\mathbb{M}}(\overline{a}/M) : \overline{a} \equiv_\emptyset \overline{m}\} \subset S_\lambda(\mathbb{M}/M).$$

Then,

$$S(\overline{m}) = \cap\{[\varphi] : \mathbb{M} \models \varphi[\overline{m}], \varphi \text{ an } L\text{-formula}\}.$$

Hence, $S(\overline{m})$ is closed in $S_\lambda(\mathbb{M}/M)$ under the Stone topology. In particular, $S(\overline{m})$ is compact. We make a series of observations now.

1. Suppose $\sigma, \sigma' \in Aut(\mathbb{M})$ are such that $tp^{\mathbb{M}}(\sigma(\overline{m})/M) = tp^{\mathbb{M}}(\sigma'(\overline{m})/M)$. Then, there exists a $\tau \in Aut_M(\mathbb{M})$ such that $\tau(\sigma'(\overline{m})) = \sigma(\overline{m})$. Hence,

$$\sigma^{-1} \circ \tau \circ \sigma' = (\sigma^{-1} \circ \tau \circ \sigma) \circ (\sigma^{-1} \circ \sigma') \in Aut_M(\mathbb{M}) \subset Autf_L(\mathbb{M}).$$

Since $Autf_L(\mathbb{M})$ is a normal subgroup of $Aut(\mathbb{M})$, $\sigma^{-1} \circ \tau \circ \sigma \in Autf_L(\mathbb{M})$. It follows that $\sigma^{-1} \circ \sigma' \in Autf_L(\mathbb{M})$. Thus, we have a well-defined map

$$\rho_{\overline{m}} : S(\overline{m}) \to Gal_L(\mathbb{M})$$

defined by

$$\rho_{\overline{m}}(tp^{\mathbb{M}}(\sigma(\overline{m})/M)) = [\sigma], \ \sigma \in Aut(\mathbb{M}).$$

2. Next take small $M \preceq N \preceq \mathbb{M}$. Take enumerations $M = \overline{m}$ and $N = \overline{m}\,\overline{n}$. We have the restriction map $r : S(\overline{m}\,\overline{n}) \to S(\overline{m})$ defined by

$$r(tp^{\mathbb{M}}(\sigma(\overline{m}\,\overline{n})/N) = tp^{\mathbb{M}}(\sigma(\overline{m})/M)), \ \sigma \in Aut(\mathbb{M}).$$

This map is continuous and onto. Further, $\rho_{\overline{m}} \circ r = \rho_N$.

We equip $Gal_L(\mathbb{M})$ with the largest topology making $\rho_{\overline{m}}$ continuous for some fixed $\overline{m} = M \preceq \mathbb{M}$, $|M| < \kappa$. This topology is independent of M. To see this, take any two small elementary substructures M and M' of $\mathbb{M}$. Since M and M' are elementarily equivalent, by Proposition 2.8.1, there is a common elementary extension N of M and M' which is a small elementary substructure of $\mathbb{M}$. Thus, without any loss of generality, we assume $M \preceq N$. Fix an enumeration $\overline{m}\,\overline{n}$ of N. Let $D \subset Gal_L(\mathbb{M})$. Then, $\rho_{\overline{m}\,\overline{n}}^{-1}(D) = r^{-1}(\rho_{\overline{m}}^{-1}(D))$. It is now easy to see that $\rho_{\overline{m}}^{-1}(D)$ is closed if and only if $\rho_{\overline{m}\,\overline{n}}^{-1}(D)$ is closed. Following usual notation from topology, for any set $X \subset Gal_L(\mathbb{M})$, $\overline{X}$ will denote the closure of X.

Since $S(\overline{m})$ is compact, we have the following theorem.

Theorem 6.5.1 $Gal_L(\mathbb{M})$ *is compact.*

Proposition 6.5.2 $\pi : Aut(\mathbb{M}) \to Gal_L(\mathbb{M})$ *is continuous.*

Proof Define $\eta_{\overline{m}} : Aut(\mathbb{M}) \to S(\overline{m})$ by

$$\eta_{\overline{m}}(\sigma) = tp(\sigma(\overline{m})/M), \ \sigma \in Aut(\mathbb{M}).$$

Then, $\pi = \rho_{\overline{m}} \circ \eta_{\overline{m}}$. So, our result will be proved if we show that $\eta_{\overline{m}}$ is continuous.

Take any $\sigma_0 \in Aut(\mathbb{M})$ and assume that $tp(\sigma_0(\overline{m})/M) \in [\varphi[\overline{x}]]$, where φ is an L_M-formula. Let $\overline{m} = (m_\beta)_{\beta<\lambda}$ and suppose variables having a free occurrence in φ are among $x_{\beta_0}, \cdots, x_{\beta_n}$. Then, for any $\sigma \in Aut(\mathbb{M})$ such that $\sigma(m_{\beta_i}) = \sigma_0(m_{\beta_i})$, $0 \leq i \leq n$, $\eta_{\overline{m}}(\sigma) \in [\varphi]$. $\qquad\square$

Proposition 6.5.3 *For $D \subset Gal_L(\mathbb{M})$, the following conditions are equivalent:*

1. *D is closed.*
2. *$\{\sigma(\overline{m}) : [\sigma] \in D\}$ is type-definable over M for every small elementary substructure $\overline{m} = M$ of $\mathbb{M}$.*
3. *$\{\sigma(\overline{a}) : [\sigma] \in D\}$ is type-definable for every small sequence $\overline{a}$ in $\mathbb{M}$.*
4. *$\{\sigma(\overline{m}) : [\sigma] \in D\}$ is type-definable for some small elementary substructure $\overline{m} = M$ of $\mathbb{M}$.*
5. *$\{\sigma(\overline{m}) : [\sigma] \in D\}$ is type-definable over M for some small elementary substructure $\overline{m} = M$ of $\mathbb{M}$.*

Proof Let $D \subset Gal_L(\mathbb{M})$ be closed and $\overline{m} = M$ a small elementary substructure of $\mathbb{M}$. Get a set $p(\overline{x})$ of L_M-formula such that $\rho_{\overline{m}}^{-1}(D) = \cap_{\varphi \in p}[\varphi]$. Then, for any $\sigma \in Aut(\mathbb{M})$,

$$\begin{aligned} [\sigma] \in D &\Leftrightarrow tp^{\mathbb{M}}(\sigma(\overline{m})/M) \in \rho_{\overline{m}}^{-1}(D) \\ &\Leftrightarrow \forall \varphi \in p(\mathbb{M} \models \varphi[\sigma(\overline{m})]) \\ &\Leftrightarrow \sigma(\overline{m}) \in p(\mathbb{M}). \end{aligned}$$

Thus, we have proved that (1) implies (2). Next assume (2) and take any small sequence $\overline{a}$, say of length $\lambda < \kappa$. Take a small elementary substructure $M = \overline{m} = \overline{a}\,\overline{b}$ of $\mathbb{M}$, say of length $\mu < \kappa$. By (2), there is a set $p(\overline{x}\,\overline{y})$, $\overline{x}$ of length λ, $\overline{x}\,\overline{y}$ of length μ, of L_M-formulas such that $\{\sigma(\overline{m}) : [\sigma] \in D\} = p(\mathbb{M})$. Set $q(\overline{x}) = \{\varphi[\overline{x}] : \varphi \in p\}$. Then, $q(M) = \{\sigma(\overline{a}) : [\sigma] \in D\}$. So, (2) implies (3)

Clearly, (3) implies (4). To see that (4) implies (5), get M as in (4). Since $Aut_M(\mathbb{M}) \subset Autf_L(\mathbb{M})$ and E_L is the orbit equivalence relation under the action of $Autf_L(\mathbb{M})$, the set $\{\sigma(\overline{m}) : [\sigma] \in D\}$ is invariant over M. Hence, by Proposition 4.6.3, $\{\sigma(\overline{m}) : [\sigma] \in D\}$ is type-definable over M.

Now assume (5). Get a set $p(\overline{x})$ of L_M-formulas such that $p(\mathbb{M}) = \{\sigma(\overline{m}) : [\sigma] \in D\}$. Then, it is easily seen that $\rho_{\overline{m}}^{-1}(D) = \cap_{\varphi \in p}[\varphi]$. Hence, D is closed. $\qquad\square$

The next result is technical but quite useful.

Proposition 6.5.4 *Let $\overline{m} = M$ be a small elementary substructure of $\mathbb{M}$, $p(\overline{x}) = tp(\overline{m})$ and G a closed subgroup of $Gal_L(\mathbb{M})$. Then, there is an $\emptyset$-type-definable, bounded, equivalence relation R on $p(\mathbb{M})$ such that*

$$[\sigma] \in G \Leftrightarrow \sigma(\overline{m}) \, R \, \overline{m}.$$

Proof Since G is closed in $Gal_L(\mathbb{M})$, by Proposition 6.5.3, there is a set of L-formulas $q(\overline{x}, \overline{y})$ such that

$$q(\mathbb{M}, \overline{m}) = \{\sigma(\overline{m}) : [\sigma] \in G\}.$$

Set $R' = q(\mathbb{M})$. We show that $R = R'|p(\mathbb{M})$ is a bounded equivalence relation on $p(\mathbb{M})$. Since R is clearly invariant, this will complete the proof.

Since $e = [id] \in G$, $(\overline{m}, \overline{m}) \models q$. Since q is without any parameter, $(\sigma(\overline{m}), \sigma(\overline{m})) \models q$ for every $\sigma \in Aut(\mathbb{M})$. This shows that R is reflexive.

Now take any two $\sigma, \tau \in Aut(\mathbb{M})$ such that $(\sigma(\overline{m}), \tau(\overline{m})) \models q$. Then, $(\tau^{-1}(\sigma(\overline{m})), \overline{m}) \models q$. Hence, $[\tau^{-1} \circ \sigma] \in G$. Since G is a subgroup, $[\sigma^{-1} \circ \tau] \in G$. Hence, $((\sigma^{-1} \circ \tau)(\overline{m}), \overline{m}) \models q$. This implies that $(\tau(\overline{m}), \sigma(\overline{m})) \models q$. So, R is symmetric.

We leave the routine proof of the transitivity of R for the reader.

R must be bounded because $\neg R(\tau(\overline{m}), \sigma(\overline{m}))$ implies that σ and τ are in different cosets of $Autf_L(\mathbb{M})$.   □

We give below another useful description of the topology on $Gal_L(\mathbb{M})$.

Theorem 6.5.5 *A set $D \subset Gal_L(\mathbb{M})$ is closed if and only if whenever $\mathcal{U}$ is an ultrafilter on an index set I, $\mathbb{M}' = \mathbb{M}^{\mathcal{U}}$, $\{\sigma_i : i \in I\} \subset Aut(\mathbb{M})$ such that for each $i \in I$, $[\sigma_i] \in D$, $\alpha_{\mathbb{M}}^{\mathbb{M}'}((\Pi_i \sigma_i)^{\mathcal{U}}) \in D$.*

Proof 'If' part: Take any small $\overline{m} = M \preceq \mathbb{M}$. Set $X = \rho_{\overline{m}}^{-1}(D)$. Take $q(\overline{x}) = tp$ $(\sigma(\overline{m})/M) \in \overline{X}$, $\sigma \in Aut(\mathbb{M})$. We are required to show that $[\sigma] \in D$.

Let I denote the set of all finite subsets of $q(\overline{x})$. Suppose $i = \{\varphi_1, \cdots, \varphi_k\} \subset q$. Then, $q \in \cap_{j=1}^{k}[\varphi_j]$. Since q is in the closure of X, there is a $\sigma_i \in Aut(\mathbb{M})$ with $[\sigma_i] \in D$ such that $\sigma_i(\overline{m}) \models i$.

For each $\varphi \in q$, set $B_\varphi = \{i \in I : \varphi \in i\}$. Clearly, $\{B_\varphi : \varphi \in q\}$ has finite intersection property. Let $\mathcal{U}$ be an ultrafilter on I containing $\{B_\varphi : \varphi \in q\}$.

Set $\sigma' = (\Pi_i \sigma_i)^{\mathcal{U}}$. Then, by Łoś Theorem 2.1.3, $tp(\sigma'(\overline{m})/M) = q = tp(\sigma(\overline{m})/M)$. This implies that $\alpha_{\mathbb{M}}^{\mathbb{M}'}(\sigma') = [\sigma]$ by the definition of $\alpha_{\mathbb{M}}^{\mathbb{M}'}$. By our hypothesis, $[\sigma] = \alpha_{\mathbb{M}}^{\mathbb{M}'}(\sigma') \in D$.

'Only if' part: Now assume that D is closed. Fix any small $\overline{m} \preceq \mathbb{M}$. By Proposition 6.5.3, there is a set $p(\overline{x})$ of L_M-formulas such that $p(\mathbb{M}) = \{\sigma(\overline{m}) : [\sigma] \in D\}$.

Take any index set I, an ultrafilter $\mathcal{U}$ on I and for each $i \in I$ a $\sigma_i \in Aut(\mathbb{M})$ such that $[\sigma_i] \in D$. Set $\mathbb{M}' = \mathbb{M}^{\mathcal{U}}$ and $\sigma' = (\Pi_i \sigma_i)^{\mathcal{U}}$. We are required to show that $\alpha_{\mathbb{M}}^{\mathbb{M}'}(\sigma') \in D$.

Since each $\sigma_i(\overline{m}) \models p$, $\sigma'(\overline{m}) \models p$. Get $\tau \in Aut(\mathbb{M})$ such that $tp(\tau(\overline{m})/M) = tp(\sigma'(\overline{m})/M)$. Since $p(\overline{x}) \subset tp(\sigma'(\overline{m})/M)$, $\tau(\overline{m}) \models p(\overline{x})$. Hence, $[\tau] \in D$. Since $tp(\sigma'(\overline{m})/M) = tp(\tau(\overline{m})/M)$, $\alpha_{\mathbb{M}}^{\mathbb{M}'}(\sigma') = [\tau]$. Therefore, $\alpha_{\mathbb{M}}^{\mathbb{M}'}(\sigma') \in D$.   □

Using the fact that for an ultrafilter $\mathcal{U}$ on a set I, whenever a subset A of I is not in $\mathcal{U}$, $A^c \in \mathcal{U}$, we easily get the following result.

Corollary 6.5.6 *A subset U of $Gal_L(\mathbb{M})$ is open if and only if whenever $\{\sigma_i : i \in I\} \subset Aut(\mathbb{M})$ is such that $\alpha_{\mathbb{M}}^{\mathbb{M}'}((\Pi_i \sigma_i)^{\mathcal{U}}) \in V$, V open in $Gal_L(\mathbb{M})$, $\{i \in I : [\sigma_i] \in V\} \in \mathcal{U}$.*

Proposition 6.5.7 *Let $A \subset Gal_L(\mathbb{M}) = Aut(\mathbb{M})/Autf_L(\mathbb{M})$. Then, $[\sigma] \in \overline{A}$ if and only if there exist a non-empty set I, an ultrafilter $\mathcal{U}$ on I, for each $i \in I$ a $[\sigma_i] \in A$ such that $\alpha_{\mathbb{M}}^{\mathbb{M}'}((\pi_{i \in I} \sigma_i)^{\mathcal{U}}) = [\sigma]$, where $\mathbb{M}' = \mathcal{M}^{\mathcal{U}}$.*

Proof Fix a small elementary substructure $\overline{m} = M$ of $\mathbb{M}$.

Let B be the set of all elements of the form $\alpha_{\mathbb{M}}^{\mathbb{M}'}((\pi_{i \in I} \sigma_i)^{\mathcal{U}})$, where $\mathcal{U}$ is an ultrafilter on an index set I, for each $i \in I$, $[\sigma_i] \in A$ and $\mathbb{M}' = \mathcal{M}^{\mathcal{U}}$. By Theorem 6.5.5, $A \subset B \subset \overline{A}$. Therefore, it is sufficient to show that B is closed.

Take any index set I, an ultrafilter $\mathcal{U}$ on I and $[\sigma_i] \in B$, $i \in I$. Then, for each $i \in I$, there is an ultrafilter $\mathcal{U}_i$ on an index set J_i and for each $j \in J_i$ a $[\sigma_i^j] \in A$ such that

$$[\sigma_i] = \alpha_{\mathbb{M}}^{\mathbb{M}_i}((\pi_{j \in J_i} \sigma_i^j)^{\mathcal{U}_i}),$$

where $\mathbb{M}_i = \mathbb{M}^{\mathcal{U}_i}$. We consider $\mathbb{M}$ as an elementary substructure of $\mathbb{M}_i$ canonically. By the definition of $\alpha_{\mathbb{M}}^{\mathbb{M}_i}$, we have

$$tp(\sigma_i(\overline{m})/M) = tp((\pi_{j \in J_i} \sigma_i^j)^{\mathcal{U}_i}(\overline{m})/M).$$

By Łoś theorem, it follows that for every L_M-formula $\varphi[\overline{x}]$,

$$\models \varphi[\sigma_i(\overline{m})] \Leftrightarrow \{j \in J_i : \models \varphi[\sigma_i^j(\overline{m})]\} \in \mathcal{U}_i. \tag{$*$}$$

Next, consider $\mathbb{M}' = \mathbb{M}^{\mathcal{U}}$. Let

$$[\sigma] = \alpha_{\mathbb{M}}^{\mathbb{M}'}((\sigma_i)_{i \in I}^{\mathcal{U}}).$$

This implies that

$$tp(\sigma(\overline{m})/M) = tp((\sigma_i)_{i \in I}^{\mathcal{U}}(\overline{m})/M).$$

By Łoś theorem, for every L_M-formula $\varphi[\overline{x}]$,

$$\models \varphi[\sigma(\overline{m})] \Leftrightarrow \{i \in I : \models \varphi[\sigma_i(\overline{m})]\} \in \mathcal{U}. \tag{$**$}$$

Now set $J = \cup_i(\{i\} \times J_i)$ and

$$\mathcal{V} = \{V \subset J : \{i \in I : \{j \in J_i : (i, j) \in V\} \in \mathcal{U}_i\} \in \mathcal{U}\}.$$

It is routine to check that $\mathcal{V}$ is an ultrafilter on J.

Set $\mathbb{M}'' = \mathbb{M}^{\mathcal{V}}$. Using $(*)$ and $(**)$, it is now easy to check that

$$[\sigma] = \alpha_{\mathbb{M}}^{\mathbb{M}''}((\pi_{(i,j) \in J} \sigma_i^j)^{\mathcal{V}}) \in B.$$

By Theorem 6.5.5, it follows that B is closed. $\qquad \square$

Proposition 6.5.8 *Translations on $Gal_L(\mathbb{M})$ are homeomorphisms.*

Proof Let $D \subset Gal_L(\mathbb{M})$ be closed and $\tau \in Aut(\mathbb{M})$. We show that $[\tau] \cdot D$ is closed in $Gal_L(\mathbb{M})$. Fix any small elementary substructure $\overline{m} = M$ of $\mathbb{M}$. By Proposition 6.5.3, there is a set of L-formulas $p(\overline{x}, \overline{y})$ such that

$$p(\mathbb{M}, \overline{m}) = \{\sigma(\overline{m}) : [\sigma] \in D\}.$$

Then,

$$p(\mathbb{M}, \tau(\overline{m})) = \{\tau(\sigma(\overline{m})) : [\sigma] \in D\}.$$

Hence, $[\tau] \cdot D$ is closed in $Gal_L(\mathbb{M})$ by Proposition 6.5.3.

We have proved that for every $\tau \in Aut(\mathbb{M})$, $[\sigma] \to [\tau] \cdot [\sigma]$ is continuous. Hence, it is a homeomorphism. Similarly, we show that for every $\tau \in Aut(\mathbb{M})$, $[\sigma] \to [\sigma] \cdot [\tau]$ is a homeomorphism. $\qquad\square$

Proposition 6.5.9 *Inversion on $Gal_L(\mathbb{M})$ is a homeomorphism.*

Proof Let $D \subset Gal_L(\mathbb{M})$ be closed. Fix any small elementary substructure $\overline{m} = M$ of $\mathbb{M}$. By Proposition 6.5.3, there is a set of L-formulas $p(\overline{x}, \overline{y})$ such that

$$p(\mathbb{M}, \overline{m}) = \{\sigma(\overline{m}) : [\sigma] \in D\}.$$

Consider

$$q(\overline{x}, \overline{y}) = \{\varphi[\overline{y}, \overline{x}] : \varphi[\overline{x}, \overline{y}] \in p\}.$$

It is easy to see that $q(\mathbb{M}, \overline{m}) = \{\sigma(\overline{m}) : [\sigma]^{-1} \in D\}$. The result follows. $\qquad\square$

The following surprising result will be used to show that $\overline{\{e\}}$ is a normal subgroup of $Gal_L(\mathbb{M})$.

Lemma 6.5.10 *If $[\sigma] \in \overline{\{[\tau]\}}$, then $[\tau] \in \overline{\{[\sigma]\}}$. In particular, if $[\sigma] \in \overline{\{[\tau]\}}$, then $\overline{\{[\tau]\}} = \overline{\{[\sigma]\}}$.*

Proof Set $[\tau] \leq [\sigma]$ if $[\tau] \in \overline{\{[\sigma]\}}$. Since $Gal_L(\mathbb{M})$ is compact, by Zorn's lemma, $Gal_L(\mathbb{M})$ has a minimal element, say $[\sigma_0]$ i.e. $[\tau] \leq [\sigma_0]$ implies $[\sigma_0] \leq [\tau]$.

Take any other $[\sigma] \in Gal_L(\mathbb{M})$ and assume that $[\tau] \leq [\sigma]$, i.e. $[\tau] \in \overline{\{[\sigma]\}}$. Then, by Proposition 6.5.8, $[\sigma_0 \circ \sigma^{-1} \circ \tau] \in \overline{\{[\sigma_0]\}}$. Since $[\sigma_0]$ is minimal, we have $[\sigma_0] \leq [\sigma_0 \circ \sigma^{-1} \circ \tau]$. Hence, by Proposition 6.5.8, $[\sigma] \in \overline{\{[\tau]\}}$. So, every element of $Gal_L(\mathbb{M})$ is a minimal element. The result is clear now. $\qquad\square$

Proposition 6.5.11 $\overline{\{e\}}$ *is a normal subgroup of $Gal_L(\mathbb{M})$.*

Proof Let $[\sigma], [\tau] \in \overline{\{e\}}$. Hence, $[\sigma] \cdot [\tau] \in \overline{\{[\sigma]\}} = \overline{\{e\}}$. Next, $[\tau] \in \overline{\{e\}}$ implies that $e \in \overline{\{[\tau^{-1}]\}}$. Then, $[\tau^{-1}] \in \overline{\{e\}}$.

Next, take $[\sigma] \in \overline{\{e\}}$ and $\tau \in Aut(\mathbb{M})$. Then, $[\sigma \circ \tau] \in \overline{\{e\}} \cdot [\tau] = \overline{\{[\tau]\}}$. Hence, $[\tau^{-1} \circ \sigma \circ \tau] \in [\tau^{-1}] \cdot \overline{\{[\tau]\}} = \overline{\{e\}}$. $\qquad\square$

The next technical result will be used to show that $Gal_L(\mathbb{M})$ is a topological group. Since we have already proved that inversion on $Gal_L(\mathbb{M})$ is continuous, we shall need to show only that the product on $Gal_L(\mathbb{M})$ is jointly continuous.

Lemma 6.5.12 *Let $\sigma, \tau \in Aut(\mathbb{M})$ be such that whenever $U \ni [\sigma]$ and $V \ni [\tau]$ are open, $U \cap V \neq \emptyset$. Then, $\overline{\{[\sigma]\}} = \overline{\{[\tau]\}}$.*

Proof Suppose $\overline{\{[\sigma]\}} \neq \overline{\{[\tau]\}}$. To complete the proof, we shall obtain open $U \ni [\sigma]$ and $V \ni [\tau]$ such that $U \cap V = \emptyset$.

By Lemma 6.5.10, $[\tau] \notin \overline{\{[\sigma]\}}$. Fix a small $\overline{m} = M \preceq \mathbb{M}$ and set $p(\overline{x}) = tp(\overline{m})$. By Proposition 6.5.4, there is an $\emptyset$-type-definable bounded equivalence relation R on $p(\mathbb{M})$ such that

$$[\sigma] \in \overline{\{e\}} \Leftrightarrow \sigma(\overline{m}) \, R \, \overline{m}.$$

Let $q(\overline{x}, \overline{y})$ be a set of L-formulas such that $R = q(\mathbb{M})$. Without any loss of generality, we assume that q is closed under finite conjunctions.

By Proposition 6.5.8, $\overline{\{[\sigma]\}} = [\sigma] \cdot \overline{\{e\}}$. Since $[\tau] \notin \overline{\{[\sigma]\}}$, we have $[\sigma^{-1}] \cdot [\tau] = [\sigma^{-1} \circ \tau] \notin \overline{\{e\}}$. Hence, there is a $\varphi[\overline{x}, \overline{y}] \in q$ such that $\models \neg\varphi[(\sigma^{-1} \circ \tau)(\overline{m}), \overline{m}]$. Since q is without parameters, $\models \neg\varphi[\tau(\overline{m}), \sigma(\overline{m})]$.

We claim that there exists a $\psi \in q$ such that $\psi[\overline{x}, \overline{y}] \wedge \psi[\overline{y}, \overline{z}] \wedge \neg\varphi[\overline{x}, \overline{z}]$ is not satisfiable by any 3-tuple $(\overline{a}, \overline{b}, \overline{c})$ in $p(\mathbb{M})$. Suppose not. Then,

$$r(\overline{x}, \overline{y}, \overline{z}) = p(\overline{x}) \cup p(\overline{y}) \cup p(\overline{z}) \cup \{\psi[\overline{x}, \overline{y}] \wedge \psi[\overline{y}, \overline{z}] \wedge \neg\varphi[\overline{x}, \overline{z}] : \psi \in q\}$$

is a type over empty set in $\mathbb{M}$. Hence, by saturability of $\mathbb{M}$, it is realised in $\mathbb{M}$. This implies that R is not transitive which is a contradiction.

Put

$$r(\overline{z}) = \{\exists\overline{y}(\neg\psi[\tau(\overline{m}), \overline{y}] \wedge \xi[\overline{z}, \overline{y}]) : \xi \in q\}$$

and

$$s(\overline{z}) = \{\exists\overline{x}(\neg\psi[\overline{x}, \sigma(\overline{m})] \wedge \xi[\overline{x}, \overline{z}]) : \xi \in q\}.$$

Set

$$C_1 = \{\gamma \in Aut(\mathbb{M}) : \gamma(\overline{m}) \models r\}$$

and

$$C_2 = \{\gamma \in Aut(\mathbb{M}) : \gamma(\overline{m}) \models s\}.$$

We now show that whenever $\gamma \in C_1$ and $\sigma \in Autf_L(\mathbb{M})$, $\sigma \circ \gamma \in C_1$. Towards showing this note that $\gamma(\overline{m}) \models r$ implies that

$$r'(\overline{y}) = \{\neg\psi[\tau(\overline{m}), \overline{y}]\} \cup \{\xi[\gamma(\overline{m}), \overline{y}]) : \xi \in q\}$$

is a type over a small set. Hence, by saturability of $\mathbb{M}$, there is an $\overline{a} \models r'$. Note that $E_L|p(\overline{m}) \subset R$. Since E_L is the orbit equivalence relation under the action of $Autf_L(\mathbb{M})$, for every $\sigma \in Autf_L(\mathbb{M})$, $(\sigma \circ \gamma)(\overline{m})R\gamma(\overline{m})$. Hence, $(\sigma \circ \gamma)(\overline{m})R\overline{a}$, i.e. $((\sigma \circ \gamma)(\overline{m}), \overline{a}) \models q$. It follows that $\sigma \circ \gamma \in C_1$. Similarly, we show that whenever $\gamma \in C_2$ and $\sigma \in Autf_L(\mathbb{M})$, $\sigma \circ \gamma \in C_2$. These together with the definitions of C_1

and C_2 show that $\pi(C_1)$ and $\pi(C_2)$ are closed in $Gal_L(\mathbb{M})$, where $\pi : Aut(\mathbb{M}) \to Gal_L(\mathbb{M})$ is the quotient map.

Since $\models \neg\varphi[\tau(\overline{m}), \sigma(\overline{m})]$, for every $\gamma \in Aut(\mathbb{M})$,

$$\models \neg\psi[\tau(\overline{m}), \gamma(\overline{m})] \vee \neg\psi[\gamma(\overline{m}), \sigma(\overline{m})].$$

Suppose $\models \neg\psi[\tau(\overline{m}), \gamma(\overline{m})]$. As $(\gamma(\overline{m}), \gamma(\overline{m})) \models q$, we see that $\gamma \in C_1$. If $\neg\psi[\gamma(\overline{m}), \sigma(\overline{m})]$, by the same reason, $\gamma \in C_2$. Thus, $Aut(\mathbb{M}) = C_1 \cup C_2$.

Using $\psi \in q$, it easily seen that $\tau \notin C_1$ and $\sigma \notin C_2$. Take $U = \pi(C_1)^c$ and $V = \pi(C_2)^c$. $\qquad\square$

Theorem 6.5.13 $Gal_L(\mathbb{M})$ *is a topological group.*

Proof We need to prove that the map $([\sigma], [\tau]) \to [\sigma] \cdot [\tau] = [\sigma \circ \tau]$ from $Gal_L(\mathbb{M}) \times Gal_L(\mathbb{M})$ to $Gal_L(\mathbb{M})$ is continuous at each $([\sigma_0], [\tau_0])$. Since the translation is a homeomorphism on $Gal_L(\mathbb{M})$, without any loss of generality, we assume that $[\sigma_0] = [\tau_0] = e$.

Let $U \ni e$ be an open set. We are required to show that there exists an open set $V \ni e$ such that whenever $[\sigma], [\tau] \in V$, $[\sigma \circ \tau] \in U$. Suppose such an open set V does not exist. For each open $V \ni e$, get $[\sigma_V], [\tau_V] \in V$ such that $[\sigma_V \circ \tau_V] \notin U$.

Let I be the set of all open neighbourhood of e. For each $V \in I$, let $A_V = \{W \in I : W \subset V\}$. Then, the family of subsets $\{A_V : V \in I\}$ of I has the finite intersection property. Let $\mathcal{U}$ be an ultrafilter on I containing each A_V, $V \in I$.

Set $\mathbb{M}' = \mathbb{M}^{\mathcal{U}}$ and

$$[\sigma] = \alpha_{\mathbb{M}}^{\mathbb{M}'}((\pi_{V \in I}\sigma_V)^{\mathcal{U}}) \wedge [\tau] = \alpha_{\mathbb{M}}^{\mathbb{M}'}((\pi_{V \in I}\tau_V)^{\mathcal{U}}).$$

Claim. $\overline{\{e\}} = \overline{\{[\sigma]\}} = \overline{\{[\tau]\}}$.

Take open $U' \ni [\sigma]$ and $V' \ni e$. To prove $\overline{\{e\}} = \overline{\{[\sigma]\}}$, by Lemma 6.5.12, it is sufficient to show that $U' \cap V' \neq \emptyset$. By Corollary 6.5.6, $A = \{V \in I : [\sigma_V] \in U'\} \in \mathcal{U}$. Hence, $A \cap A_{V'} \neq \emptyset$, containing W, say. Then, $[\sigma_W] \in U' \cap W \subset U' \cap V'$. Similarly, we prove that $\overline{\{e\}} = \overline{\{[\tau]\}}$.

By Theorem 6.5.5,

$$[\sigma \circ \tau] = \alpha_{\mathbb{M}}^{\mathbb{M}'}((\pi_{V \in I}\sigma_V \circ \tau_V)^{\mathcal{U}}) \notin U.$$

By Proposition 6.5.11, it follows that $[\sigma \circ \tau] \in \overline{\{e\}}$. Hence, by Lemma 6.5.10, $e \in \overline{\{[\sigma \circ \tau]\}}$. We have arrived at a contradiction because $[\sigma \circ \tau] \notin U$ and U is an open neighbourhood of e. $\qquad\square$

Proposition 6.5.14 *Let* $\mathbb{M}' \succeq \mathbb{M}$ *be* $|\mathbb{M}|^+$*-saturated and* $|\mathbb{M}|^+$*-strongly homogeneous. Then,*

$$\beta_{\mathbb{M}}^{\mathbb{M}'} : Gal_L(\mathbb{M}') \to Gal_L(\mathbb{M})$$

is a homeomorphism.

Proof $\underline{\beta_M^{M'} \text{ is continuous}}$: Let $D \subset Gal_L(\mathbb{M})$ be closed. Fix any small $\overline{m} = M \preceq \mathbb{M}$. By Proposition 6.5.3, there is a set $p(\overline{x})$ of L_M-formulas such that

$$p(\mathbb{M}) = \{\sigma(\overline{m}) : [\sigma] \in D\}.$$

To show that $\beta_M^{M'}$ is continuous, we show that

$$p(\mathbb{M}') = \{\sigma'(\overline{m}) : [\sigma'] \in (\beta_M^{M'})^{-1}(D)\}.$$

Take $[\sigma'] \in (\beta_M^{M'})^{-1}(D)$. Assume that $\beta_M^{M'}([\sigma']) = [\tau] \in D$. Hence, $\tau(\overline{m}) \models p$. Since $\alpha_M^{M'}(\sigma') = [\tau]$,

$$tp(\sigma'(\overline{m})/M) = tp(\tau(\overline{m}/M)).$$

Since p is a set of L_M-formulas, $\sigma'(\overline{m}) \models p$.

We now show the reverse inclusion. Let $\mathbb{M}' \ni \overline{a} \models p$. By Proposition 4.3.7, there is $\overline{b} \in \mathbb{M}$ such that $tp(\overline{a}/M) = tp(\overline{b}/M)$. Hence, there is a $[\sigma] \in D$ such that $\overline{b} = \sigma(\overline{m})$. There is $g \in Aut_M(\mathbb{M}')$ such that $g(\sigma(\overline{m})) = \overline{a}$. Since $\mathbb{M}'$ is $|\mathbb{M}|^+$-strongly homogeneous, there is an extension $\sigma'' \in Aut(\mathbb{M}')$ of σ. Note that $\beta_M^{M'}([\sigma'']) = [\sigma]$. Set $\sigma' = g \circ \sigma''$. Then, $[\sigma'] = [\sigma'']$. Since $\sigma'(\overline{m}) = g(\sigma''(\overline{m})) = g(\sigma(\overline{m})) = \overline{a}$, our proof is complete.

$\underline{(\beta_M^{M'})^{-1} \text{ is continuous}}$: Let $D \subset Gal_L(\mathbb{M}')$ be closed. Fix any small $\overline{m} = M \preceq \mathbb{M}$. By Proposition 6.5.3, there is a set $p(\overline{x})$ of L_M-formulas such that

$$p(\mathbb{M}') = \{\sigma'(\overline{m}) : [\sigma'] \in D\}.$$

To complete the proof, we show the following:

$$p(\mathbb{M}) = \{\sigma(\overline{m}) : [\sigma] \in \beta_M^{M'}(D)\}.$$

Take $[\sigma] \in \beta_M^{M'}(D)$. Since $\mathbb{M}'$ is $|\mathbb{M}|^+$-strongly homogeneous, there is an extension $\sigma' \in Aut(\mathbb{M}')$ of σ. Clearly, $tp(\sigma'(\overline{m})/M) = tp(\sigma(\overline{m})/M)$. This implies that $\beta_M^{M'}([\sigma']) = [\sigma]$. Since $\beta_M^{M'}$ is one-to-one, we must have $[\sigma'] \in D$. Inclusion from right to left follows.

For the reverse inclusion, let $\mathbb{M}' \supset \mathbb{M} \ni \overline{a} \models p$. Then, there is a $[\sigma'] \in D$ such that $\sigma'(\overline{m}) = \overline{a}$. Set $\alpha_M^{M'}(\sigma') = [\sigma]$. Then, $tp(\sigma(\overline{m})/M) = tp(\sigma'(\overline{m})/M) = tp(\overline{a})/M$. Hence, there is a $g \in Aut_M(\mathbb{M})$ such that $g(\sigma(\overline{m})) = \overline{a} = \sigma'(\overline{m})$. Since $[\sigma] = [g \circ \sigma] \in \beta_M^{M'}(D)$, our proof is complete. $\qquad \square$

This theorem tells that for any two monster models $\mathbb{M}$ and $\mathbb{N}$, $Gal_L(\mathbb{M})$ and $Gal_L(\mathbb{N})$ are homeomorphic. In fact, there is an isomorphism from $Gal_L(\mathbb{M})$ onto $Gal_L(\mathbb{N})$ which is a homeomorphism. Thus, $Gal_L(\mathbb{M})$ is invariant of T, i.e. it does not depend on the monster model $\mathbb{M}$. From now on, $Gal_L(T)$ will stand for $Gal_L(\mathbb{M})$ for some monster model $\mathbb{M}$ of T. $Gal_L(T)$ is called the *Galois group of T*. This group was introduced by Lascar in [33].

6.6 The Group $Gal_0(T)$

We set
$$Gal_0(T) = \overline{\{e\}},$$

$$Gal_c(T) = Gal_L(T)/Gal_0(T)$$

and
$$Autf_{KP}(\mathbb{M}) = \pi^{-1}(Gal_0(T)),$$

where $\pi : Aut(, \mathbb{M}) \to Gal_L(T)$ is the quotient map. The group $Gal_c(T)$ is called the *closed Galois group* of T.

We make a series of simple observations.

1. In Proposition 6.5.11, we proved that $Gal_0(T)$ is a normal subgroup of $Gal_L(T)$.
2. Since $Gal_0(T)$ is a normal subgroup of $Gal_L(T)$, $Autf_{KP}(\mathbb{M})$ is a normal subgroup of $Aut(\mathbb{M})$.
3. Since $Autf_L(\mathbb{M})$ is a normal subgroup of $Aut(\mathbb{M})$, $Autf_L(\mathbb{M})$ is a normal subgroup of $Autf_{KP}(\mathbb{M})$.
4. $Gal_0(T)$ is isomorphic to $Autf_{KP}(\mathbb{M})/Autf_L(\mathbb{M})$. This is because $\pi : Autf_{KP}(\mathbb{M}) \to Gal_0(T)$ is an epimorphism with kernel $Autf_L(\mathbb{M})$.
5. $Gal_c(T)$ is a compact topological group.
6. Since $\{e'\}$, e' the identity of $Gal_c(T)$, is closed, $Gal_c(T)$ is Hausdorff.

Theorem 6.6.1 *Let $\lambda < \kappa$ and $Y \subset \mathbb{M}^\lambda$ a type-definable, E_L-invariant set. Then, Y is $Autf_{KP}(\mathbb{M})$-invariant.*

Proof Set
$$S = \{\sigma \in Aut(\mathbb{M}) : \sigma(Y) = Y\}.$$

We need to show that $Autf_{KP}(\mathbb{M}) \subset S$. For $\overline{a} \in Y$, let
$$S(\overline{a})^+ = \{\sigma \in Aut(\mathbb{M}) : \sigma(\overline{a}) \in Y\}$$

and
$$S(\overline{a})^- = \{\sigma \in Aut(\mathbb{M}) : \sigma^{-1}(\overline{a}) \in Y\}.$$

Then, $S = \cap_{\overline{a} \in Y}(S(\overline{a})^+ \cap S(\overline{a})^-)$. Since E_L is the orbit equivalence on $\mathbb{M}^\lambda$ under the action of $Autf_L(\mathbb{M})$ and since Y is assumed to be E_L-invariant, for every $\overline{a} \in Y$, $S(\overline{a})^+ \cap S(\overline{a})^- \supset Autf_L(\mathbb{M})$.

Next assume that $\sigma \in S$ and $\tau \in Autf_L(\mathbb{M})$. Then, $\tau(\sigma(Y)) = \tau(Y) = Y$. Thus, $Autf_L(\mathbb{M}) \circ S = S$. By the same argument, we see that for every $\overline{a} \in Y$, $Autf_L(\mathbb{M}) \circ S(\overline{a})^+ = S(\overline{a})^+$ and $S(\overline{a})^- \circ Autf_L(\mathbb{M}) = Autf_L(\mathbb{M}) \circ S(\overline{a})^- = S(\overline{a})^-$. It follows that

$$S = \cap_{\overline{a} \in Y}(Autf_L(\mathbb{M}) \circ S(\overline{a})^+ \cap Autf_L(\mathbb{M}) \circ S(\overline{a})^-).$$

We shall complete the proof by showing that for every $\bar{a} \in Y$,

$$Autf_{KP}(\mathbb{M}) \subset Autf_L(\mathbb{M}) \circ S(\bar{a})^+$$

and

$$Autf_{KP}(\mathbb{M}) \subset Autf_L(\mathbb{M}) \circ S(\bar{a})^-.$$

Fix an $\bar{a} \in Y$. Note that

$$Autf_L(\mathbb{M}) \circ S(\bar{a})^+ = \pi^{-1}(\pi(S(\bar{a})^+))$$

and

$$Autf_L(\mathbb{M}) \circ S(\bar{a})^- = \pi^{-1}(\pi(S(\bar{a})^-)).$$

The proof will be complete if we show that $\pi(S(\bar{a})^+), \pi(S(\bar{a})^-) \supset Gal_0(T)$. Since $e \in \pi(S(\bar{a})^+), \pi(S(\bar{a})^-)$, our contention will follow if we show that $\pi(S(\bar{a})^+)$, $\pi(S(\bar{a})^-)$ are closed in $Gal_L(T)$. Note that $\pi(S(\bar{a})^-) = \pi(S(\bar{a})^+)^{-1}$. Hence, we need to show only $\pi(S(\bar{a})^+)$ is closed in $Gal_L(T)$. We are going to use Theorem 6.5.5.

Take any small $M \preceq \mathbb{M}$. Since Y is E_L-invariant, it is invariant over M. Further, it is type-definable. Hence, by Proposition 4.6.3, there is a type $p(\bar{x})$ over M such that $Y = p(\mathbb{M})$.

Take an ultrafilter $\mathcal{U}$ on an index set I. Set $\mathbb{M}' = \mathbb{M}^{\mathcal{U}}$. We treat $\mathbb{M}'$ as an elementary extension of $\mathbb{M}$ canonically. For each $i \in I$, choose $\sigma_i \in Aut(\mathbb{M})$ such that $\sigma_i(\bar{a}) \in Y$, i.e. $\sigma_i(\bar{a}) \models p$. Set $\sigma = (\pi_{i \in I} \sigma_i)^{\mathcal{U}}$. Suppose

$$[\tau] = \alpha_{\mathbb{M}}^{\mathbb{M}'}(\sigma). \tag{$*$}$$

We need to show that $\tau(\bar{a}) \models p$.

Take a small elementary substructure N of $\mathbb{M}$ containing $\bar{a}$. Enumerate $N = \bar{n} = \overline{ab}$. By $(*)$,

$$tp^{\mathbb{M}'}(\sigma(\overline{ab})/M) = tp^{\mathbb{M}}(\tau(\overline{ab})/M).$$

By Łoś theorem, for every L_M-formula, $\psi[\overline{xy}]$, we have

$$\mathbb{M} \models \psi[\tau(\overline{ab})] \Leftrightarrow \mathbb{M}' \models \psi[\sigma(\overline{ab})] \Leftrightarrow \{i \in I : \mathbb{M} \models \psi[\sigma_i(\overline{ab})]\} \in \mathcal{U}.$$

In particular, for every $\varphi[\bar{x}] \in p$,

$$\mathbb{M} \models \varphi[\tau(\bar{a})] \Leftrightarrow \mathbb{M}' \models \varphi[\sigma(\bar{a})].$$

Since $\{i \in I : \varphi[\sigma_i(\bar{a})]\} = I \in \mathcal{U}$, $\mathbb{M} \models \varphi[\tau(\bar{a})]$ and our proof is complete. □

6.7 E_{KP} as an Orbit Space

Fix a $\lambda < \kappa$. Let F denote the orbit equivalence relation on $\mathbb{M}^\lambda$ under the action of $Autf_{KP}(\mathbb{M})$. We make a series of simple observations first.

1. Since each Kim–Pillay strong type is type-definable and E_L-invariant, by Theorem 6.6.1, $Autf_{KP}(\mathbb{M})$ stabilises all Kim–Pillay strong types. Thus, $F \subset E_{KP}$.
2. By Theorem 6.3.4, E_L is the orbit equivalence relation under the action of $Autf_L(\mathbb{M})$ and E_L is bounded. Since $Autf_L(\mathbb{M}) \subset Autf_{KP}(\mathbb{M})$, we see that F is bounded.
3. Since $Autf_{KP}(\mathbb{M})$ is a normal subgroup of $Aut(\mathbb{M})$, it is easily checked that F is invariant.

Theorem 6.7.1   *F is type-definable over empty set.*

Proof Take any small $M \preceq \mathbb{M}$. Fix an enumeration $\overline{m}$ of M. Set

$$A(\overline{m}) = \{\overline{a} : \overline{a} \equiv_\emptyset \overline{m}\}$$

and

$$S(\overline{m}) = \{tp(\overline{a}/M) : \overline{a} \in A(\overline{m})\}.$$

Note that $A(\overline{m}) = \{\sigma(\overline{m}) : \sigma \in Aut(\mathbb{M})\}$. Recall that we have an onto map

$$\rho_{\overline{m}} : S(\overline{m}) \to Gal_L(T)$$

defined by

$$\rho_{\overline{m}}(tp(\sigma(\overline{m})/M)) = [\sigma], \ \sigma \in Aut(\mathbb{M}).$$

We set

$$\Theta : A(\overline{m}) \times A(\overline{m}) \to S(\overline{m}) \times S(\overline{m})$$

by

$$\Theta(\sigma(\overline{m}), \tau(\overline{m})) = (tp(\sigma(\overline{m})/M), tp(\tau(\overline{m})/M)), \ \sigma, \tau \in Aut(\mathbb{M}),$$

$$\Psi = \rho_{\overline{m}} \times \rho_{\overline{m}} : S(\overline{m}) \times S(\overline{m}) \to Gal_L(T) \times Gal_L(T)$$

and

$$\Phi : Gal_L(T) \times Gal_L(T) \to Gal_L(T)$$

defined by

$$\Phi([\sigma], [\tau]) = [\tau^{-1} \circ \sigma], \ \sigma, \tau \in Gal_L(T).$$

Step 1. $E = F|A(\overline{m})$ is type-definable over empty set.
Set $A = \Theta^{-1}(\Psi^{-1}(\Phi^{-1}(Gal_0(T))))$. We shall prove Step 1 by showing that A is type-definable and $A = E$.

Since $Gal_0(T)$ is closed, $\Psi^{-1}(\Phi^{-1}(Gal_0(T)^c))$ is open in $S(\overline{m}) \times S(\overline{m})$. Hence, there exist families of L_M-formulas $\{\varphi_i[\overline{x}] : i \in I\}$ and $\{\psi_i[\overline{y}] : i \in I\}$ such that

$$\Psi^{-1}(\Phi^{-1}(Gal_0(T)^c)) = \cup_{i \in I}([\varphi_i] \times [\psi_i]).$$

Set

$$q(\overline{x}, \overline{y}) = \{\neg\varphi_i[\overline{x}] \vee \neg\psi_i[\overline{y}] : i \in I\}.$$

It is fairly routine to check that $A = q(\mathbb{M})$.

Now we show that $E = A$. Fix $\overline{a}, \overline{b} \in A(\overline{m})$.

Suppose $\overline{a} E \overline{b}$. Get $\sigma \in Autf_{KP}(\mathbb{M})$ and $\tau \in Aut(\mathbb{M})$ such that $\overline{b} = \tau(\overline{m})$ and $\overline{a} = \sigma(\overline{b}) = \sigma(\tau(\overline{m}))$. Hence,

$$\Phi(\Psi(\Theta(\overline{a}, \overline{b}))) = [\tau^{-1} \circ \sigma \circ \tau] \in Gal_0(T)$$

because $Autf_{KP}(\mathbb{M})$ is a normal subgroup of $Aut(\mathbb{M})$. Thus, $(\overline{a}, \overline{b}) \in A$, proving that $E \subset A$.

Conversely, assume that $(\overline{a}, \overline{b}) \in A$. Set $\overline{a} = \sigma(\overline{m})$ and $\overline{b} = \tau(\overline{m})$. We have $[\tau^{-1} \circ \sigma] \in Gal_0(T)$. This implies that $\tau^{-1} \circ \sigma \in Autf_{KP}(\mathbb{M})$. Therefore,

$$\sigma \circ \tau^{-1} = \sigma \circ (\tau^{-1} \circ \sigma) \circ \sigma^{-1} \in Autf_{KP}(\mathbb{M}).$$

Since $\overline{b} = \tau \circ \sigma^{-1}(\overline{a})$, $A \subset E$.

By observation (3) above, E is invariant. Hence, by Proposition 4.6.3, E is type-definable over empty set, say by a set of L-formulas $p(\overline{x}, \overline{y})$.

Claim. For any $\overline{a}, \overline{b} \in \mathbb{M}^\lambda$,

$$\overline{a} F \overline{b} \Leftrightarrow \exists \overline{c}(\overline{a}\,\overline{m} \equiv_\emptyset \overline{b}\,\overline{c} \wedge \overline{c} \models p(\overline{x}, \overline{m})).$$

Assuming the claim, we complete the proof first. Consider

$$q(\overline{x}, \overline{y}, \overline{z}) = \{\varphi[\overline{x}, \overline{m}] \leftrightarrow \varphi[\overline{y}, \overline{z}] : \varphi \text{ an } L\text{-formula}\} \cup p(\overline{z}, \overline{m}).$$

By our claim, $F = proj(q(\mathbb{M}))$. Hence, by Proposition 4.6.2, F is type-definable.

Proof of the claim. Suppose $\overline{a} F \overline{b}$. Get $\sigma \in Autf_{KP}(\mathbb{M})$ such that $\overline{b} = \sigma(\overline{a})$. Take $\overline{c} = \sigma(\overline{m})$. Then, $tp(\overline{a}, \overline{m}/\emptyset) = tp(\overline{b}, \overline{c}/\emptyset)$. Further, $\overline{c} \models p(\overline{x}, \overline{m})$. Thus, we have proved the implication from left to right.

Conversely, let $\overline{a}\,\overline{m} \equiv_\emptyset \overline{b}\,\overline{c} \wedge \overline{c} \models p(\overline{x}, \overline{m})$. Get $\sigma \in Aut(\mathbb{M})$ such that $\sigma(\overline{a}\,\overline{m}) = \overline{b}\,\overline{c}$. Since p type defines $F|A(\overline{m})$, $\overline{c} F \overline{m}$. Hence, there exists a $\tau \in Autf_{KP}(\mathbb{M})$ such that $\tau(\overline{m}) = \overline{c} = \sigma(\overline{m})$ implying $\tau^{-1} \circ \sigma(\overline{m}) = \overline{m}$. It follows that $\tau^{-1} \circ \sigma \in Autf_L(\mathbb{M}) \subset Autf_{KP}(\mathbb{M})$. Since $\tau \in Autf_{KP}(\mathbb{M})$, $\sigma \in Autf_{KP}(\mathbb{M})$. Thus, $\overline{a} F \overline{b}$.

Since F is type-definable and invariant, by Proposition 4.6.3, F is type-definable over empty set. $\square$

Theorem 6.7.2 $E_{KP} = F$, *the orbit equivalence relation under the action of* $Autf_{KP}(\mathbb{M})$.

Proof In the beginning of this section, we observed that F is bounded and $F \subset E_{KP}$. In the last theorem, we proved that F is type-definable over empty set. Since E_{KP} is the smallest, $\emptyset$-type-definable, bounded equivalence relation, $E_{KP} \subset F$. Hence, $E_{KP} = F$.   □

6.8   Connection with Descriptive Set Theory

Throughout this section, T is a countable complete theory, λ a countable ordinal and parameter sets are countable. Further, by small substructures of the monster, we shall mean countable substructures.

We let $S(\overline{x}, \overline{y}/A)$ denote the set of all complete types over A in variables $\overline{x}, \overline{y}$ of equal length and of length a countable ordinal λ. We equip $S(\overline{x}, \overline{y}/A)$ with the Stone topology. Then, $S(\overline{x}, \overline{y}/A)$ is a compact, metrisable, zero-dimensional space.

We refer the reader to [58, Sect. 3.6] for the definition Borel sets in a metrisable space of additive (multiplicative) class α, $1 \le \alpha < \omega_1$.

Let $r : S(\overline{x}, \overline{y}/A) \to S(\overline{x}, \overline{y}/\emptyset)$ denote the canonical restriction map. The map r is continuous and onto. By (Theorem 5.2.11, [58]), we have the following.

Proposition 6.8.1 *For every countable ordinal α and every $B \subset S(\overline{x}, \overline{y}/\emptyset)$, $r^{-1}(B)$ is Borel of additive (multiplicative) class α if and only if B is Borel of additive (multiplicative) class α.*

Now let E be a bounded, invariant, equivalence relation on $\mathbb{M}^\lambda$, where as usual $\mathbb{M}$ is a monster model of T. Define

$$E_A = \{tp(\overline{a}, \overline{b}/A) : \overline{a} E \overline{b}\} \subset S(\overline{x}, \overline{y}/A).$$

Note that $E_A = r^{-1}(E_\emptyset)$. We call E Borel of additive (multiplicative) class α if $E_\emptyset$ is Borel of additive (multiplicative) class α. By the last proposition, we have the following result.

Proposition 6.8.2 *The following statements are equivalent:*

1. *E is Borel of additive (multiplicative) class α.*
2. *E_A is Borel of additive (multiplicative) class α for all countable A.*
3. *E_A is Borel of additive (multiplicative) class α for some countable A.*

Next, fix a small $M \preceq \mathbb{M}$. We define a binary relation E^M on $S(\overline{x}/M)$ as follows: We make an observation first. Take any $\overline{a}, \overline{a}', \overline{b}, \overline{b}' \in \mathbb{M}^\lambda$ such that $tp(\overline{a}/M) = tp(\overline{a}'/M)$ and $tp(\overline{b}/M) = tp(\overline{b}'/M)$. Then, $\overline{a} \equiv_M \overline{a}'$. Hence, $\overline{a} E_L \overline{a}'$. But E_L is the smallest bounded, invariant, equivalence relation. Therefore, $\overline{a} E \overline{a}'$. By the same argument, $\overline{b} E \overline{b}'$. It follows that

$$\bar{a} E \bar{b} \Leftrightarrow \bar{a}' E \bar{b}'.$$

Let $p(\bar{x}) = tp(\bar{a}/M)$ and $q(\bar{x}) = tp(\bar{b}/M)$. We define

$$p E^M q \Leftrightarrow \bar{a} E \bar{b}.$$

By the above remark, this is well defined.

It is easy to check that E^M is an equivalence relation on $S(\bar{x}/M)$.

Now consider a map $h : S(\bar{x}, \bar{y}/M) \to S(\bar{x}/M) \times S(\bar{x}/M)$ defined by

$$h(tp(\bar{a}, \bar{b}/M)) = (tp(\bar{a}/M), tp(\bar{b}/M)), \quad \bar{a}, \bar{b} \in \mathbb{M}^\lambda.$$

It is easy to check that h is well defined and onto. Clearly, $h^{-1}(E^M) = E_M$.

h is continuous: Let $\varphi[\bar{x}]$, $\psi[\bar{y}]$ be L_M-formulas and $\bar{a}, \bar{b} \in \mathbb{M}^\lambda$ be such that

$$(tp(\bar{a}/M), tp(\bar{b}/M)) \in [\varphi] \times [\psi].$$

Then, $h([\xi]) \subset [\varphi] \times [\psi]$, where $\xi[\bar{x}, \bar{y}] = \varphi[\bar{x}] \wedge \psi[\bar{y}]$.

By (Theorem 5.2.11, [58]), it follows the following proposition.

Proposition 6.8.3 *For every $\alpha < \omega_1$, E_M is Borel of additive (multiplicative) class α if and only if E^M is Borel of additive (multiplicative) class α.*

Let X and Y be Polish spaces, and E and F be equivalence relations on X and Y, respectively. We say that Borel cardinality of E is less than or equal to F, written $E \leq_B F$, if there is a Borel function $f : X \to Y$ such that

$$\forall x, y \in X(xEy \Leftrightarrow f(x)Ff(y)).$$

We say E and F have the same Borel cardinality, written $|E| =_B |F|$, if both $E \leq_B F$ and $F \leq_B E$ hold. It is clear that $=_B$ is reflexive, symmetric and transitive.

Theorem 6.8.4 *Let $M, N \preceq \mathbb{M}$ be countable. Then, the Borel cardinalities of E^M and E^N are equal.*

Proof Since M and N are countable and elementarily equivalent, there is a common countable elementary extension of M and N. Hence, without any loss of generality, we can assume that $M \preceq N$.

We have a continuous, onto restriction map $r : S(\bar{x}/N) \to S(\bar{x}/M)$. For any $\bar{a}, \bar{b} \in \mathbb{M}^\lambda$, we have

$$tp(\bar{a}/N)E^N tp(\bar{b}/N) \Leftrightarrow \bar{a} E \bar{b} \Leftrightarrow tp(\bar{a}/M)E^M tp(\bar{b}/M).$$

This shows that $E^N \leq_B E^M$.

By Novikov's selection theorem (Theorem 5.7.1, [58]), there is Borel map $s :$ $S(\overline{x}/M) \to S(\overline{x}/N)$ such that $r \circ s = id$. We have

$$s(tp(\overline{a}/M))E^N s(tp(\overline{b}/M)) \Leftrightarrow tp(\overline{a}/M) = r(s(tp(\overline{a}/M)))E^M r(s(tp(\overline{b}/M))) = tp(\overline{b}/M).$$

This shows that $E^M \leq_B E^N$. $\qquad\qquad\qquad\qquad\qquad\qquad\qquad\qquad\qquad$ □

The connection with descriptive set theory is important to understand the spaces of strong types and associated Galois groups as mathematical objects. The idea of measuring the complexity of bounded, invariant equivalence relations via Borel cardinalities was formulated in the paper of Krupinski, Pillay and Solecki [30]. In this paper, it was conjectured that E_L restricted to a Kim–Pillay strong type is either trivial or non-smooth, i.e. there is no Borel function inducing the equivalence relation. This conjecture was proved by Kaplan, Miller and Simon in [24]. This result was extended by Kaplan and Miller in [23] and by Krupinski and Rzepecki in [29]. Finally, a very general trichotomy theorem for arbitrary strong types was proved by Krupinski, Pillay and Rzepecki in [28]. For more recent results, see [26, 27]. For relevance of these in stability theory and model theory, see [46, 50].

Chapter 7
Model Theory of Valued Fields

Abstract This chapter is devoted to the model theory of valued fields, which is due to Ax and Kochen. We also present Ax–Kochen's solution of Artin's conjecture that for every prime p, the field of p-adic real numbers $\mathbb{Q}_p$ is a $C_2(d)$ field for every $d \geq 1$ (See [2–4]). This was probably the first occasion when model theoretic methods were used to solve an outstanding conjecture in mathematics. This chapter requires a good knowledge of valued fields. It is a specialised topic not commonly covered in graduate courses. In Appendix C, we have given a self-contained account of the theory of valued fields that we require. The reader not familiar with valued fields should go through Sect. C.1 before proceeding with this chapter.

7.1 The Language for Valued Fields

In Sect. C.1, it is shown that there is a one-to-one correspondence between valuation subrings V and divisibility relations $|$ on a field $\mathbb{F}$. Further, the value group can be taken to be $\mathbb{F}^\times / V^\times$ with valuation the quotient map $v : \mathbb{F}^\times \to \mathbb{F}^\times / V^\times$.

We are going to take the language for valued fields to be the extension of the language of rings by a divisibility relation symbol $|$. There are several reasons to start with a divisibility relation:

1. It makes sense on a commutative ring with identity.
2. It is easy to characterize substructures of a field with divisibility relations. They are precisely integral domains with divisibility relations. (See Proposition C.1.16.)
3. If V is the corresponding valuation subring, we can express $x \in V$ by the formula $1|x$ and $x \in V^\times$ by $1|x \wedge x|1$.
4. If v is a compatible valuation, then we can express many statements involving v: $v(x) \geq 0 \leftrightarrow 1|x$, $v(x) \leq v(y) \leftrightarrow x|y$, $v(x) < v(y) \leftrightarrow y \nmid x$, $v(x) = 0 \leftrightarrow (x|1 \wedge 1|x)$, etc., in terms of the divisibility relation.

From now on, the language of a valued field is an extension of the language of rings with a new binary relation symbol $|$. This language is further extended by definitions by adding unary predicate symbols V and $V^\times$ defined by

$$V(x) \leftrightarrow 1|x \text{ and } V^\times(x) \leftrightarrow (1|x \wedge x|1)$$

© Springer Nature Singapore Pte Ltd. 2017
H. Sarbadhikari and S.M. Srivastava, *A Course on Basic Model Theory*,
DOI 10.1007/978-981-10-5098-5_7

respectively. The above remarks show that if v is a compatible valuation, then various statements involving the valuation v can be expressed in our language. It is easily seen that the classes of all integral domains with a divisibility relation, all valued fields, all algebraically closed valued fields and all algebraically closed valued fields of a fixed characteristic are elementary. Also note that $char(\mathbb{F}^{\sim}) = p, p > 0$, is expressed by the formula $\neg p|1$ and $char(\mathbb{F}^{\sim}) = 0$ by the set of formulas $\{p|1 : p$ a positive prime$\}$.

We are now going to show that the truth of first-order statements in the residue field or in the value group of a valued field can be decided in the valued field itself.

By induction on the length of a formula φ in the language of rings, we define a formula φ_r in the language of valued fields as follows:

$$(t_1 = t_2)_r = (V(t_1) \wedge V(t_2) \wedge \neg V^{\times}(t_1 - t_2)),$$

$$(\neg\varphi)_r = \neg\varphi_r,$$

$$(\varphi \vee \psi)_r = \varphi_r \vee \psi_r.$$

and

$$(\exists x\varphi)_r = \exists x(V(x) \wedge \varphi_r).$$

By induction on the length of a formula $\varphi[x_0, \cdots, x_{n-1}]$ in the language of rings, it is easy to check that for all valued fields $(\mathbb{F}, V)$ and all $a_0, \cdots, a_{n-1} \in V$

$$\mathbb{F}^{\sim} \models \varphi[[a_0], \cdots, [a_{n-1}]] \Leftrightarrow (\mathbb{F}, V) \models \varphi_r[a_0, \cdots, a_r].$$

By induction on the length of a formula φ in the language of ordered groups, we define a formula φ_g in the language of valued fields as follows:

$$(t_1 = t_2)_g = \exists x(V^{\times}(x) \wedge x \cdot t_1 = t_2),$$

$$(t_1 < t_2)_g = \exists x(\neg V^{\times}(x) \wedge V(x) \wedge x \cdot t_1 = t_2),$$

$$(\neg\varphi)_g = \neg\varphi_g,$$

$$(\varphi \vee \psi)_g = \varphi_g \vee \psi_g,$$

and

$$(\exists x\varphi)_g = \exists x(x \neq 0 \wedge \varphi_g).$$

By induction on the length of a formula $\varphi[x_0, \cdots, x_{n-1}]$ in the language of ordered groups, it is easy to check that for every valued field $(\mathbb{F}, V)$ with value group Γ and every $a_0, \cdots, a_{n-1} \in \mathbb{F}^{\times}$,

$$\Gamma \models \varphi[v(a_0), \cdots, v(a_{n-1})] \Leftrightarrow (\mathbb{F}, V) \models (\varphi[a_0, \cdots, a_{n-1}])_g.$$

As a corollary, we get

Theorem 7.1.1 *If $(\mathbb{F}, V)$ is κ-saturated, so are its residue field $\mathbb{F}^{\sim}$ and the value group Γ.*

In Sect. C.6, we defined that a valued field $(\mathbb{F}, V)$ is Henselian if for every algebraic extension $\mathbb{L}$ of $\mathbb{F}$, V has a unique extension to $\mathbb{L}$. As an easy consequence of Theorem C.6.2 giving several characterizations of Henselian valued fields, we get the following result.

Proposition 7.1.2 *The class of all Henselian valued fields is elementary.*

Proof For each $n > 1$, let H_n denote the formula

$$\wedge_{i=0}^{n-1} V(x_i) \rightarrow \forall x[(V(x) \wedge \neg V^{\times}(x^n + \sum_{i=0}^{n-1} x_i x^i)$$
$$\wedge V^{\times}(nx^{n-1} + \sum_{i=1}^{n-1} i x_i x^{i-1}))$$
$$\rightarrow \exists y(V(y) \wedge \neg V^{\times}(x - y) \wedge y^n + \sum_{i=0}^{n-1} x_i y^i = 0)].$$

Then, the class of all Henselian valued fields is the set of all models of the theory of valued fields extended by axioms $\{H_n : n > 1\}$.   □

7.2 Ultraproduct of Valued Fields

Example 7.2.1 Take a family of valued fields $\{(\mathbb{F}_i, v_i, \Gamma_i) : i \in I\}$ and $\mathcal{U}$ an ultra-filter on I. Now set $\Gamma_i' = \Gamma_i \cup \{\infty_i\}$ with $a <_i \infty_i$ for all $a \in \Gamma_i$. By Theorem 2.1.3, $(\times_i \Gamma_i')/\mathcal{U} = \Gamma(\mathcal{U}) \cup \{\infty\}$ where $\infty = [(\infty_i)]$ and $[(a_i)] < \infty$ for all $[(a_i)] \in (\times_i \Gamma_i)/\mathcal{U}$.

We now define a valuation $v^{\mathcal{U}}$ on the ultraproduct $\mathbb{F}(\mathcal{U}) = (\times_i \mathbb{F}_i)/\mathcal{U}$ with value group $\Gamma(\mathcal{U}) = (\times_i \Gamma_i)/\mathcal{U}$ as follows:

$$v^{\mathcal{U}}([(a_i)]) = [(v_i(a_i))].$$

Using Theorem 2.1.3, it is quite easy to check that this defines a valuation on the ultraproduct

Proposition 7.2.2 *The residue field of the ultraproduct $\mathbb{F}(\mathcal{U})$ equals $(\times_i \mathbb{F}_i^{\sim})/\mathcal{U}$, the ultraproduct of the residue fields of $\mathbb{F}_i$s.*

Proof The valuation subring of $\mathbb{F}(\mathcal{U})$ is given by

$$V(\mathcal{U}) = \{[(a_i)] \in \mathbb{F}(\mathcal{U}) : v^{\mathcal{U}}([(a_i)]) \geq 0\}.$$

By Theorem 2.1.3, for each $[(a_i)] \in V(\mathcal{U})$, $\{i \in I : a_i \in V_i\} \in \mathcal{U}$. Hence, there exists (b_i) such that $[(b_i)] = [(a_i)]$ and $b_i \in V_i$ for all $i \in I$.

We consider the map q that assigns $[(a_i)]$ to $[([b_i])] \in (\times_i \mathbb{F}_i^\sim)/\mathcal{U}$. Using Theorem Theorem 2.1.3 it is easy to check that q is well defined and is an epimorphism. We now compute the kernel of q. Let $b_i \in V_i, i \in I$. By Theorem 2.1.3,

$$[([b_i])] = 0 \Leftrightarrow \{i \in I : b_i \in M_i\} \in \mathcal{U}.$$

But $\{i \in I : b_i \in M_i\} = \{i \in I : v_i(b_i) > 0\}$. Hence, by Theorem 2.1.3,

$$q([(a_i)]) = 0 \Leftrightarrow v^\mathcal{U}([(a_i)]) > 0.$$

Thus, the kernel of q is the unique maximal ideal of $V^\mathcal{U}$. This proves the result. $\square$

The next theorem is a direct consequence of Theorem 4.3.13

Theorem 7.2.3 *Let $\mathcal{U}$ be a free ultrafilter on the set of all primes $\mathbb{P}$. Then $\times_p \mathbb{Q}_p/\mathcal{U}$ and $\times_p \mathbb{F}_p((X))/\mathcal{U}$ are both $\aleph_1$-saturated.*

Proposition 7.2.4 *If $A = \{i \in I : (\mathbb{F}_i, V_i) \text{ is Henselian}\} \in \mathcal{U}$, then $(\mathbb{F}(\mathcal{U}), V(\mathcal{U}))$ is Henselian.*

Proof Take a polynomial

$$f(X) = X^n + a^{n-1}X^{n-1} + a^{n-2}X^{n-2} + \cdots + a^1 X + a^0 \in V(\mathcal{U})[X]$$

with $a^{n-1} \notin M(\mathcal{U})$ and $a^{n-2}, \cdots, a^0 \in M(\mathcal{U})$. We need to show that f has a root in $\mathbb{F}(\mathcal{U})$.

Let $a^j = [(a_i^j)], 0 \le j \le n - 1$. Then, by Theorem 2.1.3, $A_{n-1} = \{i \in I : a_i^{n-1} \in V^i \setminus M^i\}$ and $A_j = \{i \in I : a_i^j \in M_i\}, j = 0, \cdots, n - 2$, are all in $\mathcal{U}$. Therefore, $B = A \cap \cap_{j=0}^{n-1} A_j \in \mathcal{U}$. Fix an $i \in B$. Since $\mathbb{F}_i$ is Henselian, there is a root $b_i \in \mathbb{F}_i$ of $f_i(X) = x^n + a_i^{n-1}X^{n-1} + a_i^{n-2}X^{n-2} + \cdots + a_i^0$. Set $b_i = 0$ for $i \in I \setminus B$. Then $f([(b_i)]) = 0$ by Theorem 2.1.3. $\square$

Corollary 7.2.5 *Let $\mathbb{P}$ denote the set of all primes and $\mathcal{U}$ a free ultrafilter on $\mathbb{P}$. Then $\times_p \mathbb{Q}_p/\mathcal{U}$ and $\times_p \mathbb{F}_p((X))/\mathcal{U}$ are $\aleph_1$-saturated Henselian valued fields with same residue field $\times_p \mathbb{F}_p/\mathcal{U}$ of characteristic 0 and same value group $\mathbb{Z}^\mathbb{P}/\mathcal{U}$.*

We call a field $\mathbb{F}$ a $C_i(d)$ field, if every homogeneous polynomial of total degree d in more than d^i variables has a non-zero root. It is easy to see that the the class of all $C_i(d)$ fields is elementary.

Let $p(X_1, \cdots, X_{d^i+1}, \cdots, X_n)$ be a homogeneous polynomial of total degree d. Consider

$$q(X_1, \cdots, X_{d^i+1}) = p(X_1, \cdots, X_{d^i+1}, 0, \cdots, 0).$$

Then q is a homogeneous polynomial of degree d in $d^i + 1$ variables. If $q = 0$, then $(1, \cdots, 1, 0, \cdots, 0)$ is a non-zero root of p. Otherwise, if $\bar{a}$ is a non-zero root of $q(X_1, \cdots, X_{d^i+1}), (\bar{a}, 0, \cdots, 0)$ is a non-zero root of $p(X_1, \cdots, X_{d^i+1}, \cdots, X_n)$. Hence, $\mathbb{F}$ is a $C_i(d)$-field if and only if every homogeneous polynomial $p(X_1, \cdots, X_{d^i+1})$ of degree d over $\mathbb{F}$ in $d^i + 1$ variables has a non-zero root.

Proposition 7.2.6 *Let* $\{\mathbb{F}_i : i \in I\}$ *be a family of fields and* $\mathcal{U}$ *an ultrafilter on* I. *Then, the ultraproduct* $\mathbb{F}(\mathcal{U})$ *is a* $C_i(d)$ *field if and only if* $\{l \in I : \mathbb{F}_l$ *is a* $C_i(d)$ *field*$\}$ $\in \mathcal{U}$.

Proof Let

$$A = \{l \in I : \mathbb{F}_l \text{ is a } C_i(d) \text{ field}\} \in \mathcal{U}.$$

Suppose $p(X_1, \cdots, X_m)$ is a homogeneous polynomial over $\mathbb{F}(\mathcal{U})$ of degree $m > d^i$. Let $c^1 = [(c_l^1)], \cdots, c^k = [(c_l^k)]$ be all the coefficients of p in some order. For each $l \in A$, let p_l be the homogeneous polynomial over $\mathbb{F}_l$ obtained from p by replacing each c^j by c_l^j. Let $a_l \in \mathbb{F}_l$ be a non-zero root of p_l. For $l \notin I$, take $a_l = 0$. By Theorem 2.1.3, $[(a_l)] \neq 0$ and is a root of p.

Now assume that $A \notin \mathcal{U}$. Since $\mathcal{U}$ is an ultrafilter, $A^c \in \mathcal{U}$. For each $l \in A^c$, get a homogeneous polynomial $p_l(X_1, \cdots, X_{d^i+1})$ over $\mathbb{F}_l$ that has no non-zero root. Assume that all monomials in $degree(d)$ occur in each p_l, may be some with coefficient 0. Let $c_l^1, \cdots, c_l^k$ be all the coefficients of p_l in some order. Define $c_l^j = 0$ for $l \in A$. Set $c^j = [(c_l^j)] \in \mathbb{F}(\mathcal{U})$.

Let p be the corresponding homogeneous polynomial over $\mathbb{F}(\mathcal{U})$ obtained from p_l with c_l^j replaced by c^j. If possible, suppose p has a non-zero root $a = ([(a_l^1)], \cdots, [(a_l^m)])$, where $m = d^i + 1$. Since $a \neq 0$, by Theorem 2.1.3, $\cup_{j=1}^m \{l \in I : a_l^j \neq 0\} \in \mathcal{U}$. Since $\mathcal{U}$ is maximal,

$$A_j = \{l \in I : a_l^j \neq 0\} \in \mathcal{U}$$

for some $1 \leq j \leq m$. So, $A_j \cap A^c \neq \emptyset$. Take an $l \in A_j \cap A^c$. Then p_l has a non-zero root which is a contradiction. □

7.3 Ax–Kochen Theorem on Artin's Conjecture

It was known that for every prime p, the field of formal Laurentz series $\mathbb{F}_p((X))$ over $\mathbb{F}_p$ is a $C_2(d)$ field for every $d \geq 1$. Since $\mathbb{Q}_p$ and $\mathbb{F}_p((X))$ have many algebraic properties in common, Artin conjectured that for all primes p, $\mathbb{Q}_p$ is a $C_2(d)$ field for every $d \geq 1$. This is not exactly correct. However, using model-theoretic methods, Ax–Kochen showed that for every $d \geq 1$, $\mathbb{Q}_p$ is a $C_2(d)$ field for all but finitely many primes p. In this section we are going to prove Ax–Kochen theorem.

Lemma 7.3.1 *Let* G *be a torsion-free abelian group having a non-trivial cyclic subgroup* H *such that* $[G : H] < \infty$. *Then* G *is cyclic.*

Proof We prove the result by induction on $[G : H]$. Suppose $[G : H] = n > 1$ and the result holds for all integers less than n. Let g be a generator of H. Take a $x \in G \setminus H$. By our hypothesis, there exists $a > 1$ such that $x^a \in H$. Suppose $b \in \mathbb{Z}$ is such that $x^a = g^b$. Let $d = (a, b)$. Set $y = x^{a/d} g^{-b/d}$. Then $y^d = e_g$. Since G is torsion free, $y = e_g$. Replacing a by a/d and b by b/d if necessary, without any loss of generality, we assume that $(a, b) = 1$

Now get $p, q \in \mathbb{Z}$ such that $pa + qb = 1$. Set $h = g^p x^q$. Then

$$h^a = g^{ap} x^{aq} = g^{ap+bq} = g.$$

And

$$h^b = g^{bp} x^{bq} = x^{ap} x^{bq} = x^{ap+bq} = x.$$

Thus, $H = \langle g \rangle$ is a proper subgroup of the cyclic subgroup $H' = \langle h \rangle$ of G. Therefore, $[G : H'] < n$. Hence, G is cyclic by the induction hypothesis. $\square$

Theorem 7.3.2 *Let $(\mathbb{K}_1, V_1, v_1)$ and $(\mathbb{K}_2, V_2, v_2)$ be Henselian-valued fields with same residue fields of characteristic 0 and the same value group Γ. Assume that $\mathbb{K}_2$ is $\aleph_1$-saturated. Suppose $\mathbb{F}$ is a countable, Henselian subfield of $\mathbb{K}_1$ with respect to v_1 with $v_1(\mathbb{F})$ pure in Γ (i.e., $\Gamma/v_1(\mathbb{F})$ is torsion-free) and $\sigma : \mathbb{F} \to \mathbb{K}_2$ a value preserving embedding such that $[a] = [\sigma(a)]$ for every $a \in \mathbb{F} \cap V_1$. Then for every $b_1, \cdots, b_k \in \mathbb{K}_1$, there exists a countable subfield $\mathbb{F}' \supset \mathbb{F}(b_1, \cdots, b_k)$ of $\mathbb{K}_1$ which is Henselian with respect to v_1 such that $v_1(\mathbb{F}')$ is pure in Γ and to which σ admits a value preserving extension $\sigma' : \mathbb{F}' \to \mathbb{K}_2$ such that $[a] = [\sigma'(a)]$ for every $a \in \mathbb{F}' \cap V_1$.*

Proof By Proposition C.7.9, we get a countable subfield $\mathbb{F}'$ of $\mathbb{K}_1$ such that $\mathbb{F}' \supset \mathbb{F}(b_1, \cdots, b_k)$ and $v_1(\mathbb{F}')$ is pure in Γ. Since $\mathbb{K}_1$ is Henselian and since Henselization of a countable-valued field is countable and an immediate extension (Proposition C.7.2), without any loss of generality, we assume that $\mathbb{F}'$ is Henselian with respect to v_1. Set $\Gamma' = v_1(\mathbb{F}')$.

By Zorn's lemma, there exists a maximal $(\mathbb{H}, \sigma')$ where $\mathbb{F} \subset \mathbb{H} \subset \mathbb{F}'$ and $\sigma' : \mathbb{H} \to \mathbb{K}_2$ a value preserving embedding extending σ such that $[a] = [\sigma'(a)]$ for every $a \in \mathbb{H} \cap V_1$ and $v_1(\mathbb{H})$ pure in Γ. If possible, suppose $\mathbb{H}$ is not Henselian with respect to v_1. Then, $\sigma'(\mathbb{H})$ is not Henselian. Let $\mathbb{H}^h \subset \mathbb{K}_1$ be the Henselization of $\mathbb{H}$ and $\sigma'(\mathbb{H})^h \subset \mathbb{K}_2$ be that of $\sigma'(\mathbb{H})$. By the uniqueness of Henselization of valued fields (see Sect. C.7), it follows that there is a value preserving isomorphism $\tau : \mathbb{H}^h \to \sigma'(\mathbb{H})^h \subset \mathbb{K}_2$ extending σ' such that $[a] = [\tau(a)]$ for every $a \in \mathbb{H}^h \cap V_1$. This contradicts the maximality of $(\mathbb{H}, \sigma')$. So, $\mathbb{H}$ is Henselian with respect to v_1.

To complete the proof, we show that $\mathbb{H} = \mathbb{F}'$. We first show that $\mathbb{F}'$ is an immediate extension of $\mathbb{H}$.

Step 1. $\mathbb{F}'^\sim = \mathbb{H}^\sim$.

If possible, suppose $\mathbb{F}'^\sim \neq \mathbb{H}^\sim$. Let $[x] \in \mathbb{F}'^\sim \setminus \mathbb{H}^\sim$.

Let $[x]$ be transcendental over $\mathbb{H}^\sim$. Then, x is transcendental over $\mathbb{H}$. Let $y \in V_2$ be such that $[y] = [x]$. Since the residue fields of $\mathbb{K}_1$ and $\mathbb{K}_2$ are the same, such a y exists in $\mathbb{K}_2$. If possible, suppose $[y]$ is algebraic over $\sigma'(\mathbb{H})^\sim$. Let

$$[y]^m + [\sigma'(a_{m-1})][y]^{m-1} + \cdots + [\sigma'(a_1)][y] + [\sigma'(a_0)] = 0.$$

Since for every $a \in \mathbb{H} \cap V_1$, $[a] = [\sigma'(a)]$ and $[y] = [x]$, we then have

$$[x]^m + [a_{m-1}][x]^{m-1} + \cdots + [a_1][x] + [a_0] = 0.$$

This contradicts that $[x]$ is transcendental over $\mathbb{H}^\sim$. So, $[y]$ is transcendental over $\sigma'(\mathbb{H})^\sim$. Also note that $v_1(x) = v_2(y) = 0$. Therefore, by Theorem C.2.2,

$$v_1(\textstyle\sum_{i=0}^n a_i x^i) = \min\{v_1(a_i) : 0 \le i \le n\}$$
$$= \min\{v_2(\sigma'(a_i)) : 0 \le i \le n\}$$
$$= v_2(\textstyle\sum_{i=0}^n \sigma'(a_i) y^i)$$

This implies that the isomorphism $\tau : \mathbb{H}(x) \to \sigma'(\mathbb{H})$ that extends σ' and sends x to y is value preserving.

To arrive at a contradiction, we now need to show that $[f/g] = [\tau(f/g)]$ for every $f/g \in \mathbb{H}(x) \cap V_1$. First take a polynomial

$$f(x) = \sum_{i=0}^n a_i x^n \in \mathbb{H}[x] \cap V_1.$$

Then $v_1(f) = \min\{v_1(a_i) : 0 \le i \le n\} \ge 0$. Hence, each $a_i \in V_1$. Since $z \to [z]$ is a homomorphism, it now follows that $[f] = [\tau(f)]$. It also follows that whenever $f, g \in \mathbb{H}[x] \cap V_1^\times$, $[f/g] = [\tau(f/g)]$.

Next let $f/g \in \mathbb{H}(x) \cap M_1$. Since τ is value preserving, $\tau(f/g) \in \sigma'(\mathbb{H})(y) \cap M_2$. Hence, $[f/g] = 0 = [\tau(f/g)]$.

Now take $\frac{f}{g} \in \mathbb{H}(x) \cap V_1^\times$. Let $f(x) = \sum_{i=0}^n a_i x^i$ and $g(x) = \sum_{j=0}^n b_j x^j$. Hence, by Theorem C.2.2,

$$v_1(f) = \min\{v_1(a_i) : 0 \le i \le n\} = \min\{v_1(b_j) : 0 \le j \le m\} = v_1(g).$$

Let $v_1(a_I) = \min\{v_1(a_i) : 0 \le i \le n\}$ and $v_1(b_J) = \min\{v_1(b_j) : 0 \le j \le m\}$. Then

$$[f/g] = [a_I/b_J][f'/g'],$$

where $f', g' \in \mathbb{H}[x] \cap V_1^\times$. Now note that $[f'/g'] = [\tau(f'/g')]$. So,

$$[\tau(f/g)] = [\sigma'(a_I/b_J)][\tau(f'/g')] = [a_I/b_J][f'/g'] = [f/g].$$

Thus we have proved $[x]$ has to be algebraic over $\mathbb{H}^\sim$.

We now show that $[x]$ is not algebraic over $\mathbb{H}^\sim$ either. Suppose not. Let

$$f(X) = X^m + a_{m-1} X^{m-1} + \cdots + a_1 X + a_0 \in (\mathbb{H} \cap V_1)[X]$$

be such that $f^\sim$ is the minimal polynomial of $[x]$ over $\mathbb{H}^\sim$.

In case 2 of Proposition C.7.9, using Hensel's lemma we showed that f has a root in $\mathbb{F}'$ with residue class $[x]$. Without any loss of generality, we take x to be such a root. Clearly, $v_1(x) = 0$. In case 2 of Proposition C.7.9, we showed that $v_1(\mathbb{H}(x)) = v_1(\mathbb{H}[x]) = v_1(\mathbb{H})$.

Arguing as above, using Hensel's lemma, $\sigma'(f)$ has a root with residue class $[x]$. We take y to be such a root. So, we have a canonical isomorphism $\tau : \mathbb{H}(x) \rightarrow \sigma'(\mathbb{H})(y)$. Since $\mathbb{H}$ is Henselian, there is exactly one valuation on every algebraic extension of $\mathbb{H}$ extending v_1. Hence, τ is also value preserving.

We have arrived at a contradiction in this case too.

Step 2. $v_1(\mathbb{F}') = v_1(\mathbb{H})$.

Suppose not. Get $x \in \mathbb{F}'$ such that $v_1(x) \notin v_1(\mathbb{H})$. Replacing x by x^{-1} if necessary, without any loss of generality, we assume that $x \in V_1$. Since $v_1(x) \notin v_1(\mathbb{H})$, $0 < v_1(x) < \infty$. Also note that $[x] = 0$.

Since $v_1(\mathbb{H})$ is pure in Γ, for no non-zero integer l, $lv_1(x) \in v_1(\mathbb{H})$. So, for $0 \leq i \neq j < \infty$ and $a_i, a_j \in \mathbb{H}$, $v_1(a_i x^i) \neq v_1(a_j x^j)$. Hence, by Lemma C.1.9,

$$v_1\left(\sum_{i=0}^{m} a_i x^i\right) = \min\{v_1(a_i x^i) : 0 \leq i \leq m\}$$

whenever $a_0, \cdots, a_m \in \mathbb{H}$. Further, if not all $a_0, \cdots, a_m$ are 0,

$$v_1\left(\sum_{i=0}^{m} a_i x^i\right) = \min\{v_1(a_i) + iv_1(x) : 0 \leq i \leq m\} < \infty.$$

Hence, x is not algebraic over $\mathbb{H}$.

Since $\mathbb{K}_1$ and $\mathbb{K}_2$ have the same value group, there exists a $y \in \mathbb{K}_2$ such that $v_2(y) = v_1(x) > 0$. In particular, $[y] = 0 = [x]$. Since $v_1(\mathbb{H}) = v_2(\sigma'(\mathbb{H}))$, by the above argument, for $0 \leq i \neq j < \infty$ and $b_i, b_j \in \sigma'(\mathbb{H})$, $v_2(b_i y^i) \neq v_2(b_j y^j)$. As before, this proves that y is transcendental over $\sigma'(\mathbb{H})$ and $v_1(f(x)) = v_2(\sigma'(f)(y))$ for every $f \in \mathbb{H}[x]$. Thus, the isomorphism σ'' from $\mathbb{H}(x)$ to $\sigma'(\mathbb{H})(y)$ sending x to y and $a \in \mathbb{H}$ to $\sigma'(a)$ is value preserving. As in case of Step 1, it follows that $[\frac{f}{g}] = [\sigma''(\frac{f}{g})]$ for every $\frac{f}{g} \in \mathbb{H}(x) \cap V_1$.

However, $v_1(\mathbb{H}(x))$ may not be pure in Γ'. Take

$$\mathbb{H}^* = \{z \in \mathbb{F}' : z \text{ is algebraic over } \mathbb{H}(x)\}.$$

We first show that $v_1(\mathbb{H}^*)$ is pure in Γ'. Suppose there exists $a \in \mathbb{F}'^{\times}$, $b \in \mathbb{H}^*$ and $m > 1$ such that $mv_1(a) = v_1(b)$. Then $v_1(\frac{a^m}{b}) = 0$. Since the residue fields of $\mathbb{H}$ and $\mathbb{F}'$ are the same (Step 1), there exists a $c \in \mathbb{H} \cap V_1$ such that $v_1(c) = 0$ and $[\frac{a^m}{b}] = [c]$.

Now consider the polynomial

$$f(Z) = Z^m - \frac{a^m}{bc} \in (\mathbb{F}' \cap V_1)[Z].$$

Then

$$v_1(f(1)) = v_1(c - \frac{a^m}{b}) - v_1(c) = v_1(c - \frac{a^m}{b}) > 0,$$

and

$$v_1(f'(1)) = v_1(m) = 0$$

because $char(\mathbb{K}_1^\sim) = 0$. Since $\mathbb{F}'$ is Henselian, by Hensel's lemma, f has a root, say z, in $\mathbb{F}'$ with residue 1. In particular, $v_1(z) = 0$. Now $(\frac{a}{z})^m = bc \in \mathbb{H}^*$. Thus, $\frac{a}{z} \in \mathbb{F}'$ and is algebraic over $\mathbb{H}^*$. So, $\frac{a}{z} \in \mathbb{H}^*$. But then $v_1(a) = v_1(\frac{a}{z}) \in v_1(\mathbb{H}^*)$. Thus, we have proved that $v_1(\mathbb{H}^*)$ is pure in Γ'.

To arrive at a contradiction, we shall extend σ' to a value preserving embedding τ to $\mathbb{H}^*$ such that $[\tau(a)] = [a]$ for every $a \in \mathbb{H}^* \cap V_1$. We shall use $\aleph_1$-saturability of $\mathbb{K}_2$ to achieve this. Enumerate $\mathbb{H}^* = \{a_n : n \in \mathbb{N}\}$. We take a type over a countable $A \subset \mathbb{K}_2$ consisting of the following formulas:

$$X_i + X_j = X_k \text{ if } a_i + a_j = a_k,$$

$$X_i \cdot X_j = X_k \text{ if } a_i \cdot a_j = a_k,$$

$$X_n = \sigma'(a_n) \text{ if } a_n \in \mathbb{H}$$

$$v_2(X_n) = v_1(a_n)$$

and

$$[X_n] = [a_n] \text{ if } v_1(a_n) \geq 0.$$

Suffices to show that every finite set of these formulas is realized in $\mathbb{K}_2$. This will follow if we show the following: Let $\mathbb{H}(x) \subset \mathbb{H}' \subset \mathbb{H}^*$, $[\mathbb{H}' : \mathbb{H}(x)] < \infty$. Then σ' admits a value preserving extension τ to $\mathbb{H}'$.

Let $\mathbb{H}' = \mathbb{H}(x, \alpha)$ be a finite extension of $\mathbb{H}(x)$. By Chevalley's fundamental inequality, $[v_1(\mathbb{H}') : v_1(\mathbb{H}(x))] < \infty$. Since $v_1(\mathbb{H})$ is pure in Γ',

$$v_1(\mathbb{H}(x)) = v_1(\mathbb{H}) \oplus \mathbb{Z}v_1(x).$$

Also

$$v_1(\mathbb{H}') = v_1(\mathbb{H}) \oplus G$$

with G an infinite torsion-free abelian group with $[G : \mathbb{Z}v_1(x)] < \infty$. Hence, $G = \mathbb{Z}v_1(y)$ for some $y \in \mathbb{H}'$ by Lemma 7.3.1. Since $\mathbb{H}$ is Henselian with residue field of characteristic 0, it is algebraically maximal by Theorem C.7.8. So, y is transcendental over $\mathbb{H}$. Hence,

$$v_1(\mathbb{H}(y)) = v_1(\mathbb{H}) \oplus \mathbb{Z}v_1(y) = v_1(\mathbb{H}')$$

by Theorem C.2.3. Thus, by Step 1, $\mathbb{H}'$ is an immediate extension of $\mathbb{H}(y)$. Note that $\mathbb{H}(y)$ must contain x. So, $\mathbb{H}(x, \alpha)$ is an algebraic extension of $\mathbb{H}(y)$. Hence, $\mathbb{H}'$ is contained in the Henselization of $\mathbb{H}(y)$. Replacing x by y in the argument contained in the first part of step 1, we see σ' can be extended to a value preserving extension τ to the Henselization $(\mathbb{H}(y))^h$ of $\mathbb{H}(y)$. We have thus contradicted the maximality of $(\mathbb{H}, \sigma')$

We have now proved that $\mathbb{F}'$ is an immediate extension of $\mathbb{H}$. If possible, suppose there exists an $x \in \mathbb{F}' \setminus \mathbb{H}$. Without any loss of generality, we assume that $v_1(x) \geq 0$. Since $\mathbb{H}$ is Henselian with residue field of characteristic 0, it is algebraically maximal by Theorem C.7.8. So, x is transcendental over $\mathbb{H}$. Enumerate $\mathbb{H} = \{a_n : n \in \mathbb{H}\}$ with $a_0 = 0$.

Since the value groups of $\mathbb{F}'$ and $\mathbb{H}$ are the same (Step 2) and $x \notin \mathbb{H}$, for every n, there is a $b_n \in \mathbb{H}^\times$ such that

$$v_1(x - a_n) = v_1(b_n).$$

Claim 1. There exists a $y \in \mathbb{K}_2$ such that for every n,

$$v_1(b_n) = v_1(x - a_n) = v_2(y - \sigma'(a_n)). \qquad (*)$$

Since $\mathbb{K}_2$ is $\aleph_1$-saturated, it is sufficient to show that for every m, there is a $y \in \mathbb{K}_2$ satisfying $(*)$ for every $n \leq m$. Let

$$v_1(x - a_p) = \min\{v_1(x - a_i) : 0 \leq i \leq m\}.$$

Since the residue fields of $\mathbb{F}'$ and $\mathbb{H}$ are the same, there exists a $c \in \mathbb{H} \cap V_1$ such that $[\frac{x - a_p}{b_p}] = [c]$. Hence,

$$v_1(x - a_p - b_p c) > v_1(b_p) = v_1(x - a_p) \geq v_1(x - a_i)$$

for every $0 \leq i \leq m$. Set $d = a_p - b_p c \in \mathbb{H}$. In particular, $v_1(x - d) > v_1(x) \geq 0$. Hence, $v_1(d) \geq 0$ and $[x] = [d]$.
    For $0 \leq i \leq m$,

$$v_1(d - a_i) = v_1((x - a_i) - (x - d_i)) = v_1(x - a_i) = v_1(b_i).$$

Now take $y = \sigma'(d)$. Thus, by $\aleph_1$-saturability of $\mathbb{K}_2$ there is a $y \in \mathbb{K}_2$ satisfying $(*)$ for all n.

Since $y \notin \sigma'(\mathbb{H})$ and $\sigma'(\mathbb{H})$ is Henselian and finitely ramified, it is algebraically maximal by Theorem C.7.8. Hence, y will necessarily be transcendental over $\sigma'(\mathbb{H})$.

Claim 2. For every $f(x) \in \mathbb{H}[x]$, $v_1(f(x)) = v_2(\sigma'(f)(y))$.

Then the isomorphism τ from $\mathbb{H}(x)$ to $\sigma'(\mathbb{H})(y)$ sending x to y and $a \in \mathbb{H}$ to $\sigma'(a)$ will necessarily be a value preserving extension of σ'. This will contradict the maximality of $(\mathbb{H}, \sigma')$ and will finally prove our theorem.

We prove claim 2 by induction on $d = degree(f)$. Claim 1 proves the hypothesis for $d = 1$. Let the statement be true for all $p(x) \in \mathbb{H}[x]$ of degree less than d and $f(x) \in \mathbb{H}[x]$ be an irreducible polynomial of degree d. Set

$$\mathbb{H}_1 = \mathbb{H}[x]/(f).$$

Since $\mathbb{H}_1$ is just the space of all polynomials in $\mathbb{H}[x]$ of degree less than d, we can restrict v_1 to it. Now let $g, h \in \mathbb{H}[x]$ be of degree less than d. Write

$$gh = qf + r$$

with $degree(r) < d$. If always $v_1(r(x)) = v_1(gh(x))$, then $v_1|\mathbb{H}_1$ would be a valuation. It will follow that $\mathbb{H}$ has a proper, algebraic immediate extension. But such an extension does not exist because $\mathbb{H}$ is Henselian. Hence, there exist $g, h \in \mathbb{H}_1$ such that

$$v_1(r(x)) \neq v_1(gh(x)) = v_1(g(x)) + v_1(h(x)).$$

Hence,

$$\begin{aligned}
v_1(f(x)) &= -v_1(q(x)) + v_1(r(x) - gh(x)) \\
&= -v_1(q(x)) + \min\{v_1(r(x)), v_1(gh(x))\}.
\end{aligned}$$

By induction hypothesis

$$v_2(\sigma'(r)(y)) \neq v_2(\sigma'(g)(y)) + v_2(\sigma'(h)(y)) = v_2(\sigma'(gh)(y)).$$

Therefore,

$$\begin{aligned}
v_2(\sigma'(qf)(y)) &= v_2(\sigma'(r)(y) - \sigma'(gh)(y)) \\
&= \min\{v_2(\sigma'(r)(y)), v_2(\sigma'(gh)(y))\}
\end{aligned}$$

So,

$$\begin{aligned}
v_2(\sigma'(f)(y)) &= -v_2(\sigma'(q)(y)) + \min\{v_2(\sigma'(r)(y)), v_2(\sigma'(gh)(y))\} \\
&= -v_1(q(x)) + \min\{v_1(r(x), v_1(gh(x))\} = v_1(f).
\end{aligned}$$

The first equality holds by the induction hypothesis. □

Theorem 7.3.3 (Ax–Kochen) *Let $d \geq 1$. Then*

$$A_d = \{p \in \mathbb{P} : \mathbb{Q}_p \text{ is not a } C_2(d) \text{ field}\},$$

where $\mathbb{P}$ denotes the set of all prime numbers, is finite.

Proof If possible, suppose A_d is infinite. Let $\mathcal{U}$ be a free ultrafilter on $\mathbb{P}$ containing A_d. Consider

$$\mathbb{K}_1 = \times_p \mathbb{Q}_p / \mathcal{U} \text{ and } \mathbb{K}_2 = \times_p \mathbb{F}_p((X)) / \mathcal{U}.$$

Then both $\mathbb{K}_1$ and $\mathbb{K}_2$ are $\aleph_1$-saturated Henselian fields (Theorem 4.3.13) with same residue field $\times_p \mathbb{F}_p / \mathcal{U}$ of characteristic 0 and same value group $\Gamma = \times_p \mathbb{Z} / \mathcal{U}$ which is torsion free (Corollary 7.2.5). Since $A_d \in \mathcal{U}$, $\mathbb{K}_1$ is not a $C_2(d)$ field (Proposition 7.2.6). It is well known that $\mathbb{F}_p((X))$ is a $C_2(d)$ field for all primes p. Hence, $\mathbb{K}_2$ is a $C_2(d)$ field (Proposition 7.2.6).

From the common properties of $\mathbb{K}_1$ and $\mathbb{K}_2$ listed above and using $\mathbb{K}_2$ is a $C_2(d)$ field, we are going to prove that $\mathbb{K}_1$ is a $C_2(d)$ field. Thus we shall arrive at a contradiction.

Let v_i denote the valuation on $\mathbb{K}_i$, $i = 1, 2$. Note that both v_1 and v_2 are trivial on the prime field $\mathbb{Q}$. So, $\mathbb{Q}$ is Henselian with respect to both v_1 and v_2, the identity isomorphism $id : (\mathbb{Q}, v_1) \to (\mathbb{Q}, v_2)$ is value preserving with $\Gamma / v_1(\mathbb{Q}) = \Gamma$ torsion free.

Let $f \in \mathbb{K}_1[X_1, \cdots, X_{d^2+1}]$ be a homogeneous polynomial of degree d. Assume that all monomials in $X_1, \cdots, X_{d^2+1}$ of degree d occur in f, possibly some with coefficient 0. We now proceed to show that f has a non-zero root.

Let $a_1, \cdots, a_n$ be all the coefficients of f in some order. By Theorem 7.3.2, there is a countable Henselian subfield $\mathbb{F}_1 \supset \mathbb{Q}(a_1, \cdots, a_n)$ such that $\Gamma / v_1(\mathbb{F}_1)$ is torsion free and there exists a value-preserving extension $\sigma : \mathbb{F}_1 \to \mathbb{K}_2$ of id.

Set $\mathbb{F}_2 = \sigma(\mathbb{F}_1)$ and $g = \sigma(f)$. Then g is a homogeneous polynomial over $\mathbb{K}_2$ of degree d in $X_1, \cdots, X_{d^2+1}$. Since $\mathbb{K}_2$ is a $C_2(d)$ field, g has a non-zero solution, say $x_1, \cdots, x_{d^2+1}$. By the last theorem, there is an embedding $\tau : \mathbb{F}_2(x_1, \cdots, x_{d^2+1}) \to \mathbb{K}_1$ extending σ^{-1}. Then $\tau(x_1), \cdots, \tau(x_{d^2+1})$ is a non-zero root of f. We have arrived at a contradiction now. $\square$

7.4 Quantifier Elimination and Model Completeness of Valued Fields

We saw earlier that given any valuation subring V of a field $\mathbb{F}$, there exists an extension of V to the algebraic closure $\overline{\mathbb{F}}$. Further, such an extension is essentially unique, i.e. if V_1 and V_2 are two extensions of V to $\overline{\mathbb{F}}$, there is an $\mathbb{F}$-automorphism α of $\overline{\mathbb{F}}$ such that $\alpha(V_1) = V_2$. It follows that if $|$ is a valuation divisibility relation on D, then there is (essentially a unique) valuation divisibility relation on $\overline{\mathbb{F}}$ extending $|$.

We are now in a position to prove quantifier elimination for theory of algebraically closed fields with nontrivial valuation divisibility relation.

Theorem 7.4.1 *The theory of algebraically closed fields with non-trivial valuation divisibility relation admits quantifier elimination.*

Proof Let $(\mathbb{F}_1, |_1)$, $(\mathbb{F}_2, |_2)$ be two algebraically closed fields with non-trivial valuation divisibility relations, and $(D, |)$ a common substructure. Let $\mathbb{F}$ be the quotient field of D. Note that there is an $\mathbb{F}$-isomorphism f of the algebraic closure of $\mathbb{F}$ in $\mathbb{F}_1$ onto the algebraic closure of $\mathbb{F}$ in $\mathbb{F}_2$ that takes $|_1$ to $|_2$. It follows that we can assume that $(\mathbb{F}, |)$ is a common substructure of $(\mathbb{F}_1, |_1)$ and $(\mathbb{F}_2, |_2)$ with $\mathbb{F}$ also algebraically closed.

Let $\varphi[x, \bar{a}]$ be an open formula with parameters $\bar{a} \in D \subset \mathbb{F}$. Suppose there exists a $t \in \mathbb{F}_1$ such that $\mathbb{F}_1 \models \varphi[t, \bar{a}]$. We need to produce a $s \in \mathbb{F}_2$ such that $\mathbb{F}_2 \models \varphi[s, \bar{a}]$. If $t \in \mathbb{F}$, we simply take $s = t$. So, we assume that $t \in \mathbb{F}_1 \setminus \mathbb{F}$. Then t is transcendental over $\mathbb{F}$. Set $\mathbb{F}' = \mathbb{F}(t)$ and $|' = |_1|\mathbb{F}'$. Let $\kappa = |\mathbb{F}|$. Since there exists a κ^+-saturated elementary extension of $\mathbb{F}_2$, without any loss of generality, we assume that $\mathbb{F}_2$ is κ^+-saturated.

Let V, V' and V_2 denote the valuation subrings of $\mathbb{F}$, $\mathbb{F}'$ and $\mathbb{F}_2$ respectively. We shall produce an $\mathbb{F}$-monomorphism from $(\mathbb{F}', V')$ into $(\mathbb{F}_2, V_2)$ respectively. This will complete our proof.

Let Γ, Γ' and Γ_2 denote the value groups, v, v' and v_2 the valuations and $\mathbb{F}^{\sim} = V/M$, $\mathbb{F}'^{\sim} = V'/M'$ and $\mathbb{F}_2^{\sim} = V_2/M_2$ denote the residue fields of $(\mathbb{F}, V)$, $(\mathbb{F}', V')$ and $(\mathbb{F}_2, V_2)$ respectively.

Case 1 : $\mathbb{F}^{\sim} \neq \mathbb{F}'^{\sim}$. Note that there is a canonical embedding of $\mathbb{F}^{\sim}$ into $\mathbb{F}'^{\sim}$. Take $[x] \in \mathbb{F}'^{\sim} \setminus \mathbb{F}^{\sim}$. So, $x \notin \mathbb{F}$. In particular, x is transcendental over $\mathbb{F}$. Also, $v'(x) = 0$. Hence, by Theorem C.2.2, for every $a_0, \cdots, a_m \in \mathbb{F}$, $v'(\sum_i a_i x^i) = \min_i v(a_i)$.

Using saturability of $\mathbb{F}_2$, we now produce a $x_2 \in \mathbb{F}_2$ transcendental over $\mathbb{F}$. Towards showing this consider

$$\Phi(x) = \{1|x \wedge x|1\} \cup \{1|a - x \wedge a - x|1 : a \in V^{\times}\}.$$

Since $\mathbb{F}$ is algebraically closed, $\mathbb{F}^{\sim}$ is algebraically closed, and so infinite. Hence, $\Phi[x]$ is finitely satisfiable in $\mathbb{F}$. So, $\Phi(x)$ is satisfiable in $\mathbb{F}_2$, say by x_2. Then, $[x_2] \in \mathbb{F}_2^{\sim} \setminus \mathbb{F}^{\sim}$. As before, it follows that $x_2 \notin \mathbb{F}$ and so is transcendental over $\mathbb{F}$. Hence, there is an $\mathbb{F}$-isomorphism g from $\mathbb{F}(x)$ onto $\mathbb{F}(x_2)$ that sends x to x_2. Like before, since $[x_2]$ is transcendental over $\mathbb{F}^{\sim}$, we see that

$$v_2(\sum_i a_i x_2^i) = \min_i v(a_i).$$

It follows that g preserves the valuation divisibility relation.

Since $x \in \mathbb{F}(t)$, there exist polynomials $p(X), q(X) \in \mathbb{F}[X]$ with $q \neq 0$ such that $x = \frac{p(t)}{q(t)}$. So, $p(t) - xq(t) = 0$. Thus, t is algebraic over $\mathbb{F}(x)$. Hence, $\mathbb{F}' = \mathbb{F}(t)$ is algebraic over $\mathbb{F}(x)$. Since $\mathbb{F}_2$ is algebraically closed, we now have an $\mathbb{F}$-monomorphism from $\mathbb{F}'$ into $\mathbb{F}_2$. Since extensions of a valuation subring to the algebraic closure is essentially unique, we can easily modify the $\mathbb{F}$-monomorphism so that the valuation divisibility relations are also preserved.

Case2 : $\Gamma \neq \Gamma'$ Recall that we can assume $\Gamma = \mathbb{F}^\times / V^\times$ and $\Gamma' = \mathbb{F}'^\times / V'^\times$. Clearly, there is a natural, order preserving embedding of Γ into Γ'. Now take $x \in V'$ such that $v'(x) \in \Gamma' \setminus \Gamma$. So, $x \notin \mathbb{F}$. By Theorem C.2.3, for every $a_0, \cdots, a_n \in \mathbb{F}$,

$$v'(\sum_i a_i x^i) = \min_i \{v(a_i) + i v'(x)\}.$$

Now consider the following:

$$\Phi(v_0) = \{\neg v_0 | a : v(a) < v'(x) \ \& \ a \in \mathbb{F}\} \cup \{\neg a | v_0 : v'(x) < v(a) \ \& \ a \in \mathbb{F}\}.$$

Since Γ_2 is divisible and $|_2$ is non-trivial, $\Phi(v_0)$ is finitely satisfiable in $\mathbb{F}_2$. Hence, by saturability, there exists an $x_2 \in \mathbb{F}_2$ such that for all $a \in \mathbb{F}$, $v(a) < v'(x) \rightarrow v(a) < v_2(x_2)$ and $v'(x) < v(a) \rightarrow v_2(x_2) < v(a)$. Since $v(x) \notin \Gamma$, it follows that $v_2(x_2) \notin \Gamma$. As before, for $a_0, \cdots, a_n \in \mathbb{F}$,

$$v_2(\sum_i a_i x_2^i) = \min_i \{v(a_i) + i v_2(x)\}.$$

Since $\mathbb{F}$ is algebraically closed, x_2 is transcendental over $\mathbb{F}$. By the above observation the canonical $\mathbb{F}$-isomorphism from $\mathbb{F}(x)$ onto $\mathbb{F}_2$ preserves the valuations. The proof in this case is completed as in case 1.

Case3 : $\mathbb{F}^\sim = \mathbb{F}'^\sim \ \& \ \Gamma = \Gamma'$. In this case, we prove that there exists a $t_2 \in \mathbb{F}_2$ such that

$$\forall a \in \mathbb{F}(v'(t - a) = v_2(t_2 - a)).$$

Assuming this we complete the proof first. Let $g : \mathbb{F}(t) \rightarrow \mathbb{F}_2(t_2)$ be the $\mathbb{F}$-isomorphism with $g(t) = t_2$. Now take $a_0, \cdots, a_n \in \mathbb{F}$ with $n > 0$ and $a_n \neq 0$. Since $\mathbb{F}$ is algebraically closed, write

$$\sum_i a_i X^i = a_n (X - b_1) \cdots (X - b_n),$$

$b_1, \cdots, b_n \in \mathbb{F}$.
   Then

$$v'(\sum_i a_i t^i) = v(a_n) + \sum_i v'(t - b_i) = v(a_n) + \sum_i v_2(t_2 - b_i) = v_2(g(\sum_i a_i t^i)).$$

It follows that g preserves the valuations too and the proof is complete.

It remains to show the existence of a $t_2 \in \mathbb{F}_2$ satisfying the above condition. Since $\Gamma = \Gamma'$, for every $a \in \mathbb{F}$, we get a $t_a \in \mathbb{F}$ such that $v'(t - a) = v(b_a)$. Now consider

$$\Phi(v_0) = \{v_0 - a | t_a \wedge t_a | v_0 - a : a \in \mathbb{F}\}.$$

We now show that Φ is finitely satisfiable in $\mathbb{F}$. Then, by saturability, it is satisfiable in $\mathbb{F}_2$, say by t_2. Clearly t_2 satisfies the above condition.

Take $a_1, \cdots, a_n \in \mathbb{F}$. Let k be such that for $a = a_k$, $v(t_a) = \max_i v(t_{a_i})$. Since $v'(t - a) = v(t_a)$, $(t - a)t_a^{-1} \in V'^\times$. Since $\mathbb{F}^\sim = \mathbb{F}'^\sim$, there is a $c \in V$ such that

$$[(t - a)t_a^{-1}] = [c].$$

Set $d = a + ct_a \in \mathbb{F}$. Then

$$v'(\frac{t - d}{t_a}) > 0,$$

i.e.,

$$v'(t - d) > v(t_a).$$

It follows that for all $1 \leq i \leq n$,

$$v'(t - d) > v(t_a) \geq v(t_{a_i}).$$

Hence,

$$v(d - a_i) = v'((t - a_i) - (t - d)) = v'(t - a_i) = v(t_{a_i}).$$

Our proof is complete now. □

Corollary 7.4.2 *The theory T of algebraically closed, non-trivial valued fields is model complete. It is the model companion of the theory of valued fields.*

Proof The first part is a direct consequence of quantifier elimination for T. Let $(\mathbb{F}, V)$ be a valued field. If $V \neq \mathbb{F}$, then recall that V can be extended to the algebraic closure $\overline{\mathbb{F}}$ which is then an algebraically closed non-trivial valued field extending $(\mathbb{F}, V)$. If $V = \mathbb{F}$, first get a non-trivial valuation on $\mathbb{F}(X)$ and take its extension to the algebraic closure of $\mathbb{F}(X)$. □

Corollary 7.4.3 *The theory T of non-trivial, algebraically closed valued fields with fixed characteristic and fixed characteristic of the residue field is complete.*

Proof By quantifier elimination proved in Theorem 7.4.1, it is sufficient to show the existence of a prime structure in all possible cases. If $char(\mathbb{F}) = char(\mathbb{F}^\sim) = 0$, then $\mathbb{Q}$ with trivial valuation is a prime structure. If $char(\mathbb{F}) = char(\mathbb{F}^\sim) = p > 0$, then $\mathbb{F}_p$ with trivial valuation is a prime structure. If $char(\mathbb{F}) = 0$ and $char(\mathbb{F}^\sim) = p > 0$, then $\mathbb{Q}$ with p-adic valuation is a prime structure. □

Appendix A
Set Theory

In this chapter, we present the results and concepts from naive set theory that we shall need. Some of the proofs are omitted. For cardinals and ordinals, the reader may see ([58], Chap. 1) and for infinite combinatorics ([22], Chap. 9).

A.1 Ordinal Numbers

A *well-ordered set* is a linearly ordered set $(W, <)$ such that every non-empty subset A of W has a (unique) least element. *Well-ordering principle*, in notation WOP, is the statement "every set can be well-ordered". Throughout this book, we have assumed WOP.

For any $u \in W$,

$$W(u) = \{v \in W : v < u\}$$

is called an *initial segment* of W.

Proposition A.1.1 *If $(W, <)$ is a well-ordered set and $u \in W$, then there is no order-preserving injection $f : W \to W(u)$.*

Proof Suppose to the contrary an order-preserving injection $f : W \to W(u)$ exists. Set

$$u_0 = u \; \& \; \forall n(u_{n+1} = f(u_n)).$$

Then $A = \{u_0, u_1, \ldots\} \subset W$ is a non-empty set with no least element. This is a contradiction. $\qquad\square$

We have the following two methods of transfinite induction.

Theorem A.1.2 *Let $(W, <)$ be a well-ordered set.*

1. **(Proof by transfinite induction.)** *Suppose for each $u \in W$, P_u is a statement such that whenever P_v holds for all $v < u$, P_u holds. Then for all $u \in W$, P_u holds.*

© Springer Nature Singapore Pte Ltd. 2017
H. Sarbadhikari and S.M. Srivastava, *A Course on Basic Model Theory*,
DOI 10.1007/978-981-10-5098-5

2. (**Definition by transfinite induction**.) *Let X be a set and $F : \mathcal{I}(W) \to X$ a function, where $\mathcal{I}(W)$ is the set of all functions from an initial segment of W to X. Then there is a unique function $G : W \to X$ such that*

$$\forall u \in W(G(u) = F(G|W(u))).$$

The class of all *ordinal numbers* (or simply ordinals) is a class ON such that each $\alpha \in ON$ is a well-ordered set and every well-ordered set is order isomorphic to a unique $\alpha \in ON$. That such a class exists follows from Zermelo–Fraenkel axioms. Ordinal numbers will generally be denoted by α, β and γ with or without suffixes or prefixes.

For ordinals α, β, we write $\alpha < \beta$ if α is order isomorphic to an initial segment of β. This initial segment is necessarily unique. We have the following trichotomy theorem for ordinals.

Proposition A.1.3 *1. For ordinals α, β, exactly one of $\alpha < \beta$, $\alpha = \beta$, $\beta < \alpha$ holds.*
2. Every set of ordinal numbers is well-ordered by $<$.

In view of this proposition, we identify each ordinal α with $\{\beta \in ON : \beta < \alpha\}$ with the ordering as defined above.

Note that two finite well-ordered sets W_1, W_2 are order isomorphic if and only if they have the same number of elements. The ordinals corresponding to finite well-ordered sets are denoted in increasing order by $0, 1, 2, \ldots$. The set $\{0, 1, 2, \ldots\}$ is denoted by ω. Note that by our convention, $0 = \emptyset$ and $n = \{o, \ldots, n-1\}, n \in \omega$.

Let α, β be ordinals. Choose well-ordered sets $(W_1, <_1), (W_2, <_2)$ order isomorphic to α, β respectively with $W_1 \cap W_2 = \emptyset$. Set $W = W_1 \cup W_2$. For $u, v \in W$, define $u < v$ by

1. $u, v \in W_1$ and $u <_1 v$.
2. $u, v \in W_2$ and $u <_2 v$.
3. $u \in W_1$ and $v \in W_2$.

The ordinal corresponding to $(W, <)$ is denoted by $\alpha + \beta$. An ordinal of the form $\alpha + 1$ is called a *successor ordinal*. An ordinal which is not a successor ordinal is called a *limit ordinal*. It is not hard to prove that every ordinal α has a unique representation $\alpha = \beta + n$, β a limit ordinal and $n \in \omega$. In this case, we call α an *even ordinal* if n is even and an *odd ordinal* if n is odd.

A.2 Axiom of Choice

The well-ordering principle WOP is a non-constructive principle which merely asserts the existence of a well-ordering of an arbitrary set without specifying any. In mathematics, there are two more such non-constructive principles which are commonly used. We state these and show that they are equivalent statements in Zermelo–

Fraenkel set theory. It has been shown that they are undecidable in Zermelo–Fraenkel set theory.

Axiom of Choice. (AC) *For every family* $\{X_i : i \in I\}$ *of non-empty sets, there is a function* $f : I \to \cup_{i \in I} X_i$ *such that for every* $i \in I, f(i) \in X_i$.

A function f satisfying this condition is called a *choice function* for $\{X_i : i \in I\}$. The set of all choice function for $\{X_i : i \in I\}$ is denoted by $\times_{i \in I} X_i$. If each $X_i = X$, we denote this set by X^I.

Let $(\mathbb{P}, \leq)$ be a partially ordered set. A *chain* in $\mathbb{P}$ is a subset C of $\mathbb{P}$ such that $\leq |C$ is a linear order on C.

(**Zorn's Lemma**.) (ZL) *Let* $(\mathbb{P}, \leq)$ *be a non-empty partially ordered set such that every chain in* $\mathbb{P}$ *is bounded above. Then* $(\mathbb{P}, \leq)$ *has a maximal element.*

The following is a theorem of Zermelo–Fraenkel set theory.

Theorem A.2.1 *The following statements are equivalent.*

(a) Zorn's Lemma.
(b) Well-ordering principle.
(c) Axiom of choice.

Proof (a) implies (b): Let X be a set. If $X = \emptyset$, then empty relation well orders X. Assume that $X \neq \emptyset$. Set

$$\mathbb{P} = \{(A, <) : A \subset X \ \& \ < \text{ a well-order on } A\}.$$

Then $\mathbb{P}$ is non-empty. Define

$$(A, <) \prec (B, <_1) \Leftrightarrow (A, <) \text{ is an initial segment of } (B, <_1).$$

If $\{(A_i, <_i) : i \in I\}$ is a chain C in $\mathbb{P}$, then $\cup C$ is an upper bound of C. Hence, $\mathbb{P}$ has a maximal element, say $(A, <)$. We claim that $A = X$. For otherwise, take an $x \in X \setminus A$. Now extend $<$ to a well-order on $A \cup \{x\}$ by declaring x larger than every $a \in A$. This contradicts the maximality of $(A, <)$.

(b) implies (c): Given a family of non-empty sets $\{X_i : i \in I\}$, set $X = \cup_i X_i$. By (b), there is a well-order $<$ on X. Define $f(i)$ to be the least element of $X_i, i \in I$.

(c) implies (a): Let $(\mathbb{P}, \leq)$ be a non-empty partially ordered set such that every chain C in $\mathbb{P}$ has an upper bound in $\mathbb{P}$.

Let $\mathcal{C}$ denote the set of all chains in $\mathbb{P}$. For each chain C, define

$$C' = \{p \in \mathbb{P} : \forall x \in C(x < p)\}$$

and set

$$\mathcal{C}_0 = \{C \in \mathcal{C} : C' \neq \emptyset\}.$$

By the axiom of choice, there is a function $f : C_0 \to \mathbb{P}$ such that $f(C) \in C'$ for every $C \in C_0$. For each chain C, define its *successor* $s(C)$ as follows:

$$s(C) = \begin{cases} C & \text{if } C' = \emptyset, \\ C \cup \{f(C)\} & \text{otherwise.} \end{cases}$$

We need to show that there exists a chain C such that $s(C) = C$. (Then an upper bound of C (that exists by the hypothesis) will be a maximal element.)

We call a family $\mathcal{M}$ of chains a *normal family* if the following three conditions are satisfied.

(1) The empty chain $\emptyset \in \mathcal{M}$.
(2) If $\{C_i : i \in I\} \subset \mathcal{M}$ and $C = \cup_i C_i$ is a chain, then $C \in \mathcal{M}$.
(3) $s(C) \in \mathcal{M}$ whenever $C \in \mathcal{M}$.

The set of all chains $\mathcal{C}$ is a normal family. Further, it is easy to check that the intersection of a set of normal families of chains is normal.

Let $\mathcal{N}$ denote the intersection of all normal families of chains. Then $\mathcal{N}$ is a normal family of chains that is contained in all normal families.

Main Observation. For every chain $C, D \in \mathcal{N}$ either $C \subset D$ or $D \subset C$.

Assume this for the time being. Let

$$C_0 = \cup_{C \in \mathcal{N}} C.$$

By the above observation, C_0 is a chain. Since $\mathcal{N}$ is normal, $C_0 \in \mathcal{N}$. Moreover, it is the largest element of $\mathcal{N}$. By the same reason, $s(C_0) \in \mathcal{N}$. Thus, $s(C_0) \subset C_0 \subset s(C_0)$ and the proof of Zorn's lemma is complete.

We now proceed to prove the main observation. Call a chain $C \in \mathcal{N}$ *good* if for every $D \in \mathcal{N}$, either $C \subset D$ or $D \subset C$. We need to prove that every $C \in \mathcal{N}$ is good. The following is the crucial property of good sets:

Fact. If C is good, for every $N \in \mathcal{N}$ either $N \subset C$ or $s(C) \subset N$.

Assuming this fact, we complete the proof of the main observation first. Towards proving this, consider

$$\mathcal{M}_1 = \{C \in \mathcal{N} : C \text{ is good}\}.$$

We have the following:

(i) Since the empty chain $\emptyset$ is contained in all chains, $\emptyset \in \mathcal{M}_1$.
(ii) Suppose $\{C_i : i \in I\} \subset \mathcal{M}_1$ and $C = \cup_i C_i$ is a chain. Take any $D \in \mathcal{N}$. If each $C_i \subset D$, $C \subset D$. Otherwise, there is a $C_i \not\subset D$. But $C_i \in \mathcal{M}_1$. So, $D \subset C_i \subset C$.
(iii) Now let $C \in \mathcal{M}_1$, i.e., C is good and $D \in \mathcal{N}$. Then by the above fact, either $D \subset C \subset s(C)$ or $s(C) \subset D$.

These prove that $\mathcal{M}_1 \subset \mathcal{N}$ is normal. Hence, $\mathcal{M}_1 = \mathcal{N}$. This proves the main observation.

It remains to prove the fact which we do now. Let $C \in \mathcal{N}$ be a good chain. Consider

$$\mathcal{M}_2 = \{N \in \mathcal{N} : N \subset C \vee s(C) \subset N\}.$$

Suffices to show that $\mathcal{M}_2$ is normal.

(a) Clearly, the empty chain $\emptyset \subset C$. Hence, $\emptyset \in \mathcal{M}_2$.
(b) Suppose $\{\mathcal{N}_i : i \in I\} \subset \mathcal{M}_2$ and $N = \cup_i N_i$ is a chain. Then $N \in \mathcal{N}$. If each $N_i \subset C, N \subset C$. Otherwise, there is a $N_i \not\subset C$. Since $N_i \in \mathcal{M}_2, s(C) \subset N_i \subset N$.
(c) Now let $N \in \mathcal{M}_2$. If $s(C) \subset N, s(C) \subset s(N)$. So assume that $N \subset C$. Since C is good and $s(N) \in \mathcal{N}$, either $s(N) \subset C$ or $C \subset s(N)$. In the first case $s(N) \in \mathcal{M}_2$. In the second case, we have $N \subset C \subset s(N)$. But $s(N)$ differs from N by at most one point. So, either $C = N \subset s(N)$ or $C = s(N)$ implying $s(N) \subset C$.

We have now proved that $\mathcal{M}_2$ is normal. Thus, we have completed the proof of Zorn's lemma. $\quad\dashv$

Remark A.2.2 The proof of Zorn's lemma from the axiom of choice presented above is due to Hausdorff.

We shall be using minor variants of these three equivalent principles.

A.3 Cardinal Numbers

Let X and Y be any two sets. We shall $|X| \leq |Y|$ if there is an injection $f : X \to Y$. In this case, we say that the *cardinality of X is less than or equal to that of Y*. We shall write $|X| = |Y|$ if there is a bijection $f : X \to Y$. In this case, we say that the *cardinality of X is equal to that of Y*. We shall write $|X| < |Y|$ if $|X| \leq |Y|$ & $|X| \neq |Y|$.

Theorem A.3.1 *For sets X and Y, we have*

1. $|X| < |\mathcal{P}(X)| = |2^X|$, *where* $\mathcal{P}(X)$ *denotes the power set of X. (Recall that* 2^X *is the set of all indicator functions on X.)*
2. (**Cantor–Dedekind–Schroder–Bernstein Theorem.**)

$$(|X| \leq |Y| \wedge |Y| \leq |X|) \Rightarrow |X| = |Y|.$$

3. (**AC**) *X is infinite if and only if* $|X| = |Y|$ *for some proper subset Y of X.*
4. (**AC**) $|X| \leq |Y|$ *or* $|Y| \leq |X|$.

5. **(AC)** *If X is infinite,*
$$|X| = |X \times \{0, 1\}| = |X \times X|.$$

6. **(AC)** *If $\{X_i : i \in I\}$ and $\{Y_i : i \in I\}$ such that $|X_i| < |Y_i|$ for all $i \in I$, then*
   $|\cup_i X_i| < |\times_i Y_i|.$

7. **(WOP)** *There is an ordinal α such that $|X| = |\alpha|$.*

A set X is called *countable* if $|X| \leq |\omega|$. Otherwise, X is called *uncountable*. It is useful to see that countable union of countable sets is countable. An ordinal α is called *countable* if α as a set is countable.

A *cardinal number* is an ordinal number λ such that for every ordinal $\beta < \lambda$, $|\beta| < |\lambda|$. By **WOP**, for each set X, there is a cardinal number λ such that $|X| = |\lambda|$. In this case, we shall write $|X| = \lambda$ and call λ the cardinality of X. Cardinal numbers will be denoted by λ, μ, ν and κ with or without suffixes.

If λ and μ are cardinal numbers, we define $\lambda \cdot \mu = |\lambda \times \mu|$ and $\lambda^\mu = |\lambda^\mu|$, where λ^μ on the right-hand side stands for the set of all functions from μ to λ. Let $\{\lambda_i : i \in I\}$ is a family of cardinal numbers. Suppose $\{X_i : i \in I\}$ is a family of pairwise disjoint sets such that for each $i \in I$, $|X_i| = \lambda_i$. Then we define $\sum_i \lambda_i = |\cup_i X_i|$.

Since every set of ordinal numbers is well-ordered, every set of cardinal numbers is well-ordered. Therefore, for any set of cardinals $\{\lambda_i : i \in I\}$, $\sup_i \lambda_i$ makes sense. For any cardinal λ, λ^+ denotes the least cardinal greater than λ. We call λ^+ the *successor* of λ. Such cardinals are also called *successor cardinals*. Other cardinals are called *limit cardinals*. From the last theorem, we easily get

Theorem A.3.2 *1. For every cardinal λ, $\lambda < 2^\lambda$.*
2. $(\lambda \leq \mu \wedge \mu \leq \lambda) \Rightarrow \lambda = \mu$.
3. If $\lambda \leq \mu$ and μ is infinite, then

$$\mu = \lambda + \mu = \lambda \cdot \mu.$$

4. If λ, μ, ν are cardinals, then

$$\lambda^\mu \cdot \lambda^\nu = \lambda^{\mu+\nu} \ \& \ (\lambda^\mu)^\nu = \lambda^{\mu \cdot \nu}.$$

Note that ω is a cardinal number. Indeed, it is the least infinite cardinal number. By transfinite induction, we define a class of cardinals $\{\aleph_\alpha : \alpha \in ON\}$ as follows:

$$\aleph_0 = \omega,$$

$$\aleph_{\alpha+1} = \aleph_\alpha^+,$$

and for limit α,

$$\aleph_\alpha = \sup\{\aleph_\beta : \beta < \alpha\}.$$

These $\aleph$'s are all the infinite cardinals. There are two very famous hypotheses on these $\aleph$'s.

Continuum Hypothesis. (CH) $2^{\aleph_0} = \aleph_1$.

Generalised Continuum Hypothesis. (GCH) $\forall \alpha \in ON(2^{\aleph_\alpha} = \aleph_{\alpha+1})$.

For an ordinal α, we define the *cofinality* of α, denoted by $cf(\alpha)$ to be the least ordinal β such that there is a map $f : \beta \to \alpha$ with $\alpha = \sup\{f(\gamma) : \gamma < \beta\}$. Clearly, $cf(\alpha)$ is a cardinal number. An infinite cardinal κ is called a *regular cardinal* if $cf(\kappa) = \kappa$. Otherwise, κ is called a *singular cardinal*. Every successor cardinal κ is regular, so is $\aleph_0$. $\aleph_\omega$ is a singular cardinal.

It is known that $|\mathbb{R}| = 2^{\aleph_0}$. So, $2^{\aleph_0}$ is also referred to as the *continuum* and is denoted by $\mathfrak{c}$.

A.4 Ultrafilters

Let I be a non-empty set. A *filter on I* is a family $\mathcal{F}$ of subsets of I satisfying the following conditions:

(∗) $\emptyset \notin \mathcal{F}$ and $I \in \mathcal{F}$.
(∗) $A, B \in \mathcal{F} \Rightarrow A \cap B \in \mathcal{F}$.
(∗) If $A \in \mathcal{F}$ and $B \supset A$, $B \in \mathcal{F}$.

It is clear that if $\mathcal{F}$ is a filter, it satisfies *finite intersection property*, i.e. for every finite $\mathcal{F}' \subset \mathcal{F}, \cap \mathcal{F}' \neq \emptyset$.

Remark A.4.1 Let $\mathcal{B}$ be a family of subsets of I with finite intersection property. Then
$$\mathcal{F} = \{A \subset I : \exists B_1, \ldots B_n \in \mathcal{B}(\cap_j B_j \subset A)\}$$

is a filter on I. Indeed, the filter $\mathcal{F}$ described above is the smallest filter containing $\mathcal{B}$ which we shall refer to as the filter generated by $\mathcal{B}$.

Given a filter $\mathcal{F}$ on I, consider

$$\mathbb{P} = \{\mathcal{F}' : \mathcal{F}' \supset \mathcal{F} \text{ and } \mathcal{F}' \text{ a filter on } I\}.$$

Since $\mathcal{F} \in \mathbb{P}, \mathbb{P} \neq \emptyset$. $\mathbb{P}$ is a partially ordered set, partially ordered by the inclusion $\subset$.

If $\{\mathcal{F}_a : a \in A\}$ is a chain in $\mathbb{P}$, $\cup_a \mathcal{F}_a$ is a filter on I containing each $\mathcal{F}_a$. So, by Zorn's lemma, $\mathcal{F}$ is contained in a maximal filter. Maximal filters are also called *ultrafilters*. Combining this with the above remark, we have

Proposition A.4.2 *Every family $\mathcal{B}$ of subsets of I with finite intersection property is contained in an ultrafilter on I.*

Proposition A.4.3 *Let $\mathcal{F}$ be a filter on I. The following conditions are equivalent.*

(a) *$\mathcal{F}$ is an ultrafilter.*
(b) *If $B \subset I$ is such that $B \cap A \neq \emptyset$ for every $A \in \mathcal{F}$, $B \in \mathcal{F}$.*
(c) *For $B_1, \ldots, B_n \subset I$, $\cup_n B_n \in \mathcal{F} \Leftrightarrow \exists 1 \le i \le n (B_i \in \mathcal{F})$.*
(d) *Whenever $A \cup B \in \mathcal{F}$, A or B is in $\mathcal{F}$.*
(e) *For every $B \subset I$, $B \in \mathcal{F}$ or $I \backslash B \in \mathcal{F}$.*

Proof Assume (a). If B satisfies the hypothesis of (b), $\mathcal{F} \cup \{B\}$ satisfies the finite intersection property. Hence, there is a filter $\mathcal{F}' \supset \mathcal{F} \cup \{B\}$. The maximality of $\mathcal{F}$ implies $B \in \mathcal{F}$.

Now assume (b). Let $B_1, \ldots, B_n \subset I$ and no $B_i \in \mathcal{F}$, $1 \le i \le n$. By (b), there exist $A_1, \ldots A_n \in \mathcal{F}$ such that $B_i \cap A_i = \emptyset$, $1 \le i \le n$. This implies that $\cup_{1 \le i \le n} B_i) \cap (\cap_{1 \le i \le n} A_i) = \emptyset$. Since $\cap_{1 \le i \le n} A_i \in \mathcal{F}$, it follows that $\cup_{1 \le i \le n} B_i \notin \mathcal{F}$. Since the reverse implication is clear, (b) implies (c).

(d) is a special case of (c) and (e) is a special case of (d).

Now assume (e). Suppose there is a filter $\mathcal{F}'$ on I containing $\mathcal{F}$ properly. Take $B \in \mathcal{F}'$ that does not belong to $\mathcal{F}$. By (e), $I \setminus B \in \mathcal{F}$. So, both $B, I \setminus B \in \mathcal{F}'$ which is not possible because $\mathcal{F}'$ is a filter. This contradiction shows that (e) implies (a). $\square$

Corollary A.4.4 *If $\mathcal{U}$ is an ultrafilter on I, then $\cap \mathcal{U}$ contains at most one point.*

Proof Suppose $x \neq y$ in I belong to all sets in $\mathcal{U}$. Then both $\{x\}$ and $\{y\}$ satisfy (b). Hence, both these sets belong to $\mathcal{U}$. This is a contradiction. $\square$

An ultrafilter $\mathcal{F}$ is called *free* or *non-principal* if $\cap \mathcal{F} = \emptyset$. Otherwise, it is called a *principal ultrafilter*. Thus, $\mathcal{U}$ is a principal ultrafilter on I if and only if $\mathcal{U} = \{A \subset I : x \in A\}$ for some $x \in I$. Clearly, a free ultrafilter does not contain a finite set. In particular, every ultrafilter on a finite set I is principal. Conversely, if I is infinite, the family of all cofinite (complement of finite) subset of I has finite intersection property. Hence, there is a free ultrafilter on I. Also, note that a free ultrafilter on I contains every cofinite subset of I.

A.5 Some Infinite Combinatorics

If $u = (u(0), \ldots, u(n - 1))$ is a finite sequence and $i < n$, then we define $u|i = (u(0), \ldots, u(i - 1))$ and call it an *initial segment* of u. By taking $i = 0$, we see that the empty sequence e either equals u or is an initial segment of u. For finite sequences u, v, we write $u \preceq v$ if either $u = v$ or u is an initial segment of v. For a finite sequence u, $|u|$ will denote its length. If $u = (u(0), \ldots, u(n - 1))$ and $v = (v(0), \ldots, v(m - 1))$ are finite sequences, then their concatenation $uv = (u(0), \ldots, u(n - 1), v(0), \ldots, v(m - 1))$. For an infinite sequence $\alpha = (\alpha(0), \alpha(1), \ldots)$ and $n \in \omega$, $\alpha|n = (\alpha(0), \ldots, \alpha(n - 1))$.

A *tree* T is a non-empty set of finite sequences such that whenever $u \in T$ all the initial segments of u are in T. Thus, the empty sequence e belongs to every tree T. If T is a tree and $u \in T$, then we define

$$T_u = \{v : uv \in T\}.$$

Note that $T = T_e$ and $T_u = \cup\{T_v : u \preceq v \wedge |v| = |u| + 1\} \cup \{u\}$. A tree T is called *finitely splitting* if for all $u \in T$, $\{v \in T : u \preceq v \wedge |v| = |u| + 1\}$ is finite. An infinite sequence $\alpha = (\alpha(0), \alpha(1), \ldots)$ such that for every $n \in \omega$, $\alpha|n \in T$ is called an *infinite branch* of T.

Theorem A.5.1 (König's infinity Lemma) *Every finitely splitting, infinite tree T has an infinite branch.*

Proof We define an infinite branch α of T by induction such that for every $n \in \omega$, $T_{\alpha|n}$ is infinite. By our hypothesis, it is not very hard to see that such an α exists. □

For any set X and any cardinal μ, $[X]^\mu$ will denote the set of all subsets of X of cardinality μ. For a map $f : [X]^\mu \to Y$, a subset $Z \subset X$ is called *homogeneous* if f is constant on $[Z]^\mu$. For cardinals κ, λ, μ and η, one writes

$$\kappa \to (\eta)^\mu_\lambda$$

if whenever $|X| \geq \kappa$, every function of $f : [X]^\mu \to \lambda$ has a homogeneous set of cardinality $\geq \eta$.

Theorem A.5.2 (Ramsey Theorem) *For every $n \geq 1$ and every infinite set X, every function f from $[X]^n$ into a finite set Y has an infinite homogeneous set, i.e. for every $m, n \geq 1$,*

$$\aleph_0 \to (\aleph_0)^n_m.$$

Proof We prove this result by induction on n. It is obvious for $n = 1$. Let the result be true for n, X be an infinite set and f a function from $[X]^{n+1}$ into a finite set Y.

Set $Z_0 = X$ and take any $z_0 \in Z_0$. Define $f_0 : [Z_0 \setminus \{z_0\}]^n \to Y$ by

$$f_0(A) = f(\{z_0\} \cup A), \quad A \in [Z_0 \setminus \{z_0\}]^n.$$

By induction hypothesis, there exists an infinite homogeneous set $Z_1 \subset Z_0 \setminus \{z_0\}$ for f_0.

Now assume that $Z_0 \supset Z_1 \supset \ldots \supset Z_k$, each infinite, and for $i < k$, $z_i \in Z_i$ have been defined. Take any $z_k \in Z_k$. Define $f_k : [Z_k \setminus \{z_k\}]^n \to Y$ by

$$f_k(A) = f(\{z_k\} \cup A), \quad A \in [Z_k \setminus \{z_k\}]^n.$$

By induction hypothesis there exists an infinite homogeneous set $Z_{k+1} \subset Z_k \setminus \{z_k\}$ for f_k.

Set $Z_\infty = \{z_0, z_1, \ldots\}$. For each k, each subset A of Z_∞ of cardinality $n + 1$ with $z_k \in A$ and $A \setminus \{z_k\} \subset \{z_i : i > k\}$ takes the same value. Since there are only finitely many possibilities, the value will be the same for infinitely many k's. The set of these z_k's is infinite and homogeneous for f.   □

We now prove yet another important result in infinite combinatorics known as Erdös–Rado theorem. For any infinite cardinal κ and any ordinal α, we define *Beth cardinals* $\beth_\alpha(\kappa)$ by transfinite induction as follows:

$$\beth_0(\kappa) = \kappa,$$

$$\beth_\alpha(\kappa) = \sup_{\beta < \alpha} \beth_\beta(\kappa) \quad \text{if } \alpha \text{ limit,}$$

$$\beth_{\alpha+1}(\kappa) = 2^{\beth_\alpha(\kappa)}.$$

The following simple identities will be used in the proof of Erdös–Rado theorem without mention.

Lemma A.5.3 *For every* $2 \le \mu \le \beth_{\alpha+1}(\kappa)$,

$$\mu^{\beth_\alpha(\kappa)} = \beth_{\alpha+1}(\kappa).$$

Proof We have

$$\beth_{\alpha+1}(\kappa) = 2^{\beth_\alpha(\kappa)} \le \kappa^{\beth_\alpha(\kappa)} \le \beth_{\alpha+1}(\kappa)^{\beth_\alpha(\kappa)} = (2^{\beth_\alpha(\kappa)})^{\beth_\alpha(\kappa)} = \beth_{\alpha+1}(\kappa).$$

The result is easily seen from here.

Theorem A.5.4 (Erdös–Rado Theorem) *For every infinite cardinal* κ *and every* $n \ge 0$,

$$\beth_n(\kappa)^+ \to (\kappa^+)^{n+1}_\kappa.$$

Proof We prove the result by induction on n. For $n = 0$ the result is trivially seen. Suppose the result is true for $n - 1$. Set $\lambda = \beth_n(\kappa)^+$ and take any $f : [\lambda]^{n+1} \to \kappa$.

For any $\alpha < \lambda$, define $f_\alpha : [\lambda \setminus \{\alpha\}]^n \to \kappa$ by

$$f_\alpha(A) = f(A \cup \{\alpha\}), \quad A \in [\lambda \setminus \{\alpha\}]^n.$$

Inductively we define

$$X_0 \subset X_1 \subset \cdots \subset X_\alpha \subset \cdots \subset \lambda, \quad \alpha < \beth_{n-1}(\kappa)^+$$

such that

1. $X_0 = \beth_n(\kappa)$;
2. $X_\alpha = \cup_{\beta < \alpha} X_\beta$, if α is limit;
3. $|X_\alpha| = \beth_n(\kappa)$, and
4. for all $Y \subset X_\alpha$ of cardinality $\leq \beth_{n-1}(\kappa)$ and all $\beta \in \lambda \setminus Y$, there is a $\gamma \in X_{\alpha+1} \setminus Y$ such that $f_\beta|[Y]^n = f_\gamma|[Y]^n$.

We only need to define $X_{\alpha+1}$ from X_α so that $X_{\alpha+1}$ satisfies (3) and (4). Take any $Y \subset X_\alpha$ of cardinality at most $\beth_{n-1}(\kappa)$. By the last Lemma A.5.3, the number of such Y is $\beth_n(\kappa)$. For each such Y, by the last Lemma A.5.3 again, there are at most $\beth_n(\kappa)$ many functions from $[Y]^n$ to κ. So,

$$|\{f_\beta|[Y]^n : Y \subset X_\alpha \wedge |Y| \leq \beth_{n-1}(\kappa) \wedge \beta \in \lambda \setminus Y\}| \leq \beth_n(\kappa).$$

Hence, $X_{\alpha+1}$ with desired properties exists.

Now set

$$X = \cup_{\alpha < \beth_{n-1}(\kappa)^+} X_\alpha.$$

If $Y \subset X$ and $|Y| \leq \beth_{n-1}(\kappa)$, then there is an $\alpha < \beth_{n-1}(\kappa)^+$ such that $Y \subset X_\alpha$. So, for every $\beta \in \lambda \setminus Y$ there is a $\gamma \in X \setminus Y$ such that $f_\beta|[Y]^n = f_\gamma|[Y]^n$.

Now choose any $\delta \in \lambda \setminus X$. Inductively we define $y_\alpha \in X$, $\alpha < \beth_{n-1}(\kappa)^+$ satisfying

$$f_{y_\alpha}|[\{y_\beta : \beta < \alpha\}]^n = f_\delta|[\{y_\beta : \beta < \alpha\}]^n.$$

($y_0 \in X$ is chosen arbitrarily.)

By induction hypothesis there is a $Z \subset \{y_\alpha : \alpha < \beth_{n-1}(\kappa)^+\}$ of cardinality $\geq \kappa^+$ which is homogeneous for f_δ.

Let $\alpha_0 < \cdots < \alpha_n$ be such that $y_{\alpha_i} \in Z$, $i \leq n$.

$$f(\{y_{\alpha_0}, \ldots, y_{\alpha_n}\}) = f_{y_{\alpha_n}}(\{y_{\alpha_0}, \ldots, y_{\alpha_{n-1}}\}) = f_\delta(\{y_{\alpha_0}, \ldots, y_{\alpha_{n-1}}\}).$$

It follows that Z is homogeneous for f. $\square$

Appendix B
Algebra

B.1 Field Extensions and Galois Group

Let $\mathbb{F}$ be a field. A field $\mathbb{K}$ is called a *field extension* of $\mathbb{F}$ if $\mathbb{F}$ a subfield of $\mathbb{K}$. If $\mathbb{K}$ is a field extension of $\mathbb{F}$, then $\mathbb{K}$ is canonically a vector space over $\mathbb{F}$. We let $[\mathbb{K} : \mathbb{F}]$ denote the dimension of $\mathbb{K}$ as a vector space over $\mathbb{F}$. We call $\mathbb{K}$ a *finite extension* of $\mathbb{F}$ if $[\mathbb{K} : \mathbb{F}] < \aleph_0$.

An element $a \in \mathbb{K}$ is called *algebraic* over $\mathbb{F}$ if there is a polynomial $f(X) \in \mathbb{F}[X]$ such that $f(a) = 0$. Otherwise, a is called *transcendental* over $\mathbb{F}$. If $a \in \mathbb{K}$ is algebraic over $\mathbb{F}$, then there is a unique monic irreducible polynomial $f(X) \in \mathbb{F}[X]$ such that $f(a) = 0$. f is called the *minimal polynomial* of a. An extension field $\mathbb{K}$ of $\mathbb{F}$ is called an *algebraic extension* of $\mathbb{F}$ if every $a \in \mathbb{K}$ is algebraic over $\mathbb{F}$. A field $\mathbb{K}$ is called *algebraically closed* if every polynomial $f(X) \in \mathbb{K}[X]$ has a root in $\mathbb{K}$.

Let $\mathbb{K}$ be an extension of $\mathbb{F}$ and $a \in \mathbb{K}$. We set

$$\mathbb{F}[a] = \{f(a) \in \mathbb{K} : f(X) \in \mathbb{F}[X]\}$$

and

$$\mathbb{F}(a) = \{f(a)/g(a) \in \mathbb{K} : f(X), g(X) \in \mathbb{F}[X] \ \& \ g(a) \neq 0\}.$$

Proposition B.1.1 *et $\mathbb{F}$ be a field and $f(X) \in \mathbb{F}[X]$ an irreducible polynomial. Let (f) denote the smallest ideal in $\mathbb{F}[X]$ containing f. Then*

1. *(f) is a maximal ideal and the quotient $\mathbb{F}[X]/(f)$ is a field and an algebraic extension of $\mathbb{F}$. Further, $[\mathbb{F}[X]/(f) : \mathbb{F}] = degree(f)$.*
2. *f has a root in $\mathbb{F}[X]/(f)$.*
3. *Every finite extension of $\mathbb{F}$ is an algebraic extension of $\mathbb{F}$.*
4. *If $\mathbb{K}$ is a finite extension of $\mathbb{F}$ and $\mathbb{L}$ a finite extension of $\mathbb{K}$, then $\mathbb{L}$ is a finite extension of $\mathbb{F}$.*
5. *Let $\mathbb{K}$ be an extension of $\mathbb{F}$ and $a \in \mathbb{K}$. Then a is algebraic over $\mathbb{F}$ if and only if $[\mathbb{F}(a) : \mathbb{F}] < \aleph_0$ and equals the degree of the minimal polynomial of a over $\mathbb{F}$. Further, in this case $\mathbb{F}(a) = \mathbb{F}[a]$.*

© Springer Nature Singapore Pte Ltd. 2017
H. Sarbadhikari and S.M. Srivastava, *A Course on Basic Model Theory*,
DOI 10.1007/978-981-10-5098-5

Theorem B.1.2 *Every field* $\mathbb{F}$ *has an algebraically closed, algebraic extension* $\mathbb{K}$. *Further, if* $\mathbb{K}'$ *is another such extension of* $\mathbb{F}$, *then there is an isomorphism* $h : \mathbb{K} \to \mathbb{K}'$ *such that* $h|\mathbb{F}$ *is identity on* $\mathbb{F}$.

We call $\mathbb{K}$ the *algebraic closure* of $\mathbb{F}$ and denote it by $\overline{\mathbb{F}}$. By the above theorem, algebraic closure of a field is unique in the above sense.

An element $a \in \overline{\mathbb{F}}$ is called *separable over* $\mathbb{F}$ if $f'(X) \neq 0$ where $f(X)$ is the minimal polynomial of a and $f'(X)$ its formal termwise derivative. This is equivalent to saying that $f'(a) \neq 0$. In this case, a is called a *simple root* of f. An algebraic extension $\mathbb{K}$ of $\mathbb{F}$ is called a *separable extension* of $\mathbb{F}$ if every $a \in \mathbb{K}$ is separable over $\mathbb{F}$. We set

$$\mathbb{F}^s = \{a \in \overline{\mathbb{F}} : a \text{ separable over } \mathbb{F}\}.$$

$\mathbb{F}^s$ is a field, called the *separable closure* of $\mathbb{F}$. If $\mathbb{K}$ is a separable extension of $\mathbb{F}$, then there is an embedding $h : \mathbb{K} \to \mathbb{F}^s$ such that $h|\mathbb{F}$ is identity on $\mathbb{F}$.

Proposition B.1.3 *Let* $a \in \overline{\mathbb{F}}$.

1. *If* $a \in \overline{\mathbb{F}}$ *is separable over* $\mathbb{F}$ *and* $f(X) \in \mathbb{F}[X]$ *its minimal polynomial, then all roots of* f *are separable over* $\mathbb{F}$.
2. *If* a *is separable over* $\mathbb{F}$, *then* $\mathbb{F}[a]$ *is a separable extension of* $\mathbb{F}$.
3. *If* $\mathbb{K}$ *is a separable extension of* $\mathbb{F}$ *and* $\mathbb{L}$ *a separable extension of* $\mathbb{K}$, *then* $\mathbb{L}$ *is a separable extension of* $\mathbb{F}$.
4. $(\mathbb{F}^s)^s = \mathbb{F}^s$.

Proposition B.1.4 *For every* $d > 1$, *there is a field* $\mathbb{F}$ *such that* $\mathbb{F}$ *is not algebraically closed but every* $f(X) \in \mathbb{F}[X]$ *of degree* $\leq d$ *(and* > 1) *has a root in* $\mathbb{F}$

Proof Take any prime $p > d$ and consider $\mathbb{K} = \mathbb{F}_p(Y)$, the field of rational functions over $\mathbb{F}_p$ in variable Y. We now show that $\mathbb{F} = \mathbb{K}^s$, the separable closure of $\mathbb{K}$, will do our job.

$\mathbb{K}$ has an element a such that for no $b \in \mathbb{K}$, $b^p = a$. For instance, $Y \in \mathbb{K}$ has this property. Since $\mathbb{K}$ is of characteristic p, for every $x, y \in \mathbb{K}$, $(x - y)^p = x^p - y^p$. This implies that all the roots of $X^p - a \in \mathbb{K}[X]$ are equal. In particular, it has no root in $\mathbb{F} = \mathbb{K}^s$. Thus, we have shown that $\mathbb{F}$ is not algebraically closed.

Let $f(X) \in \mathbb{F}[X]$ with $1 < degree(f) \leq d$. Suppose $c \in \overline{\mathbb{F}} = \overline{\mathbb{K}}$ is a root of f. Without any loss of generality assume that $f(X) \in \mathbb{F}[X]$ is the monic minimal polynomial of c. Since $\mathbb{F}$ is of characteristic $p > d$, the formal derivative f' of f satisfies $1 \leq degree(f') < degree(f)$. If possible, let $c \notin \mathbb{F} = \mathbb{K}^s$. Then $f'(c) = 0$. This is a contradiction. So, $f(X)$ has a root in $\mathbb{F}$. □

Proposition B.1.5 *If* $\mathbb{F}$ *is of characteristic* 0 *or if* $\mathbb{F}$ *is of characteristic* $p > 0$ *and for every* $x \in \mathbb{F}$ *there is a* $y \in \mathbb{F}$ *such that* $x = y^p$, *then* $\mathbb{F}^s = \overline{\mathbb{F}}$. *Otherwise,* $\mathbb{F}^s \subsetneq \overline{\mathbb{F}}$.

Proposition B.1.6 *Let* $\mathbb{F}$ *be of characteristic* $p > 0$ *and* $a \in \overline{\mathbb{F}}$. *Then there is a natural number* k *such that* a^{p^k} *is separable over* $\mathbb{F}$. *So, given* $a_1, \ldots, a_n \in \overline{\mathbb{F}}$, *there is a natural number* k *such that each* $a_i^{p^k}$, $1 \leq i \leq n$, *is separable over* $\mathbb{F}$.

An algebraic extension $\mathbb{K}$ of $\mathbb{F}$ is called a *normal extension* of $\mathbb{F}$ if whenever an irreducible polynomial $f(X) \in \mathbb{F}[X]$ has a root in $\mathbb{K}$, all its roots are in $\mathbb{K}$. So, $\overline{\mathbb{F}}$ and $\mathbb{F}^s$ are normal extensions of $\mathbb{F}$. Also, if $\mathbb{K}$ is a normal extension of $\mathbb{F}$, so is $\mathbb{K} \cap \mathbb{F}^s$.

Proposition B.1.7 *Let $\mathbb{K}$ be a field extension of $\mathbb{F}$. Then $\mathbb{K}$ is a finite normal extension of $\mathbb{F}$ if and only if $\mathbb{K}$ is the splitting field of some $f(X) \in \mathbb{F}[X]$.*

The Galois group $G(\mathbb{K}, \mathbb{F}) \subset \mathbb{K}^{\mathbb{K}}$. We equip $\mathbb{K}^{\mathbb{K}}$ with the product of discrete topology on $\mathbb{K}$ and $G(\mathbb{K}, \mathbb{F})$ with the subspace topology. If $\mathbb{F} \subset \mathbb{L} \subset \mathbb{K}$, then $G(\mathbb{K}, \mathbb{L})$ is a closed subset of $G(\mathbb{K}, \mathbb{F})$.

In the rest of this section, we assume that $\mathbb{K}$ is a normal separable extension of $\mathbb{F}$.

Proposition B.1.8 *If $\mathbb{F} \subset \mathbb{L} \subset \mathbb{K}$. Then $\mathbb{K}$ is a normal separable extension of $\mathbb{L}$.*

Proof Suppose $f(X) \in \mathbb{L}[X]$ is an irreducible polynomial which has a root, say a, in $\mathbb{K}$. Let $g(X) \in \mathbb{F}[X]$ be the minimal polynomial of a over $\mathbb{F}$. Then $f(X)$ divides $g(X)$. So, all the roots of f are roots of g. Thus, all the roots of f are in $\mathbb{K}$. $\square$

We have a map from the set of all fields $\mathbb{L}$, $\mathbb{F} \subset \mathbb{L} \subset \mathbb{K}$, to the set of all closed subgroups of $G(\mathbb{K}, \mathbb{F})$, given by $\mathbb{L} \to G(\mathbb{K}, \mathbb{L})$. The fundamental theorem of Galois theory states that this correspondence is a bijection.

We now proceed to describe the inverse of this correspondence. Let H be a closed subgroup of $G(\mathbb{K}, \mathbb{F})$. Set

$$F(H) = \{x \in \mathbb{K} : \forall \sigma \in H (\sigma(x) = x)\}.$$

Then

(a) $F(H)$ is a subfield of $\mathbb{K}$ and $\mathbb{F} \subset F(H)$.
(b) $H \subset G(\mathbb{K}, F(H))$.
(c) If $\mathbb{F} \subset \mathbb{L} \subset \mathbb{K}$, then $\mathbb{L} \subset F(G(\mathbb{K}, \mathbb{L}))$.

$F(H)$ is called the *fixed field* of H.

Theorem B.1.9 (The Fundamental Theorem of Galois Theory) *Let $\mathbb{K}$ be a normal, separable extension of $\mathbb{F}$.*

1. *If H is a closed subgroup of $G(\mathbb{K}, \mathbb{F})$, then $H = G(\mathbb{K}, F(H))$.*
2. *If $\mathbb{F} \subset \mathbb{L} \subset \mathbb{K}$, then $\mathbb{L} = F(G(\mathbb{K}, \mathbb{L}))$.*

B.2   Ring of Polynomials and Zariski Topology

A good but probably not well-known reference for commutative algebra is [16].

Let R be a commutative ring with identity. An *ideal* in R subring I of R such that whenever $a \in I$, $b \cdot a \in I$ for all $b \in R$. We assume that all ideals are proper. The ideal I is called a *maximal ideal* if it is not a proper subset of any ideal. It is called a *prime ideal* if whenever $a \cdot b \in I$, a or b is in I.

Fix an ideal I in R. For $x, y \in R$, define

$$x \sim y \Leftrightarrow x - y \in I.$$

Then $\sim$ is an equivalence relation on R. Let R/I be the set of all equivalence classes. For $x, y \in R$, define

$$(x + I) + (y + I) = x + y + I$$

and

$$(x + I) \cdot (y + I) = x \cdot y + I.$$

This makes R/I into a commutative ring with identity. The following result is quite easy to prove.

Proposition B.2.1 *Let I be an ideal in R. Then*

1. *I is a maximal ideal if and only if R/I is a field.*
2. *I is a prime ideal if and only if R/I is an integral domain.*

For an ideal I in R, we set

$$\sqrt{I} = \{x \in R : x^n \in I \text{ for some } n \geq 1\}.$$

$\sqrt{I}$ is an ideal, called the *radical* of I. The ideal I is called a *radical ideal* if $I = \sqrt{I}$.

R is called a *noetherian ring* if it has no strictly increasing infinite sequence of ideals. This is clearly equivalent to saying that every ideal in R is finitely generated. Since a field has no non-trivial proper ideal, every field is a Noetherian ring.

Proposition B.2.2 *If R is a Noetherian ring, so is the ring of polynomials $R[X]$ over R in one variable.*

By induction, now we get

Theorem B.2.3 (Hilbert Basis Theorem) *For every field $\mathbb{K}$, the ring of polynomials $\mathbb{K}[X_1, \ldots, X_n]$ is Noetherian. Hence, each ideal in $\mathbb{K}[X_1, \ldots, X_n]$ is finitely generated.*

Theorem B.2.4 (Prime Decomposition Theorem) *Let $\mathbb{K}$ be a field and $I \subset \mathbb{K}[\overline{X}]$ a radical ideal. Then there exist prime ideals $P_1, \ldots, P_k$ such that $I = \cap_{i=1}^{k} P_i$.*

Let R be a commutative ring with identity and S a subring. We call an element $x \in R$ *integral* over S if there exist $a_1, \ldots, a_n \in S$ such that $x^n + a_1 x^{n-1} + \cdots + a_{n-1}x + a_n = 0$. The ring R is called an *integral extension* of S if every element in R is integral over S.

Proposition B.2.5 *The set*

$$\{x \in R : x \text{ integral over } S\}$$

is a ring.

It is clear that every algebraic extension $\mathbb{F}$ of a field $\mathbb{K}$ is an integral extension of $\mathbb{K}$. Thus, we have the following result.

Proposition B.2.6 *If a field $\mathbb{F}$ is an algebraic extension of a field $\mathbb{K}$, then $\mathbb{F}[X_1, \ldots, X_n]$ is an integral extension of $\mathbb{K}[X_1, \ldots, X_n]$.*

Proposition B.2.7 *([16], Proposition 4.2.4.) Let R be a commutative ring with identity and S a subring of R. Assume that R is an integral extension of S. Then for every prime ideal P in S, there is a prime ideal Q in R such that $P = Q \cap S$.*

Proposition B.2.8 *Let $\mathbb{F}$ be a field, $\mathbb{K}$ a subfield and $P \subset \mathbb{K}[X_1, \ldots, X_n]$ a prime ideal. Then there is a prime ideal $Q \subset \mathbb{F}[X_1, \ldots, X_n]$ such that $P = Q \cap \mathbb{K}[X_1, \ldots, X_n]$.*

Proof Let $T \subset \mathbb{F}$ be a maximal algebraically independent subset of $\mathbb{F}$ over $\mathbb{K}$ and $\mathbb{L} \subset \mathbb{F}$ the subfield generated by $\mathbb{K} \cup T$. Since $\mathbb{F}$ is an algebraic extension of $\mathbb{L}$, by the last proposition, it is sufficient to show that there is a prime ideal $Q \subset \mathbb{L}[X_1, \ldots, X_n]$ such that $P = Q \cap \mathbb{K}[X_1, \ldots, X_n]$.

Let $D \subset \mathbb{L}$ be the subring generated by $\mathbb{K} \cup T$. Then D is an integral domain with $\mathbb{L}$ its quotient field.

Claim. The subring $P[T] \subset D[X_1, \ldots, X_n]$ generated by $P \cup T$ is a prime ideal.

We can view $D[X_1, \ldots, X_n]$ as the polynomial ring $\mathbb{K}[X_1, \ldots, X_n][T]$. Hence, our claim will be proved if we show that $\mathbb{K}[X_1, \ldots, X_n][T]/P[T]$ is an integral domain. Since T is algebraically independent over $\mathbb{K}$,

$$\mathbb{K}[X_1, \ldots, X_n][T]/P[T] = \frac{\mathbb{K}[X_1, \ldots, X_n]}{P}[T].$$

Since P is a prime ideal, $\frac{\mathbb{K}[X_1,\ldots,X_n]}{P}$ is an integral domain Hence, $\frac{\mathbb{K}[X_1,\ldots,X_n]}{P}[T]$ is an integral domain.

Since T is transcendental over $\mathbb{K}$, $P[T] \cap \mathbb{K}[X_1, \ldots, X_n] = P$. Consider the localization

$$Q = \{\frac{a}{s} \in \mathbb{L}[X_1, \ldots, X_n] : a \in P[T] \wedge s \in D \setminus \{0\}\}.$$

By ([1], Proposition 2.3.14), Q is a prime ideal in $\mathbb{L}[X_1, \ldots, X_n]$ with $Q \cap D[X_1, \ldots, X_n] = P[T]$. It follows that $Q \cap \mathbb{K}[X_1, \ldots, X_n] = P$. $\qquad\qquad\square$

Let $\mathbb{K}$ be a field. For $S \subset \mathbb{K}[\overline{X}]$, define

$$\mathcal{V}(S) = \{\overline{x} \in \mathbb{K}^n : f(\overline{x}) = 0 \,\forall f \in S\}.$$

If $I(S)$ is the ideal generated by S, then by Hilbert basis theorem, there exist finitely many $g_1, \ldots, g_k \in S$ that generate $I(S)$. In particular,

$$\mathcal{V}(S) = \mathcal{V}(I(S)) = \cap_{i=1}^{k} \{\overline{x} \in \mathbb{K}[\overline{X}] : g_i(\overline{x}) = 0\}.$$

It is now easy to see that

$$\mathcal{T} = \{\mathcal{V}(S) \subset \mathbb{K}^n : S \subset \mathbb{K}[\overline{X}]\}$$

is closed under finite unions and arbitrary intersections. Further, it contains $\emptyset$ and $\mathbb{K}^n$. Hence, $\mathcal{T}$ is the family of all closed sets of a topology on $\mathbb{K}^n$. We call this topology the *Zariski topology* on $\mathbb{K}^n$ and sets of the form $\mathcal{V}(S)$ *Zariski closed*. Following Chevalley, a subset $C \subset \mathbb{K}^n$ is called *constructible* if it belongs to the Boolean algebra generated by Zariski closed sets in $\mathbb{K}^n$.

For $X \subset \mathbb{K}^n$, set

$$\mathcal{I}(X) = \{f \in \mathbb{K}[\overline{X}] : f(\overline{x}) = 0 \,\forall \overline{x} \in X\}.$$

Then $\mathcal{I}(X)$ is a radical ideal in $\mathbb{K}[\overline{X}]$ and if $X \subset Y \subset \mathbb{K}^n$, $\mathcal{I}(Y) \subset \mathcal{I}(X)$. We also have

1. $X \subset \mathcal{V}(\mathcal{I}(X))$ for all $X \subset \mathbb{K}^n$.
2. $S \subset \mathcal{I}(\mathcal{V}(S))$ for all $S \subset \mathbb{K}[\overline{X}]$.
3. For $S \subset T \subset \mathbb{K}[\overline{X}]$, $\mathcal{V}(T) \subset \mathcal{V}(S)$.
4. $\mathcal{V}(S) = \mathcal{V}(\mathcal{I}(\mathcal{V}(S)))$ for all $S \subset \mathbb{K}[\overline{X}]$.
5. For an ideal $I \subset \mathbb{K}[\overline{X}]$, $\mathcal{V}(I) = \mathcal{V}(\sqrt{I})$.

Let $\mathbb{K}$ be any field. We equip $\mathbb{K}^n$ with Zariski topology. So, sets of the form

$$D(f) = \{\overline{x} \in \mathbb{K}^n : f(\overline{x}) \neq 0\},$$

$f \in \mathbb{K}[\overline{X}]$, form a subbase for the topology. For $Z \subset \mathbb{K}^n$, $\overline{Z}$ will denote the closure of Z in Zariski topology on $\mathbb{K}^n$. A closed set $C \subset \mathbb{K}^n$ is called *irreducible* if there do not exist non-empty, closed $C_1, C_2 \subsetneq C$ such that $C = C_1 \cup C_2$.

Let $\sigma \in Aut(\mathbb{K})$. We define $\sigma : \mathbb{K}^n \to \mathbb{K}^n$ by

$$\sigma(a_1, \ldots, a_n) = (\sigma(a_1), \ldots, \sigma(a_n)), \ (a_1, \ldots, a_n) \in \mathbb{K}^n$$

Lemma B.2.9 *For any $A \subset \mathbb{K}^n$ and any σ, $\sigma(\overline{A}) = \overline{\sigma(A)}$.*

Proof It is enough to show that $\sigma(\overline{A}) \subset \overline{\sigma(A)}$: Then

$$\sigma^{-1}(\overline{\sigma(A)}) \subset \overline{\sigma^{-1}(\sigma(A))} = \overline{A}.$$

Hence,

$$\sigma(\overline{A}) \supset \overline{\sigma(A)}.$$

To complete the proof, take any $\overline{x} = (x_1, \ldots, x_n) \in \sigma(\overline{A})$ and $\cap_{j=1}^{k} D(f_j), f_1, \ldots, f_k \in \mathbb{K}[X_1, \ldots, X_n]$, a basic open set containing $\overline{x}$. Suppose

$$f_j(X_1, \ldots, X_n) = \sum d_I^j X_1^{i_1} \ldots X_n^{i_n}.$$

We have

$$\sum d_I^j x_1^{i_1} \ldots x_n^{i_n} \neq 0, \quad 1 \le j \le k. \tag{B.1}$$

We are required to show that there exists $\overline{a} = (a_1, \ldots, a_n) \in \sigma(A)$
such that

$$\sum d_I^j a_1^{i_1} \ldots a_n^{i_n} \neq 0$$

for each $1 \le j \le k$. Take $\overline{y} = (y_1, \ldots, y_n) \in \overline{A}$ such that $\overline{x} = \sigma(\overline{y})$. By (1), we have

$$\sum \sigma^{-1}(d_I^j) y_1^{i_1} \ldots y_n^{i_n} \neq 0, \quad 1 \le j \le k.$$

Hence, there exists $\overline{z} = (z_1, \ldots, z_n) \in A$ such that

$$\sum \sigma^{-1}(d_I^j) z_1^{i_1} \ldots z_n^{i_n} \neq 0, \quad 1 \le j \le k.$$

Now take $a_i = \sigma(z_i), 1 \le i \le n$. Then $(a_1, \ldots, a_n) \in \sigma(A)$ and

$$\sum d_I^j a_1^{i_1} \ldots a_n^{i_n} \neq 0, \quad 1 \le j \le k.$$

$\square$

Lemma B.2.10 *Every non-empty closed set C has unique representation*

$$C = C_1 \cup \cdots \cup C_m,$$

where $C_1, \ldots, C_m$ are irreducible closed sets such that for every $1 \le j \le m$, $C_j \not\subset \cup_{i \neq j} C_i$.

Proof Let C be a Zariski closed set such that $C = C_0$ is not a finite union of irreducible closed sets. Then $C_0 = C_1 \cup D_1$, where C_1 and D_1 are Zariski closed sets and $C_1, D_1 \neq C_0$. One of these must not be a finite union of irreducible closed sets. Let C_1 be one such. Then $C_1 = C_2 \cup D_2$, where C_2 and D_2 are Zariski closed sets and $C_2, D_2 \neq C_1$ with C_2 not a finite union of irreducible closed sets. Proceeding similarly, we get an infinite descending sequence $C_0 \supset C_1 \supset C_2 \supset \ldots$ of Zariski closed sets. This contradicts that $\mathbb{K}[X_1, \ldots, X_n]$ is Noetherian.

Thus, $C = \cup_{i=1}^{n} C_i$, where each C_i is irreducible closed. We can get a subcollection of these so that the union is irredundant. Now let $C = \cup_{i=1}^{m} C_i = \cup_{j=1}^{n} F_j$ be two representations of C as irredundant unions of irreducible closed sets. Consider $J = \{j \leq n : C_1 \cap F_j \neq \emptyset\}$. As C_1 is irreducible, there must be $j \in J$ such that $F_j \supset C_1$. Doing the same argument with F_j, we get a $C_i \supset F_j$. Because of the irredundancy, $i = 1$. Hence, $C_1 = F_j$. By reordering F_js, we assume that $C_1 = F_1$. Proceeding thus, we easily see that $m = n$ and $\{C_1, \ldots, C_m\} = \{F_1, \ldots, F_m\}$. $\qquad \square$

We call $C_1, \ldots, C_m$ the *irreducible components* of C an the representation $C = C_1 \cup \ldots \cup C_m$ satisfying above conditions to be redundant.

Lemma B.2.11 *Let*

$$Z = \cup_{i=1}^{m} (C_i \cap U_i),$$

where $C_1, \ldots, C_m$ are irreducible closed sets, $U_1, \ldots, U_m$ Zariski open and $C_i \cap U_i \neq \emptyset$, $1 \leq i \leq m$, be a constructible set. Then Z is dense in each irreducible component of $\overline{Z}$.

Proof Set $F_i = C_i \cap U_i^c$, $1 \leq i \leq m$. Either $C_i \cap U_i = C_i$ or F_i is a proper non-empty closed subset of C_i. Then

$$C_i = (C_i \cap U_i) \cup F_i \subset \overline{C_i \cap U_i} \cup F_i \subset C_i.$$

As C_i is irreducible, it follows that $C_i = \overline{C_i \cap U_i}$. Thus, $\overline{Z} = \cup_{i=1}^{m} C_i$. Also, for each $1 \leq i \leq m$,

$$C_i \supset \overline{Z} \cap C_i \supset \overline{C_i \cap U_i} = C_i.$$

Thus, Z is dense in each irreducible component of $\overline{Z}$.

Lemma B.2.12 *Let Z_1 and Z_2 be constructible sets with the same closure, say Z. Then $Z_1 \cap Z_2 = \emptyset$.*

Proof Write

$$Z_1 = \cup_{i=1}^{m} (C_i \cap U_i),$$

and

$$Z_2 = \cup_{j=1}^{k} (D_j \cap V_j),$$

where $C_1, \ldots, C_m, D_1, \ldots, D_k$ are irreducible closed, $U_1, \ldots, U_m$, $V_1, \ldots, V_k$ Zariski open, $C_i \cap U_i \neq \emptyset \neq D_j \cap V_j$, $1 \leq i \leq m$, $1 \leq j \leq k$ Then,

$$Z = \cup_i C_i = \cup_j D_j.$$

Since $C_1, \ldots, C_m$ and $D_1, \ldots, D_k$ are irreducible, it follows that there exist i, j such that $C_i = D_j$. If $C_i = C_i \cap U_i$ or $D_j = D_j \cap V_j$, then $Z_1 \cap Z_2 \neq \emptyset$. Assume that $C_i \neq C_i \cap U_i$ and $D_j \neq D_j \cap V_j$. So, $C_i \cap U_i^c$ and $C_i \cap V_j^c$ are proper, non-empty closed subsets of C_i. Since C_i is irreducible, this implies that

$$C_i \setminus ((C_i \cap U_i^c) \cup (C_i \cap V_j^c)) = (C_i \cap U_i) \cap (D_j \cap V_j) \neq \emptyset.$$

Hence, $Z_1 \cap Z_2 \neq \emptyset$. $\qquad\square$

Finally, we need a result of André Weil from Algebraic Geometry. (See [32, 68].)

We see that $\mathbb{K}^{[n]} = \mathbb{K}[X_1, \ldots, X_n]$ is a vector space over $\mathbb{K}$ with the set of all monomials $M_\alpha(\overline{X}) = X_1^{\alpha_1} \ldots X_n^{\alpha_n}$ a basis. Let I be an ideal in $\mathbb{K}^{[n]}$. By Hilbert basis theorem it is finitely generated. Now let k be a subfield of $\mathbb{K}$. We say that I is *algebraically definable* over k if it has a basis consisting of polynomials with coefficients in k.

Note that I is a vector subspace of $\mathbb{K}^{[n]}$. So, we have a quotient space $\mathbb{K}^{[n]}/I$. Let $\{M_\beta(\overline{X})\}$ be a maximal set of monomials which are independent modulo I. So, each monomial $M_\alpha(\overline{X})$ has a unique representation

$$M_\alpha(\overline{X}) = \sum_\beta a_{\alpha\beta} M_\beta(\overline{X}) \ \ (\text{modulo } I).$$

We need one more notation. If $\sigma : \mathbb{K} \to \mathbb{K}$ is an automorphism, and $f(\overline{X}) = \sum_\alpha a_\alpha M_\alpha(\overline{X}) \in \mathbb{K}^{[n]}$, then $f^\sigma(\overline{X}) = \sum_\alpha \sigma(a_\alpha) M_\alpha(\overline{X})$. Further, $I^\sigma = \{f^\sigma : f \in I\}$.

Theorem B.2.13 (André Weil) *There exists a subfield k_0 of $\mathbb{K}$ such that*

1. *I is algebraically definable over k_0, and*
2. *If I is algebraically definable over a subfield k of $\mathbb{K}$, then $k_0 \subset k$.*

Further, for every automorphism σ of $\mathbb{K}$, $I^\sigma = I$ if and only if σ fixes k_0 pointwise.

Proof As above, let $B = \{M_\beta(\overline{X})\}$ be a basis of $\mathbb{K}^{[n]}$ modulo I and let $\{M_\gamma(\overline{X})\}$ be the set of remaining monomials. Let

$$M_\gamma(\overline{X}) = \sum_\beta a_{\gamma\beta} M_\beta(\overline{X}) \ (\text{modulo } I).$$

Take any $f \in I$. Then f has a unique representation

$$f = \sum_\gamma a_\gamma M_\gamma(\overline{X}) + \sum_\beta b_\beta M_\beta(\overline{X})$$
$$= \sum_\gamma a_\gamma (M_\gamma(\overline{X}) - \sum_\beta a_{\gamma\beta} M_\beta(\overline{X})) + \sum_\beta c_\beta M_\beta(\overline{X})$$

Because f and $M_\gamma(\overline{X}) - \sum_\beta a_{\gamma\beta} M_\beta(\overline{X})$ are in I and $\{M_\beta(\overline{X})\}$ linearly independent modulo I, each $c_\beta = 0$. Now it is easily seen that I has a basis consisting of finitely many polynomials of the form $M_\gamma(\overline{X}) - \sum_\beta a_{\gamma\beta} M_\beta(\overline{X})$. We fix a finite basis D of I consisting of polynomials of the form $M_\gamma(\overline{X}) - \sum_\beta a_{\gamma\beta} M_\beta(\overline{X})$. Let k_0 be the subfield of $\mathbb{K}$ generated by the set of all $a_{\gamma\beta}$ appearing in this finite basis of I. Thus, I is algebraically definable over k_0.

Now let k be a subfield of $\mathbb{K}$ and I have a basis $f_1, \ldots, f_m \in k[\overline{X}]$.

Let $M_0(\overline{X})$ be one of the $M_\gamma(\overline{X})$ not appearing in B. Then

$$M_0(\overline{X}) - \sum_\beta a_{0\beta} M_\beta(\overline{X}) = \sum_{i=1}^m g_i f_i,$$

where $g_i = \sum_\alpha y_{i\alpha} M_\alpha(\overline{X})$. Introduce an indeterminate $Y_{i\alpha}$ whenever $y_{i\alpha} \neq 0$ and set

$$G_i = \sum_\alpha Y_{i\alpha} M_\alpha(\overline{X}).$$

Denote $\overline{Y} = (Y_{i\alpha})$. We then have

$$\sum_{i=1}^m G_i f_i = \sum_\gamma l_\gamma(\overline{Y}) M_\gamma(\overline{X}) + \sum_\beta l_\beta(\overline{Y}) M_\beta(\overline{X}),$$

where $l_\gamma(\overline{Y})$ and $l_\beta(\overline{Y})$ are linear forms with coefficients in k.

The system of linear equations $l_0(\overline{Y}) = 1$ and $l_\gamma(\overline{Y}) = 0$ for all $\gamma \neq 0$ over k has a solution in $\mathbb{K}$. Hence, it has a solution, $\overline{y'} = (y'_{i\alpha})$ in k. Let

$$g'_i = \sum_\alpha y'_{i\alpha} M_\alpha(\overline{X}).$$

Let

$$\sum_{i=1}^m g'_i f_i = M_0(\overline{X}) - \sum_\beta z_\beta M_\beta(\overline{X}).$$

Since $\{M_\beta(\overline{X})\}$ is a basis of $\mathbb{K}^{[n]}$ modulo I, $a_{0\beta} = z_\beta \in k$. Since $M_0(\overline{X})$ among $M_\gamma(\overline{X})$ not appearing in B was arbitrary, each $a_{\gamma\beta} \in k$. Thus, $k_0 \subset k$.

Now let σ be an automorphism of $\mathbb{K}$. Suppose σ fixes k_0 pointwise. Then σ fixes a basis of I pointwise. But then $I^\sigma = I$. Conversely, let $I^\sigma = I$. Then for each γ such that $M_\gamma(\overline{X}) - \sum_\beta a_{\gamma\beta} M_\beta(\overline{X})$ appears in D,

$$M_\gamma(\overline{X}) - \sum_\beta \sigma(a_{\gamma\beta}) M_\beta(\overline{X}) \in I^\sigma = I.$$

By uniqueness of the representation, it follows that $\sigma(a_{\gamma\beta}) = a_{\gamma\beta}$ for all γ, β. Thus, σ fixes k_0 pointwise. $\qquad\qquad\square$

In terms of Zariski closed sets, this result recasts as follows.

Theorem B.2.14 *Let* $\mathbb{K} \models ACF$ *and* $Z \subset \mathbb{K}^n$ *Zariski closed. Let* I *be a radical ideal such that* $Z = \mathcal{V}(I)$ *and* k_0 *the smallest subfield such that* I *is algebraically defined over* k_0. *Then for any automorphism* σ *of* $\mathbb{K}$, $\sigma(Z) = Z$ *if and only if* σ *fixes* k_0 *pointwise.*

Proof Note that

$$
\begin{aligned}
&\sigma(Z) \\
&= \cap_{\sum_\alpha a_\alpha M_\alpha(\overline{x}) \in I} \{\sigma(\overline{x}) \in \mathbb{K}^n : \sum_\alpha a_\alpha M_\alpha(\overline{x}) = 0\} \\
&= \cap_{\sum_\alpha a_\alpha M_\alpha(\overline{x}) \in I} \{\sigma(\overline{x}) \in \mathbb{K}^n : \sigma\left(\sum_\alpha a_\alpha M_\alpha(\overline{x})\right) = \sum_\alpha \sigma(a_\alpha) M_\alpha(\sigma(\overline{x})) = 0\} \\
&= \mathcal{V}(I^\sigma)
\end{aligned}
$$

By Hilbert Nullstellensatz, $I = \mathcal{I}(Z)$ and $I^\sigma = \mathcal{I}(\sigma(Z))$. Hence, $I^\sigma = I$ if and only if $\sigma(Z) = Z$. The result now follows from the last theorem. $\square$

B.3 Real Closed Fields

We refer the reader to [6, 61] for detailed accounts of real closed fields and for real algebraic geometry.

We call an ordered field $\mathbb{F}$ *real closed* if

1. $\forall x \exists y (x = y^2 \vee x + y^2 = 0)$, i.e. for every $x \in \mathbb{F}$, either x or $-x$ has a square root and
2. every polynomial in one variable over $\mathbb{F}$ of odd degree has a root in $\mathbb{F}$.

Example B.3.1 Besides $\mathbb{R}$, the field of real algebraic numbers, denoted by $\mathbb{R}_{alg}$, is a real closed field.

Note that in a real closed field $\mathbb{F}$, if $x \in \mathbb{F}$ is a square, it must be ≥ 0, and if $x \geq 0$ there is a unique $y \geq 0$ such that $x = y^2$. We then write $\sqrt{x}$ for y. Also note that $0 \leq x \leq y \Rightarrow \sqrt{x} \leq \sqrt{y}$.

Remark B.3.2 If $\mathbb{F}$ is a real closed field, it has a unique ordering, because its non-negative elements are given by

$$
\mathbb{F}^+ = \{x \in \mathbb{F} : \exists y \in \mathbb{K}(y \neq 0 \wedge x = y^2)\}.
$$

This also shows that the ordering of $\mathbb{F}$ is definable in the language of rings. However, it may not be definable by an open formula. If $\mathbb{F} = \mathbb{R}$ and $<$ is defined by an open formula, $\mathbb{R}^+$ should be either finite or cofinite, which it is not. Also note that every definable subset of $\mathbb{F}$ is definable in the language of rings.

Let $\mathbb{F}$ be a real closed field. We can canonically topologies $\mathbb{F}^n$. Indeed, we have a $\mathbb{F}$-valued metric on $\mathbb{F}^n$.

For $\bar{x} = (x_1, \ldots, x_n) \in \mathbb{F}$, we define its *norm* by

$$|\bar{x}| = \sqrt{x_1^2 + \cdots + x_n^2}.$$

For $\bar{x}, \bar{y} \in \mathbb{F}$, we define their *inner product* by

$$\langle \bar{x}, \bar{y} \rangle = \sum_{i=1}^{n} (x_i \cdot y_i).$$

Now we follow the classical proof of Cauchy–Schwarz inequality to get

Theorem B.3.3 (Cauchy–Schwarz Inequality) *If $\mathbb{F}$ is a real closed field and $\bar{x}, \bar{y} \in \mathbb{F}$, then*

$$|\langle \bar{x}, \bar{y} \rangle| \leq |\bar{x}| \cdot |\bar{y}|.$$

Proof Take any $\lambda \in \mathbb{F}$ and set

$$z_i = x_i - \lambda \cdot y_i, \ 1 \leq i \leq n.$$

Then

$$\sum_{i=1}^{n} z_i^2 = |\bar{y}|^2 \cdot \lambda^2 - \left(2 \sum_i x_i \cdot y_i \right) \cdot \lambda + \sum_i x_i^2 \geq 0.$$

Hence,

$$\left(|\bar{y}| \cdot \lambda - \frac{\langle \bar{x}, \bar{y} \rangle}{|\bar{y}|} \right)^2 - \left(\left(\frac{\langle \bar{x}, \bar{y} \rangle}{|\bar{y}|} \right)^2 - |\bar{x}|^2 \right) \geq 0$$

for all $\lambda \in \mathbb{F}$. By taking $\lambda = \frac{\langle \bar{x}, \bar{y} \rangle}{|\bar{y}|^2}$, we get the Cauchy–Schwarz inequality as above. $\square$

For $\bar{x}, \bar{y} \in \mathbb{F}$, define

$$\rho(\bar{x}, \bar{y}) = |\bar{x} - \bar{y}|.$$

For $\bar{x}, \bar{y}, \bar{z} \in \mathbb{F}$, it is easy to check the following:

1. $\rho(\bar{x}, \bar{y}) \geq 0$.
2. $\rho(\bar{x}, \bar{y}) = 0 \Leftrightarrow \bar{x} = \bar{y}$.
3. $\rho(\bar{x}, \bar{y}) = \rho(\bar{y}, \bar{x})$.
4. $\rho(\bar{x}, \bar{z}) \leq \rho(\bar{x}, \bar{y}) + \rho(\bar{y}, \bar{z})$.

The last fact follows from

(5) (Triangle inequality) $|\bar{x} + \bar{y}| \leq |\bar{x}| + |\bar{y}|$,

which can be proved by Cauchy–Schwarz inequality as usual. By using ρ like a metric, we have usual topology on $\mathbb{F}^n$. In fact, we can regard $\mathbb{F}^n$ as an inner product space. Thus, there is a geometry based on real closed fields, which is called real algebraic geometry. This is a rich branch of mathematics and model theory plays a significant role in real algebraic geometry.

We call a field **real** if -1 is not a sum of squares. Note that every ordered field is real. Further, a field $\mathbb{F}$ is real if and only if for all $\bar{a} \in \mathbb{F}$, $\sum_i a_i^2 = 0$ implies each $a_i = 0$.

Proposition B.3.4 *Let $\mathbb{F}$ be a real field. Then the field of rational functions $\mathbb{F}(X_1, \ldots, X_n)$ is real.*

Proof Our result will be proved if we show that for polynomials $f_1(\overline{X}), \ldots, f_k(\overline{X})$ over $\mathbb{F}$, $\sum_i f_i^2 = 0$ implies each $f_i = 0$. We prove this by induction on n. In the case $n = 1$, let $f_i(X) = \sum_j a_{ij} X^j$ with $\sum_i f_i^2 = 0$. Since $\mathbb{F}$ is real, note that the leading coefficients of those f_i whose degree is the highest among those of $f_1, \ldots, f_k$ are zero. So, each f_i must be 0. For inductive step, note that

$$\mathbb{F}[X_1, \ldots, X_n] = \mathbb{F}[X_1, \ldots, X_{n-1}][X_n].$$

$\square$

Lemma B.3.5 *Let $\mathbb{F}$ be a real field and $a \neq 0$ in $\mathbb{F}$. Then $\mathbb{F}[\sqrt{a}]$ is real if and only if $-a$ is not a sum of squares in $\mathbb{F}$.*

Proof If $\mathbb{F}[\sqrt{a}]$ is real and $-a$ is a sum of squares, then $a + \sum_i b_i^2 = 0$. This implies that $a = 0$. This proves the only if part. Now assume that $\mathbb{F}[\sqrt{a}]$ is not real. Then we get $\bar{x}, \bar{y} \in \mathbb{F}$ such that $\bar{y} \neq 0$ and $\sum_i (x_i + y_i \sqrt{a})^2 = 0$. This, in particular, implies that $\sum_i x_i^2 + a \sum_i y_i^2 = 0$. Hence

$$-a = \frac{(\sum_i x_i^2)(\sum_i y_i^2)}{(\sum_i y_i^2)^2},$$

contradicting that $-a$ is not a sum of squares. $\square$

Now, by Zorn's lemma, we get the following result:

Theorem B.3.6 *Every real field $\mathbb{K}$ has a real algebraic extension $\mathbb{F}$ such that in $\mathbb{F}$ for every $a \in \mathbb{F}$, either a is a square or $-a$ is a sum of squares.*

Proof Set

$$\mathbb{P} = \{\mathbb{F} : \mathbb{F} \text{ real algebraic extension of } \mathbb{K}\}.$$

Then $\mathbb{K} \in \mathbb{P}$ showing that $\mathbb{P}$ is non-empty. We partially order $\mathbb{P}$ by inclusion $\subset$. Note that if $\{\mathbb{F}_\alpha\}$ is a chain in $\mathbb{P}$, $\cup_\alpha \mathbb{F}_\alpha$ is an upper bound of the chain in $\mathbb{P}$. So, by Zorn's lemma, $\mathbb{P}$ has a maximal element, say $\mathbb{F}$. This works. $\square$

Now assume that $\mathbb{F}$ is a real field such that for every $a \in \mathbb{F}$ either a is a square or $-a$ is a sum of squares. Since $\mathbb{F}$ is real, this implies that for every $a \neq 0$, exactly one of a and $-a$ is a sum of squares. We define

$$x < y \Leftrightarrow \exists \overline{z} \in \mathbb{F}\left(\overline{z} \neq 0 \wedge y = x + \sum_i z_i^2\right), x, y \in \mathbb{F}.$$

Then $<$ is a linear order on $\mathbb{F}$ making it into an ordered field.

If $\mathbb{K}$ is a real field with no proper real algebraic extension, then the above ordering will be called the *canonical order of* $\mathbb{K}$. The above result, in particular, gives us the following result of Artin and Schreier.

Theorem B.3.7 *Let $\mathbb{F}$ be a real field and $a \in \mathbb{F}$ not a sum of squares. Then there is an order $<$ on $\mathbb{F}$ making $\mathbb{F}$ into an ordered field and $a < 0$.*

Corollary B.3.8 *A field is orderable if and only if it is real.*

The following result is also due to Artin and Schreier.

Theorem B.3.9 (Weierstrass Nullstellensatz) *Let $\mathbb{K}$ be a real field with no proper real algebraic extension, $<$ its canonical order, $f[X] \in \mathbb{K}[X]$ and $a < b \in \mathbb{K}$ be such that $f(a) \cdot f(b) < 0$. Then there is a $c \in \mathbb{K}$ with $a < c < b$ and $f(c) = 0$.*

Proof First we observe that it is sufficient to show that f has a root in $\mathbb{K}$. Let $c_0, \ldots, c_{n-1}$ be all the roots of f less than a. Then

$$f(x) = (x - c_0) \ldots (x - c_{n-1}) \cdot g(x),$$

where g is a polynomial having no root less than a. Note that $g(a) \cdot g(b) < 0$. We now work similarly with the roots $d_0, \ldots, d_{m-1}$ of f greater than b. They are precisely the roots of g greater than b. We write

$$g(x) = (x - d_0) \ldots (x - d_{m-1}) \cdot h(x),$$

with h having no roots either less than a or greater than b. Still we have $h(a) \cdot h(b) < 0$. Any root of h is a root of f between a and b.

Suppose there is a polynomial $f \in \mathbb{K}[X]$ and $a < b$ such that $f(a) \cdot f(b) < 0$ but f has no root in $\mathbb{K}$. We choose one such f of least degree. It is easily seen that f is irreducible. Then $\mathbb{F} = \mathbb{K}[X]/(f(X))$ is a proper algebraic extension of $\mathbb{K}$, and so not real. Set $\alpha = [X] \in \mathbb{F}$. We then get non-zero polynomials $g_i(X) \in \mathbb{K}[X]$, $1 \leq i \leq k$, such that $\sum_i g_i^2(\alpha) = 0$. Hence, $\sum_i g_i^2(X) = f(X) \cdot h(X)$ for some h. We choose such an h of the least degree. Note that we can arrange things so that the degree of each g_i is less than the degree of f. This implies that the degree of h is less than the degree of f. Since $f(a) \cdot h(a), f(b) \cdot h(b) \geq 0$, and $f(a) \cdot f(b) < 0$, either $h(a) = 0$ or $h(b) = 0$ or h changes sign between a and b. Since f was one such polynomial of least degree with no root, h has a root r in $\mathbb{K}$. Since $\mathbb{K}$ is real, r is a root of each

g_i. Hence, there exist polynomials $h_i(X) \in \mathbb{K}[X]$ such that $g_i(X) = (X - r) \cdot h_i(X)$, $1 \leq i \leq k$. Since f has no root in $\mathbb{K}$, this implies that $(X - r)^2$ divides $h(X)$. Let $h(X) = (X - r)^2 \cdot f_1(X)$. Now, we get $\sum_i h_i^2(X) = f(X) \cdot f_1(X)$ contradicting that h has the least possible degree satisfying such an identity. $\qquad\square$

Artin and Schreier proved the following crucial result.

Theorem B.3.10 *Let $\mathbb{K}$ be a real field. Then the following statements are equivalent.*

(a) $\mathbb{K}$ *has no proper real algebraic extension.*
(b) $\mathbb{K}$ *is real closed.*
(c) *The ring $\mathbb{K}[i] = \mathbb{K}[X]/(X^2 + 1)$ is algebraically closed.*

dummy

Proof Suppose $\mathbb{K}$ has no proper real algebraic extension. Let $<$ denote the canonical order of $\mathbb{K}$ and $a > 0$ be in $\mathbb{K}$. Consider the polynomial

$$f(X) = X^2 - a$$

in $\mathbb{K}[X]$. Then $f(0) < 0$ and $f(1+a) > 0$. By Weierstrass Nullstellensatz (Theorem B.3.9), there is a $c \in \mathbb{K}$ such that $f(c) = 0$. Thus every positive element of $\mathbb{K}$ has a square root in $\mathbb{K}$.

Now take a monic polynomial $f(X) \in \mathbb{K}[X]$ of odd degree. Then, arguing as in the case of $\mathbb{R}$, we can find $a < b$ such that $f(a) < 0 < f(b)$. Hence, f has a root in $\mathbb{K}$ by Weierstrass nullstellensatz. Thus, (a) implies (b).

We now show that (b) implies (c). We first note that it is sufficient to prove that every $f \in \mathbb{K}[X]$ has a root in $\mathbb{K}[i]$. To see this, take any $g \in \mathbb{K}[i][X]$. Let $\bar{g}$ denote the polynomial obtained from g by replacing all its coefficients by their conjugates. Then $g \cdot \bar{g} \in \mathbb{K}[X]$. Hence, by our assumption, it has a root, say α, in $\mathbb{K}[i]$. So, α is a root of either g or $\bar{g}$. If α is not a root of g, the conjugate $\bar{\alpha}$ of α is a root of g.

Fix $f \in \mathbb{K}[X]$. Let $d = 2^m(2n + 1)$ be the degree of f. By induction on m, we show that f has a root in $\mathbb{K}[i]$. If $m = 0$, the degree of f is odd. So, by (b), it has a root in $\mathbb{K} \hookrightarrow \mathbb{K}[i]$. Now assume that the assertion is true for $m - 1$. Let $r_1, \ldots, r_d$ be all the roots of f in an algebraic closure $\overline{\mathbb{K}}$ of $\mathbb{K}$. For any $k \in \mathbb{Z}$, consider the polynomial

$$g_k(X) = \Pi_{1 \leq p < q \leq d}(X - r_p - r_q - kr_p \cdot r_q) \in \overline{\mathbb{K}}[X].$$

This is invariant under all transpositions of $r_1, \ldots, r_d$ and so is symmetric in $r_1, \ldots, r_d$. Since all the elementary symmetric polynomials in $r_1, \ldots, r_d$ are coefficients of f, they belong to $\mathbb{K}$. It is well known that every symmetric polynomial in $r_1, \ldots, r_d$ are functions of elementary symmetric polynomials. Hence, each g_k is a polynomial over $\mathbb{K}$. The degree of g_k equals $\frac{d(d-1)}{2} = 2^{m-1}n'$, n' odd. By induction hypothesis, each g_k has a root in $\mathbb{K}[i]$. By pigeonhole principle, there exist $1 \leq p < q \leq d$ and $k \neq l$ such that

$$r_p + r_q + kr_p \cdot r_q, r_p + r_q + lr_p \cdot r_q \in \mathbb{K}[i].$$

For such p, q, $r_p + r_q$ and $r_p \cdot r_q$ are in $\mathbb{K}[i]$. It follows that $r_p, r_q \in \mathbb{K}[i]$. We have shown that (b) implies (c).

To show that (c) implies (a), first note that by (c) $\mathbb{K}[i]$ is the only non-trivial algebraic extension of $\mathbb{K}$. Further, $\mathbb{K}[i]$ is clearly not real. Hence, $\mathbb{K}$ has no proper real algebraic extension. □

Corollary B.3.11 *Let* $\mathbb{F}$, $\mathbb{K}$ *be real closed fields with* $\mathbb{F}$ *a subfield of* $\mathbb{K}$. *Then every root in* $\mathbb{K}$ *of a polynomial* $f(X)$ *over* $\mathbb{F}$ *lies in* $\mathbb{F}$.

Proof Let $x \in \mathbb{K}$ be a root of a polynomial $f(X) \in \mathbb{F}[X]$. Note that the subfield generated by $\mathbb{F}$ and x is a real algebraic extension of $\mathbb{F}$. But $\mathbb{F}$ has no proper real algebraic extension. Hence, $x \in \mathbb{F}$. □

Remark B.3.12 Let $\mathbb{F}$ be a real closed field and $a \neq 0$ be in $\mathbb{F}$. Then, a is a square if and only if $-a$ is not a sum of squares.

A real algebraic extension of $\mathbb{K}$ with no proper real algebraic extension will be called a *real closure* of $\mathbb{K}$.

Remark B.3.13 Consider the real field $\mathbb{F} = \mathbb{Q}(X)$ of rational functions over $\mathbb{Q}$. Clearly, $\mathbb{F}[\sqrt{X}]$ and $\mathbb{F}[\sqrt{-X}]$ are real fields. Their real closures are not isomorphic over $\mathbb{F}$.

However, there is a unique result for ordered fields.

Proposition B.3.14 *Let* $(\mathbb{K}, 0, 1, +, \cdot, <)$ *be an ordered field,* $0 < x \in \mathbb{K}$ *which is not a square in* $\mathbb{K}$, *then there is an order on the extension field* $\mathbb{K}[\sqrt{x}]$ *extending the order* $<$ *on* $\mathbb{K}$.

Proof For $a + b\sqrt{x}$, $c + d\sqrt{x}$ in $\mathbb{K}[\sqrt{x}]$, define

$$a + b\sqrt{x} < c + d\sqrt{x}$$

if any one of the following conditions is satisfied.

  (i) $b = d$ and $a < c$.
  (ii) $b < d$ and either $a < c$ or $\frac{(a-c)^2}{(b-d)^2} < x$.
  (iii) $b > d$ and $c > a$ and $x < \frac{(a-c)^2}{(b-d)^2}$.

It is entirely routine to check that this works. □

Theorem B.3.15 *Let* $(\mathbb{K}, <)$ *be an ordered field. Then there is a real closure of* $\mathbb{K}$ *whose canonical order is compatible with* $<$. *If* $\mathbb{K}_1$, $\mathbb{K}_2$ *are real closed algebraic extensions of* $\mathbb{K}$, *then there is a unique order-preserving isomorphism* $\alpha : \mathbb{K}_1 \to \mathbb{K}_2$ *fixing* $\mathbb{K}$.

Proof Consider

$$\mathbb{P} = \{(\mathbb{F}, <') : \mathbb{F} \text{ an ordered algebraic extension of } \mathbb{K} \ \wedge \ <' \,|\mathbb{K} \, = <\},$$

partially ordered by the inclusion $\subset$. By Zorn's lemma it has a maximal element, say $(\mathbb{K}', <')$. Hence, every positive element of $\mathbb{K}'$ has a square root in $\mathbb{K}'$. Note that $\mathbb{K}'$ does not have a proper real algebraic extension because then its canonical ordering would extend $<$ since every positive element of $\mathbb{K}$ has a square root in $\mathbb{K}'$. Clearly, $\mathbb{K}'$ satisfies the desired properties. We omit the proof of the uniqueness. $\square$

If $\mathbb{K}$ is an ordered field, a real closed algebraic extension of $\mathbb{K}$ preserving $<$ is called *the real closure* of $\mathbb{K}$. Note that $\mathbb{R}_{alg}$ is the real closure of $\mathbb{Q}$.

Appendix C
Valued Fields

We are nearly self-contained in this chapter. However, the interested reader is referred to [2–4] for further study.

C.1 Basic Definitions and Examples

Let $\mathbb{F}$ be a field and $(\Gamma, +, \leq)$ an ordered abelian group. A *valuation* on $\mathbb{F}$ is a surjective map $v : \mathbb{F} \to \Gamma \cup \{\infty\}$ satisfying the following three conditions:

(i) $v(x) = \infty$ iff $x = 0$.

(ii) $v(x \cdot y) = v(x) + v(y)$.

(iii) $v(x + y) \geq \min\{v(x), v(y)\}$.

In particular, a valuation v on $\mathbb{F}$ is an epimorphism from $\mathbb{F}^\times$ to Γ. The valuation v identically equal to 0 on $\mathbb{F}$ is called *the trivial valuation* on $\mathbb{F}$. The group Γ is called *the valuation group*.

Example C.1.1 Let $\mathbb{F} = \mathbb{Q}$ be the field of all rational numbers and $p > 1$ a prime. Let $r = p^i \frac{a}{b} \neq 0$, $p \nmid a, b$. Define $v_p(r) = i$. It is easily checked that this defines a valuation on $\mathbb{Q}$. v_p is called the *p-adic valuation* on $\mathbb{Q}$.

Example C.1.2 Let $\mathbb{K}$ be a field and $\mathbb{F} = \mathbb{K}(X)$, the field of rational functions over $\mathbb{K}$. For $\frac{f}{g} \neq 0$, define $v_\infty(\frac{f}{g}) = deg(g) - deg(f)$. It is easy to check that this defines a valuation on $\mathbb{F}$. v_∞ is called the *degree valuation* on $\mathbb{K}(X)$.

Example C.1.3 Let $\mathbb{F}$ be as in Example 1.2 and $p \in \mathbb{K}[X]$ an irreducible polynomial. For $r = p^i \frac{f}{g} \neq 0$ with $p \nmid f, g$, we define $v_p(r) = i$. It is easy to check that v_p is a valuation on $\mathbb{F}$. It is called the *p-adic valuation* on $\mathbb{K}(X)$.

Example C.1.4 Let $\mathbb{F}$ be a field and $\mathbb{F}((X))$ denote the field of formal Laurent series over $\mathbb{F}$. So, an element f in $\mathbb{F}((X))$ has a formal representation

© Springer Nature Singapore Pte Ltd. 2017
H. Sarbadhikari and S.M. Srivastava, *A Course on Basic Model Theory*,
DOI 10.1007/978-981-10-5098-5

$$f(X) = \sum_{i=m}^{\infty} a_i X^i,$$

where $m \in \mathbb{Z}$ and $a_m \neq 0$. In this case we define $v(f) = m$. It is easy to check that v is a valuation on $\mathbb{F}((X))$.

For any valuation v, we have

1. $v(1) = v(1 \cdot 1) = v(1) + v(1)$. This implies that $v(1) = 0$
2. $0 = v(1) = v(-1 \cdot -1) = 2v(-1)$. Since Γ is ordered, it is torsion-free. Hence, $v(-1) = 0$. Also, $v(-x) = v(-1) + v(x) = v(x)$
3. Let $x \in \mathbb{F}^{\times}$. Then $0 = v(1) = v(x \cdot x^{-1}) = v(x) + v(x^{-1})$. So, $v(x^{-1}) = -v(x)$.

Given a valuation $v : \mathbb{F} \to \Gamma \cup \{\infty\}$, define

$$V = \{x \in \mathbb{F} : v(x) \geq 0\}.$$

Then V is a subring of $\mathbb{F}$ with identity such that for every $x \in \mathbb{F}^{\times}$, at least one of x or x^{-1} belongs to V. Such a subring is called a *valuation subring* of $\mathbb{F}$.

A *valued field* consists of a field $\mathbb{F}$ and a valuation subring V of $\mathbb{F}$.

A valuation v on $\mathbb{F}$ is called compatible with V if $V = \{x \in \mathbb{F} : v(x) \geq 0\}$. Let $(\mathbb{F}, V)$ be a valued field and v a compatible valuation. Consider

$$M = \{x \in V : v(x) > 0\}.$$

Then M is an ideal in V. If $v(x) = 0$, $v(x^{-1}) = -v(x) = 0$. Thus, $V \setminus M = V^{\times}$, the set of all units of V. Hence, M is the unique maximal ideal of V. Note that an $x \in \mathbb{F}^{\times}$ is not in V iff $x^{-1} \in M$. Clearly, $V^{\times}$ is a subgroup of $\mathbb{F}^{\times}$. Further, $\mathbb{F}^{\sim} = V/M$ is a field, called the *residue field* of $(\mathbb{F}, V)$.

For $a \in V$, $[a]$ will denote its class in $\mathbb{F}^{\sim} = V/M$ and for $f \in V[\overline{X}], f^{\sim} \in \mathbb{F}^{\sim}[\overline{X}]$ is obtained by replacing each coefficient a of f by $[a]$.

Note that $char(\mathbb{F}) = p > 0$, then $char(\mathbb{F}^{\sim}) = p$. Hence, if $char(\mathbb{F}^{\sim}) = 0$, then $char(\mathbb{F}) = 0$. However, $char(\mathbb{F}^{\sim})$ need not be 0 even if $char(\mathbb{F}) = 0$. This is shown in the next example.

Example C.1.5 Let $p > 1$ be a prime number and v_p the p-adic valuation on $\mathbb{Q}$ as defined in Example C.1.1. Then the corresponding valuation subring equals

$$V_p = \left\{ \frac{a}{b} : (a, b) = 1 \ \& \ p \nmid b \right\}.$$

Thus $V_p = \mathbb{Z}_{(p)}$, the localization of $\mathbb{Z}$ by the prime ideal (p). The maximal ideal of V_p equals $p\mathbb{Z}_{(p)}$. It follows that the residue field is of characteristic p. Further, the residue field in this case equals $\mathbb{F}_p$. To see this, consider the quotient map $q : \mathbb{Z} \to \mathbb{Z}/p\mathbb{Z} = \mathbb{F}_p$. Clearly, for every $b \notin p\mathbb{Z}$, $q(b)$ is non-zero in $\mathbb{F}_p$. So, we define $h : \mathbb{Z}_{(p)} \to \mathbb{F}_p$

by $h(a/b) = q(a).q(b)^{-1}$, where $(a, b) = 1$ and $p \nmid b$. Its kernel is clearly $p\mathbb{Z}(p)$ and range $\mathbb{F}_p$. Thus, the residue field in this case is $\mathbb{F}_p$.

Example C.1.6 Next consider the p-adic valuation on $\mathbb{K}(X)$, $p \in \mathbb{K}[X]$ irreducible. Its valuation subring equals

$$\left\{ \frac{f}{g} \in \mathbb{K}(X) : (f, g) = 1 \ \& \ p \nmid g \right\} = \mathbb{K}[X]_{(p)},$$

the localization of $\mathbb{K}[X]$ by the prime ideal (p). Its maximal ideal is $p\mathbb{K}[X]_{(p)}$. Let $q : \mathbb{K}[X] \to \mathbb{K}[X]/(p)$ be the quotient map. For $g \notin (p)$, $q(g) \neq 0$. So, we have an epimorphism $h : \mathbb{K}[X]_{(p)} \to \mathbb{K}[X]/(p)$ defined by $h(f/g) = q(f)q(g)^{-1}$ whose kernel is $p\mathbb{K}[X]_{(p)}$. Thus, the residue field in this case if $\mathbb{K}[X]/(p)$.

Example C.1.7 Consider the degree valuation v_∞ on the field of rational functions $\mathbb{K}(X)$ over a field $\mathbb{K}$. Its valuation subring

$$V_\infty = \left\{ \frac{f}{g} \in \mathbb{K}(X) : degree(g) \geq degree(f) \right\}$$

whose units are precisely those $\frac{f}{g} \in \mathbb{K}(X)$ for which $degree(g) = degree(f)$. An element in the valuation subring has the representation $\frac{\sum_{i=0}^{m} a_i X^i}{\sum_{i=0}^{m} b_i X^i}$ with $b_m \neq 0$. We define $h : V_\infty \to \mathbb{K}$ by

$$h \left(\frac{\sum_{i=0}^{m} a_i X^i}{\sum_{i=0}^{m} b_i X^i} \right) = a_m/b_m.$$

This is an epimorphism with kernel the maximal ideal of V_∞. It follows that the residue field in this case is $\mathbb{K}$.

Example C.1.8 Consider the field $\mathbb{F}((X))$ of formal Laurent series over a field $\mathbb{F}$ and the valuation defined by

$$v \left(\sum_{i=m}^{\infty} a_i X^i \right) = m,$$

where $a_m \neq 0$. Then the corresponding valuation subring is the ring of all formal power series $\mathbb{F}[[X]]$ and the maximal ideal M consists of all power series of the form $\sum_{i=1}^{\infty} a_i X^i$. The residue field in this case is easily seen to be isomorphic to $\mathbb{F}$.

Lemma C.1.9 If $x_1, \ldots, x_n \in \mathbb{F}$ and for all $i \neq j$, $v(x_i) \neq v(x_j)$. Then $v(\sum_i x_i) = \min_i v(x_i)$.

Proof Suppose k is such that $v(x_k) = \min_i v(x_i)$. Then for $i \neq k$, $x_i x_k^{-1} \in M$. Now suppose $v(\sum_i x_i) > v(x_k)$. Then $x_k^{-1} \sum_i x_i \in M$. Thus, $1 \in M$, contradicting that M is a proper ideal of V.   □

Proposition C.1.10 *Let $v : \mathbb{F} \to \Gamma \cup \{\infty\}$ be a valuation with $V = \{x \in \mathbb{F} : v(x) \geq 0\}$. Then Γ is isomorphic to $\mathbb{F}^\times / V^\times$.*

Proof Consider the epimorphism $v : \mathbb{F}^\times \to \Gamma$. Then its kernel is $V^\times$. This implies the result.   □

Remark C.1.11 We can pull back the order on Γ to $\mathbb{F}^\times / V^\times$ to make this isomorphism preserve the order too. In other words, if we define

$$[x] \leq [y] \Leftrightarrow yx^{-1} \in V$$

we see that the ordered group Γ is isomorphic to the ordered group $\mathbb{F}^\times / V^\times$. This idea gives us the next theorem.

Theorem C.1.12 *Let V be a valuation subring of a field $\mathbb{F}$. Then there exists an ordered abelian group $(\Gamma, +, \leq)$ and a valuation $v : \mathbb{F} \to \Gamma \cup \{\infty\}$ such that $V = \{x \in \mathbb{F} : v(x) \geq 0\}$.*

Proof Let $V^\times$ be the set of all units of V. This is a subgroup of the (abelian) group $\mathbb{F}^\times$. We take $\Gamma = \mathbb{F}^\times / V^\times$. We shall denote the group operation of Γ by $+$. Let $v : \mathbb{F}^\times \to \Gamma = \mathbb{F}^\times / V^\times$ be the quotient map. Set $v(0) = \infty$. Since v is a homomorphism, we have

$$v(x \cdot y) = v(x) + v(y).$$

Define $\leq$ on Γ by

$$[x] \leq [y] \Leftrightarrow y \cdot x^{-1} \in V.$$

This is well-defined: Let $x \cdot (x')^{-1} \in V^\times$ and $y' \cdot y^{-1} \in V^\times$. Then

$$y \cdot x^{-1} = (y \cdot (y')^{-1}) \cdot (y' \cdot (x')^{-1}) \cdot (x' \cdot x^{-1}).$$

So, $y' \cdot (x')^{-1} \in V$ will imply that $y \cdot x^{-1} \in V$. Clearly, $\leq$ is a linear order on Γ.
    For $x, y, z \in \mathbb{F}^\times$, note that $(y \cdot z) \cdot (x \cdot z)^{-1} = y \cdot x^{-1}$. Therefore,

$$[x] \leq [y] \Rightarrow [x] + [z] \leq [y] + [z].$$

Next, if possible, let there exist $x, y \in \mathbb{F}^\times$ such that $v(x + y) < \min\{v(x), v(y)\}$, i.e. $[x + y] < [x]$ and $[x + y] < [y]$. Then,

$$x^{-1} \cdot (x + y) = 1 + x^{-1} \cdot y \notin V$$

and

$$y^{-1} \cdot (x + y) = 1 + y^{-1} \cdot x \notin V.$$

Since $1 \in V$, then neither of $x \cdot y^{-1}$, $y \cdot x^{-1}$ is in V. This contradicts that V is a valuation ring. So, we have

$$v(x + y) \geq \min\{v(x), v(y)\}.$$

Note that $[1] = 0$. So,

$$0 \leq [x] \Leftrightarrow x \in V.$$

This implies that

$$V = \{x \in \mathbb{F} : v(x) \geq 0\}.$$

□

We call valuations $v_i : \mathbb{F}^\times \to \Gamma_i$, $i = 1, 2$, on a field $\mathbb{F}$ *isomorphic* if

$$\{x \in \mathbb{F} : v_1(x) \geq 0\} = \{x \in \mathbb{F} : v_2(x) \geq 0\},$$

i.e. they induce the same valuation subrings. From the above arguments, it follows that.

Proposition C.1.13 *Two valuations* $v_i : \mathbb{F}^\times \to \Gamma_i$, $i = 1, 2$, *on a field* $\mathbb{F}$ *are isomorphic if and only if there is an order-preserving isomorphism* $\rho : \Gamma_1 \to \Gamma_2$ *such that* $v_2 = \rho \circ v_1$.

Now we determine the set of all valuations modulo isomorphism on the field $\mathbb{Q}$ of all rational numbers and the field of all rational functions $\mathbb{F}(X)$ over a field $\mathbb{F}$.

Proposition C.1.14 *Let* v *be a non-trivial valuation on* $\mathbb{Q}$. *Then there is a prime* $p > 1$ *such that* v *is isomorphic to the p-adic valuation* v_p *on* $\mathbb{Q}$.

Proof Let V be the valuation subring and M the unique maximal ideal of V. Clearly, V contains $\mathbb{Z}$. Since v is non-trivial, $V \neq \mathbb{Q}$ and $M \neq 0$. Now note that M must contain a non-zero integer. If not, then $v(a) = 0$ for every non-zero integer a implying that v is trivial. Since M is a prime ideal of V, by the prime decomposition, there is a prime $p > 1$ in M. If $q \neq p$ is another prime, we have $ap + bq = 1$ for some $a, b \in \mathbb{Z}$. Since M is proper, it follows that $q \notin M$. So, $v(q) = 0$ if $q \neq p$ is a positive prime. Hence, if $r = p^i \frac{a}{b}$ such that $p \not| a, b$, $v(r) = v(p^i) = iv(p)$ with $v(p) > 0$. Now it is easy to see that v is isomorphic to v_p. □

Proposition C.1.15 *Let* $\mathbb{F}$ *be a field and* v *a non-trivial valuation on* $\mathbb{F}(X)$ *such that* v *is trivial on* $\mathbb{F}$. *If* $v(X) \geq 0$, *then* v *is isomorphic to* v_p *for some irreducible polynomial* $p \in \mathbb{F}[X]$. *Otherwise,* v *is isomorphic to the degree valuation* v_∞.

Proof Let V be the valuation subring corresponding to v and M its unique maximal ideal.

Suppose $v(X) \geq 0$. Then $v(f) \geq 0$ for all polynomials $f \in \mathbb{F}[X]$, implying that $\mathbb{F}[X] \subset V$. Since v is non-trivial, M contains a monic irreducible polynomial p. Arguing as in the proof of the last proposition, we see that p is the only irreducible monic polynomial in M. As in the last proposition, we see that v is isomorphic to v_p.

If $v(X) < 0$, $v(X^{-1}) > 0$. So, $v(X^m) < v(X^n)$ whenever $m > n$. Now consider $\sum_{i=0}^{n} a_i X^{i-n}$ with $a_n \neq 0$. By Lemma C.1.9 it follows that

$$v\left(\sum_{i=0}^{n} a_i X^{i-n}\right) = v(a_n) = 0.$$

This implies that $v(\sum_{i=0}^{n} a_i X^i) = v(X^n)$ provided $a_n \neq 0$. Hence,

$$v\left(\frac{\sum_{i=0}^{n} a_i X^i}{\sum_{j=0}^{m} b_j X^j}\right) = v(X^n) - v(X^m) = (m-n)v(X^{-1}),$$

with $v(X^{-1}) > 0$ where $a_n, b_m \neq 0$. Now, it easily seen that v is isomorphic to v_∞.   □

In standard algebra, one has to start with an integral domain D and consider the quotient field of D. Similar situation arises for valued fields also. What should be an additional structure on an integral domain D so that its quotient field becomes a valued field with the valuation suitably connected with the additional structure on D? We take up this problem now.

Let $(\mathbb{F}, V)$ be a valued field and v be a compatible valuation V. For $a, b \in \mathbb{F}$, define

$$a|b \Leftrightarrow \exists c \in V (ac = b).$$

Note that for $a \neq 0$, $a|b$ if and only if $ba^{-1} \in V$. We have the following:

(a) $1|0$ & $0 \nmid 1$.
(b) $\forall x (x|x)$.
(c) $\forall x, y, z ((x|y \;\&\; y|z) \Rightarrow x|z)$.
(d) $\forall x, y (x|y \vee y|x)$.
(e) $\forall x, y \forall z \neq 0 (x|y \Leftrightarrow xz|yz)$.
(f) $\forall x, y, z ((z|x \;\&\; z|y) \Rightarrow z|x+y)$.
(g) $\forall x, y (x \nmid y \Leftrightarrow v(y) < v(x))$.

Most of these are trivial to prove. To see (d), note that $x \nmid y$ implies that $yx^{-1} \notin V$. So, $xy^{-1} = z \in V$. Thus, $yz = x$ implying that $y|x$.

To see (g), note that $x \nmid y$ implies that $y|x$. So, there exists $z \in V$ such that $x = yz$. Hence $v(x) = v(y) + v(z)$ and $v(z) \geq 0$. If $v(z) = 0$, $z \in V^\times$. But then $y = xz^{-1}$ and $z^{-1} \in V$. Hence, $x|y$, a contradiction. To prove the converse, without any loss of generality, we assume that $x \neq 0$. Now $v(y) < v(x)$ implies that $v(xy^{-1}) > 0$. So, $z = xy^{-1} \in M$. In particular, z is not a unit in V and $zy = x$. So, $x \nmid y$.

Also note the following:

$$\forall x(1|x \Leftrightarrow x \in V), \tag{*}$$

and

$$\forall x(x|1 \Leftrightarrow x^{-1} \in V).$$

Any binary relation $x|y$ satisfying (a)–(f), is called a *valuation divisibility relation* on $\mathbb{F}$. If $|$ is a valuation divisibility relation on $\mathbb{F}$ and $V = \{x \in \mathbb{F} : 1|x\}$, then V is a valuation ring of $\mathbb{F}$. Thus there is a one-to-one correspondence between valuation divisibility relations on $\mathbb{F}$ and valuation subrings of $\mathbb{F}$.

Our main interest in valuation divisibility relations stems from the following useful result.

Proposition C.1.16 *Let D be an integral domain and $x|y$ a binary relation on D satisfying (a)–(f). Then there is a unique valuation divisibility relation on the quotient field $\mathbb{F}$ of D extending $x|y$ on D.*

Proof For $\frac{a}{b}, \frac{c}{d}$, define

$$\frac{a}{b}|\frac{c}{d} \Leftrightarrow ad|bc.$$

The result follows. $\qquad\qquad\qquad\qquad\qquad\qquad\qquad\qquad\qquad\qquad\qquad\qquad\qquad\qquad\qquad\square$

We close this section with a simple general result on valued fields.

Proposition C.1.17 *Let $(\mathbb{F}, v)$ be an algebraically closed valued field. Then its value group, say Γ, is divisible.*

Proof Let $a \in \Gamma$ and $m > 1$ an integer. Since v is a surjection to $\Gamma \cup \{\infty\}$, there is a $y \in \mathbb{F}$ such that $v(y) = a$. Since $\mathbb{F}$ is algebraically closed, there is an $x \in \mathbb{F}$ such that $x^m = y$. Set $b = v(x) \in \Gamma$. Then $mb = v(x^m) = v(y) = a$. Thus, we have proved that Γ is divisible. $\qquad\qquad\qquad\qquad\qquad\qquad\qquad\qquad\qquad\qquad\square$

C.2 Extensions of Valuations on Rational Function Fields

Theorem C.2.1 *Let $v : \mathbb{F} \to \Gamma \cup \{\infty\}$ be a valuation. Assume that Γ is an ordered subgroup of an ordered group Γ' and $\gamma \in \Gamma'$. For $f(X) = \sum_{i=0}^{n} a_i X^i \in \mathbb{F}[X]$, define*

$$w(f) = \min\{v(a_i) + i\gamma : 0 \le i \le n\},$$

and for $\frac{f}{g} \in \mathbb{F}(X)$, define

$$w(\frac{f}{g}) = w(f) - w(g).$$

Then w is a well-defined valuation on $\mathbb{F}(X)$ extending v with $w(X) = \gamma$. Its value group is the subgroup of Γ' generated by $\Gamma \cup \{\gamma\}$.

Proof Clearly $w(0) = \infty$, $w(a) = v(a)$ for all $a \in \mathbb{F}$ and $w(X) = \gamma$.

Take $f(X) = \sum_{i=0}^{n} a_i X^i$ and $g(X) = \sum_{i=0}^{n} b_i X^i$ in $\mathbb{F}[X] \setminus \{0\}$. For every $0 \leq i \leq n$.

$$v(a_i + b_i) + i\gamma \geq \min\{v(a_i), v(b_i)\} + i\gamma$$
$$= \min\{v(a_i) + i\gamma, v(b_i) + i\gamma\}$$
$$\geq \min\{w(f), w(g)\}$$

Therefore,

$$w(f + g) \geq \min\{w(f), w(g)\}.$$

Now $fg = \sum_{k=0}^{2n} c_k X^k$, where $c_k = \sum_{i+j=k} a_i b_j$. Let $i + j = k$. Then

$$v(a_i b_j) + k\gamma = (v(a_i) + i\gamma) + (v(b_j) + j\gamma) \geq w(f) + w(g).$$

Hence, $w(fg) \geq w(f) + w(g)$.

To prove the opposite inequality, let

$$i_0 = \min\{i : w(f) = v(a_i) + i\gamma\},$$

and

$$j_0 = \min\{j : w(g) = v(b_j) + j\gamma\}.$$

Set $k = i_0 + j_0$. Now,

$$c_k = \left(\sum_{i<i_0, i+j=k} a_i b_j \right) + a_{i_0} b_{j_0} + \left(\sum_{j<j_0, i+j=k} a_i b_j \right).$$

For $i < i_0$, $v(a_i) + i\gamma > v(a_{i_0}) + i_0\gamma$ by the definition of i_0. Therefore, for $i < i_0$ and $i + j = k$,

$$v(a_i b_j) + k\gamma = (v(a_i) + i\gamma) + (v(b_j) + j\gamma) > v(a_{i_0} b_{j_0}) + k\gamma.$$

Hence, for all $i < i_0$ and $i + j = k$,

$$v(a_i b_j) > v(a_{i_0} b_{j_0}).$$

Similarly, for all $j < j_0$ and $i + j = k$,

$$v(a_i b_j) > v(a_{i_0} b_{j_0}).$$

By Lemma C.1.9, it follows that

$$v(c_k) = v(a_{i_0} b_{j_0}).$$

Thus,

$$v(c_k) + k\gamma = v(a_{i_0}) + i_0\gamma + v(b_{j_0}) + j_0\gamma = w(f) + w(g).$$

This implies that

$$w(fg) \leq w(f) + w(g).$$

Hence, $w(fg) = w(f) + w(g)$ for all polynomials $f, g \in \mathbb{F}(X)$.
Let $\frac{f_1}{g_1} = \frac{f_2}{g_2}$. Then $f_1 g_2 = f_2 g_1$. Therefore,

$$w(f_1) + w(g_2) = w(f_2) + w(g_1).$$

This implies that w is well-defined. It is now easy to check that

$$w\left(\frac{f_1}{g_1} \frac{f_2}{g_2}\right) = w\left(\frac{f_1}{g_1}\right) + w\left(\frac{f_2}{g_2}\right).$$

Assume that $w(\frac{f_1}{g_1}) \geq w(\frac{f_2}{g_2})$. Then $w(f_1) + w(g_2) \geq w(f_2) + w(g_1)$, i.e. $w(f_1 g_2) \geq w(f_2 g_1)$. Now,

$$
\begin{aligned}
w\left(\frac{f_1}{g_1} + \frac{f_2}{g_2}\right) &= w\left(\frac{f_1 g_2 + f_2 g_1}{g_1 g_2}\right) \\
&= w(f_1 g_2 + f_2 g_1) - w(g_1 g_2) \\
&\geq \min\{w(f_1 g_2), w(f_2 g_1)\} - w(g_1 g_2) \\
&\geq w(f_2) + w(g_1) - w(g_1) - w(g_2) \\
&= \min\{w(\frac{f_1}{g_1}), w(\frac{f_2}{g_2})\}
\end{aligned}
$$

Our result is proved now. $\square$

Theorem C.2.2 Let $v : \mathbb{F} \to \Gamma \cup \{\infty\}$ be a valuation and $\mathbb{K}$ a subfield. Let $v(x) = 0$ and $[x] \in \mathbb{F}^{\sim}$ be transcendental over $\mathbb{K}^{\sim}$. Then, whenever $a_0, \ldots, a_n \in \mathbb{K}$,

$$v\left(\sum_{i=0}^{n} a_i x^i\right) = \min\{v(a_i) : 0 \leq i \leq n\}.$$

Further, $v(\mathbb{K}(x)) = v(\mathbb{K})$ *and* $\mathbb{K}(x)^{\sim} = \mathbb{K}^{\sim}([x])$.

Proof Let $f = \sum_{i=0}^{n} a_i x^i \neq 0$. Choose a $k \leq n$ such that

$$v(a_k) = \min\{v(a_i) : 0 \leq i \leq n\}.$$

Then $f = a_k \sum_{i=0}^{n} b_i x^i$, where $b_i = a_i a_k^{-1}$. So, $v(b_i) \geq 0$ for each i and $b_k \neq 0$. As $v(x) = 0$, $v(\sum_{i=0}^{n} b_i x^i) \geq 0$.

Since $[x]$ is transcendental over $\mathbb{K}^\sim$, $\sum_{i=0}^n b_i x^i$ is a unit in the valuation subring of $\mathbb{F}$. Hence, $v(\sum_{i=0}^n b_i x^i) = 0$. It follows that

$$v(f) = v(a_k) = \min\{v(a_i) : 0 \le i \le n\}.$$

It is now clear that $v(\mathbb{K}(x)) = v(\mathbb{K})$.

Let $v(f) \ge 0$. Then each $v(a_i) \ge 0$. So,

$$[f] = \sum_{i=o}^n [a_i][x]^i \in \mathbb{K}^\sim([x]).$$

This implies that $\mathbb{K}(x)^\sim \subset \mathbb{K}^\sim([x])$.

Conversely, let $f \in V[x]$, where V is the valuation subring of $\mathbb{K}$. Then $v(f) \ge 0$. Hence, $[f] \in \mathbb{K}(x)^\sim$. This shows that $\mathbb{K}^\sim([x]) \subset \mathbb{K}(x)^\sim$. Thus, our result is proved. $\qquad\square$

Theorem C.2.3 *Let v be a valuation on a field $\mathbb{K}$ with value group Γ. Let $\mathbb{K}$ be a subfield of a field $\mathbb{F}$, Γ an ordered subgroup of an ordered group Γ' such that Γ'/Γ is torsion-free, $\gamma \in \Gamma' \setminus \Gamma$ and $x \in \mathbb{F}$ transcendental over $\mathbb{K}$. Then there is a unique valuation w on $\mathbb{K}(x)$ extending v such that $w(x) = \gamma$. Further, the value group of $\mathbb{K}(x)$ is $\Gamma \oplus \mathbb{Z}\gamma$ and the residue fields of $\mathbb{K}$ and $\mathbb{K}(x)$ are the same.*

Proof Let $f = \sum_{i=0}^n a_i x^i$ with $a_0, \ldots, a_n \in \mathbb{K}$. Arguing as in the proof of Theorem C.2.1, we see that

$$w(f) = \min\{v(a_i) + i\gamma : 0 \le i \le n\} \qquad (*)$$

and for $\frac{f}{g} \in \mathbb{K}(x)$,

$$w\left(\frac{f}{g}\right) = w(f) - w(g)$$

defines a valuation on $\mathbb{K}(x)$ extending v and such that $w(x) = \gamma$.

Let w be a valuation on $\mathbb{K}(x)$ extending v and such that $w(x) = \gamma$. We now show that w satisfies $(*)$ which will prove the uniqueness part of the result.

Let $f = \sum_{i=0}^n a_i x^i \in \mathbb{K}[x]$. Take $0 \le i \ne j \le n$. By our hypothesis, $w(a_i x^i) = v(a_i) + i\gamma$ for every $0 \le i \le n$. If $a_i = a_j$,

$$w(a_i x^i) - w(a_j x^j) = (i - j)\gamma \ne 0$$

because Γ'/Γ is torsion-free. Now assume that $a_i \ne a_j$. If possible, suppose $w(a_i x^i) = w(a_j x^j)$. But then

$$(i - j)\gamma = v(a_j) - v(a_i) \in \Gamma$$

contradicting that Γ'/Γ is torsion-free. Thus, whenever $0 \le i \ne j \le n$, $w(a_i x^i) \ne w(a_j x^j)$. This implies that

$$w(f) = \min\{w(a_i x^i) : 0 \le i \le n\} = \min\{v(a_i) + i\gamma : 0 \le i \le n\}.$$

Clearly, the value group of $\mathbb{K}(x)$ equals $\Gamma \oplus \mathbb{Z}\gamma$. The sum is direct because Γ'/Γ is torsion-free.

We now show that the residue fields of $\mathbb{K}$ and $\mathbb{K}(x)$ are the same.

Claim. *If* $h \in \mathbb{K}(x) \setminus \{0\}$, *h is of the form* $ax^r(1 + u)$ *with* $a \in \mathbb{K}$, $r \in \mathbb{Z}$ *and* $u \in \mathbb{K}(x)$ *with* $w(u) > 0$.

Assuming the claim we complete the proof first. Since $w(u) > 0 = w(1)$, by Lemma C.1.9,

$$w(1 + u) = \min\{w(1), w(u)\} = w(1) = 0.$$

Therefore, $w(h) = v(a) + r \cdot \gamma$. If $w(h) = 0$, $v(a) + r \cdot \gamma = 0$. Since $r \cdot \gamma \notin \Gamma$, it follows that $v(a) = 0$ and $r = 0$ when $w(h) = 0$. Then

$$w(h - a) = w(au) = v(a) + w(u) = w(u) > 0.$$

Thus $[h] = [a]$ in the residue field of $\mathbb{K}(x)$. Since $[a]$ is in the residue field of $\mathbb{K}$, the result follows.

Proof of the claim. Let $h = \frac{f}{g}$, where $f(x) = \sum_{i=0}^{n} a_i x^i$, $g(x) = \sum_{j=0}^{m} b_j x^j$, $a_i, b_j \in \mathbb{K}$, $0 \le i \le n$ and $0 \le j \le m$. We know that there exist $0 \le i_0 \le n$ and $0 \le j_0 \le m$ such that for all $i \ne i_0$ and for all $j \ne j_0$,

$$w(f) = v(a_{i_0}) + i_0 \cdot \gamma < v(a_i) + i \cdot \gamma$$

and

$$w(g) = v(b_{j_0}) + j_0 \cdot \gamma < v(b_j) + j \cdot \gamma.$$

We have

$$f = a_{i_0} x^{i_0} \left(1 + \sum_{i \ne i_0} \frac{a_i}{a_{i_0} x^{i_0}} x^i \right) = a_{i_0} x^{i_0}(1 + u_1), \text{ say}$$

and

$$g = b_{j_0} x^{j_0} \left(1 + \sum_{j \ne j_0} \frac{b_j}{b_{j_0} x^{j_0}} x^j \right) = b_{j_0} x^{j_0}(1 + u_2), \text{ say}.$$

By the choice of i_0 and j_0, $w(u_1)$, $w(u_2) > 0$. Set

$$u = \frac{u_1 - u_2}{1 + u_2}.$$

By Lemma C.1.9, $w(1 + u_2) = 0$. Further, $w(u_1 - u_2) = \min\{w(u_1), w(u_2)\} > 0$. Hence $w(u) > 0$. The claim follows since

$$\frac{f}{g} = \frac{a_{i_0}}{b_{j_0}} x^{i_0 - j_0} (1 + u).$$

$\square$

C.3 Valuations Induced by a Norm

The p-adic valuations have an important property that they are induced by the so-called non-Archimedean absolute values. An *absolute value* or a *norm* on a field $\mathbb{K}$ is a map $|\cdot| : \mathbb{K} \to [0, \infty)$ satisfying the following three conditions:

(a) $|x| = 0 \Leftrightarrow x = 0$.
(b) $\forall x, y(|x \cdot y| = |x||y|)$.
(c) $\forall x, y(|x + y| \le |x| + |y|)$.

The absolute value is called *non-Archimedean* if instead of (c) the following stronger property is satisfied:

(c') $\forall x, y(|x + y|) \le \max\{|x|, |y|\}$.

Otherwise, the absolute value is called *Archimedean*. These properties derive their names from the following observation.

Proposition C.3.1 *An absolute value $|\cdot|$ on a field $\mathbb{K}$ is non-Archimedean if and only if $\{|n| : n \in \mathbb{N}\}$ is bounded.*

Proof Suppose $|\cdot|$ is a non-Archimedean absolute value on $\mathbb{K}$. Then $|n| \le |1|$ for every $n \in \mathbb{N}$.
  Conversely, assume that $|n| < M$ for every $n \in \mathbb{N}$. Let $x, y \in \mathbb{K}$ and n a positive integer. Then

$$|x + y|^n = |(x + y)^n| = \left| \sum_{i=0}^{n} {}_nC_i x^i y^{n-i} \right| \le (n + 1)M \max\{|x|, |y|\}^n.$$

So,

$$|x + y| \le (n + 1)^{1/n} M^{1/n} \max\{|x|, |y|\}.$$

By taking limit as $n \to \infty$, we see that $|x + y| \le \max\{|x|, |y|\}$. $\square$

Example C.3.2 The usual absolute values on $\mathbb{Q}, \mathbb{R}$ and $\mathbb{C}$ are Archimedean absolute values.

Example C.3.3 Let $p > 1$ be a prime. Consider the field $\mathbb{Q}$ of all rational numbers. For a non-zero $r = p^i \frac{a}{b}$ with $p \nmid a, b$, define

$$|r|_p = e^{-i}.$$

Then $|\cdot|_p$ is a non-Archimedean absolute value on $\mathbb{Q}$. It is called the *p-adic absolute value*.

Example C.3.4 Now consider the field of all rational functions $\mathbb{K}(X)$ over a field $\mathbb{K}$. Let $p \in \mathbb{K}[X]$ be irreducible. For a non-zero rational function $f = p^i \frac{g}{h}, p \nmid g, h$, define

$$|f|_p = e^{-i}.$$

Then $|\cdot|_p$ defines a non-Archimedean absolute value on $\mathbb{K}(X)$. It is called the *p-adic absolute value*. An important special case is obtained by taking $p = X$.

Let $|\cdot|$ be an absolute value on a field $\mathbb{K}$. Then $|1| = |1 \cdot 1| = |1|^2$. Therefore, $|1| = 1$. Further, $|-1|^2 = |-1 \cdot -1| = |1| = 1$. Therefore, $|-1| = 1$. It follows that for every $x \neq 0$, $|-x| = |x|$ and $|x^{-1}| = |x|^{-1}$.

An absolute value on a field $\mathbb{K}$ canonically induces a metric on it which is of fundamental importance. If $|\cdot|$ is an absolute value on $\mathbb{K}$, we have a metric on $\mathbb{K}$ defined by $(x, y) \rightarrow |x - y|, x, y \in \mathbb{K}$.

We can carry out the standard completion of a metric space. Given an absolute value $|\cdot|$ on a field $\mathbb{K}$, we show that the metric completion $\overline{\mathbb{K}}$ is a field and $|\cdot|$ can be extended to a complete absolute value on $\overline{\mathbb{K}}$, where an absolute value is called *complete* if it induces a complete metric. We describe this briefly below.

For Cauchy sequences $(a_n), (a'_n)$, define $(a_n) \sim (a'_n)$ if $|a_n - a'_n| \rightarrow 0$ as $n \rightarrow \infty$. The $\sim$ is an equivalence relation on the set of all Cauchy sequences. Let $[(a_n)]$ denote the equivalence class of a Cauchy sequence (a_n). Let $(a_n) \sim (a'_n)$ and $(b_n) \sim (b'_n)$. Then $(a_n + b_n)$ and $(a_n \cdot b_n)$ are Cauchy sequences and $(a_n + b_n) \sim (a'_n + b'_n)$ as well as $(a_n \cdot b_n) \sim (a'_n \cdot b'_n)$. For $a \in \mathbb{K}$, we let $\bar{a}$ denote the equivalence class of the constant sequence $(a, a, a, \ldots)$. Also, note that $||a_n| - |a'_n|| \leq |a_n - a'_n|$. These show that $\overline{\mathbb{K}}$ is a field with $[(a_n)] + [(b_n)] = [(a_n + b_n)], [(a_n)] \cdot [(b_n)] = [(a_n \cdot b_n)]$, $\bar{0}$ the additive identity and $\bar{1}$ the multiplicative identity. Further, $|[(a_n)]| = \lim_n |a_n|$ defines a complete absolute value on $\overline{\mathbb{K}}$. Also, $a \rightarrow \bar{a}$ is an isometric monomorphic embedding of $\mathbb{K}$ onto a dense subfield of $\overline{\mathbb{K}}$.

Given an absolute value $|\cdot|$ on a field $\mathbb{K}$, define

$$v(x) = -\ln(|x|), \quad x \in \mathbb{K}. \tag{*}$$

Then v defines a valuation on $\mathbb{K}$ if and only if $|\cdot|$ is non-Archimedean. Now assume that v is a valuation on a field $\mathbb{K}$ with value group and ordered subgroup of the additive group of reals. Then

$$|x| = e^{-v(x)}, \quad x \in \mathbb{K},$$

defines a non-Archimedean norm on $\mathbb{K}$. Note that p-adic valuations defined above are induced by corresponding p-adic absolute values. A valuation will be called *non-Archimedean* if it is induced by a non-Archimedean absolute value or equivalently its value group is an additive subgroup of reals.

Let v and $|\cdot|$ be related by ($*$) and V the corresponding valuation subring. Then a sequence $\{x_n\}$ is Cauchy if and only if

$$\forall M > 0 \exists N \in \mathbb{N} \forall n, m \geq N (v(x_n - x_m) \geq M).$$

The valuation v is called *complete* if $\mathbb{K}$ is complete with respect to the metric induced by corresponding absolute value. Also note that $x_n \to x$ if and only if $v(x_n - x) \to \infty$.

Let V be the corresponding valuation subring and M the maximal ideal of V. Take any $x \in \mathbb{K}$. Then

$$x + M = \{z \in \mathbb{K} : v(z - x) > 0\} = \{z \in \mathbb{K} : |z - x| < 1\}.$$

Thus, $x + M$ is the open unit ball with centre at x.

Example C.3.5 Let $p > 1$ be a prime number. The completion of $\mathbb{Q}$ with respect to the p-adic absolute value is denoted by $\mathbb{Q}_p$. It is called the *field of p-adic real numbers*. We denote its valuation by v_p itself. Its valuation subring is denoted by $\mathbb{Z}_p$. Its elements are called *p-adic integers*. Clearly, $\mathbb{Q}_p$ is of characteristic 0.

Proposition C.3.6 *Let $|\cdot|$ be a non-Archimedean absolute value on a field $\mathbb{K}$ with v corresponding valuation. Let $\overline{\mathbb{K}}$ be the metric completion of $\mathbb{K}$ with respect to $|\cdot|$, $\overline{|\cdot|}$ the canonical extension of $|\cdot|$ and $\overline{v}$ corresponding valuation. Then the residue fields and the value groups of $\mathbb{K}$ and $\overline{\mathbb{K}}$ are isomorphic.*

Proof Let V be the valuation subring of $\mathbb{K}$ and M the maximal ideal of V. Also, let $\overline{V}$ be the valuation subring of $\overline{\mathbb{K}}$ and $\overline{M}$ the maximal ideal of $\overline{V}$.

There is a canonical monomorphism $\alpha : V/M \hookrightarrow \overline{V}/\overline{M}$. Let $x \in \overline{V}$. Then $x + \overline{M}$ is an open neighbourhood of x. Since $\mathbb{K}$ is dense in $\overline{\mathbb{K}}$, there exists a $y \in \mathbb{K}$ such that $y \in x + \overline{M} \subset \overline{V}$. Since $V = \overline{V} \cap \mathbb{K}$, $y \in V$. Clearly, the residue class of y is mapped to the residue class of x. So, α is an isomorphism.

Again, there is a canonical monomorphism $\beta : \mathbb{K}^\times/V^\times \hookrightarrow \overline{\mathbb{K}}^\times/\overline{V}^\times$. Now take an $x \in \overline{\mathbb{K}}^\times$. Since $\mathbb{K}$ is dense in $\overline{\mathbb{K}}$, there is $y \in \mathbb{K}$ such that $\overline{v}(x - y) > \overline{v}(x)$. Clearly, $y \neq 0$ and $v(y) = \overline{v}(y - x + x) = \overline{v}(x)$ by Proposition 1.1. So, $yx^{-1} \in \overline{V}^\times$. Hence, $\beta([y]) = [x]$. So, β is an isomorphism. $\square$

Corollary C.3.7 *The residue field of $\mathbb{Q}_p$ is $\mathbb{F}_p$.*

Corollary C.3.8 *The residue field of the completion of $(\mathbb{F}(X), v_p)$, $p \in \mathbb{F}[X]$ irreducible, is $\mathbb{F}[X]/(p)$.*

C.4 *p*-adic Expansion and Hensel's Lemma

A valued field $(\mathbb{K}, v)$ is called a *discrete-valued field of rank* 1 if its value group is isomorphic to $\mathbb{Z}$. $(\mathbb{Q}, v_p)$, $(\mathbb{Q}_p, v_p)$, $(\mathbb{F}(X), v_\infty)$, $(\mathbb{F}(X), v_p)$, $p \in \mathbb{F}[X]$ irreducible, are discrete-valued fields of rank 1. We observed in the last section that such a v is induced by a non-Archimedean norm.

Let $(\mathbb{K}, v)$ be a discrete-valued field of rank 1. An element $\pi \in \mathbb{K}$ is called a *normalizer* if $v(\pi) = 1$. Note that any two normalizers are associates in the valuation subring V of $\mathbb{K}$. Also, if $v(x) = n \neq 0$, then x and π^n are associates in V. Note that p is a normalizer for $(\mathbb{Q}_p, v_p)$, $p > 1$ a prime; X^{-1} is a normalizer for $(\mathbb{F}(X), v_\infty)$ and p is a normalizer of $(\mathbb{F}(X), v_p)$, $p \in \mathbb{F}[X]$ irreducible.

Theorem C.4.1 *Let* $(\mathbb{K}, v)$ *be a discrete-valued field of rank 1,* π *a normalizer and* $R \subset V$ *intersects each residue class in exactly one point with* $R \cap M = \{0\}$. *Then every* $x \in \mathbb{K}^\times$ *has a unique* π-*adic representation* $x = \sum_{i=n}^\infty a_i \pi^i$ *for some integer n with* $a_n, a_{n+1}, \ldots$ *in R. Moreover, if* $\mathbb{K}$ *is complete with respect to* v, *then for every sequence* $\{a_m : m \geq n\}$ *in R, the series* $\sum_{i=n}^\infty a_i \pi^i$ *is convergent.*

Proof Let $v(x) = n$. Then $v(x\pi^{-n}) = 0$. Take $a_n \in R$ that belongs to the residue class of $x\pi^{-n}$. Since $x\pi^{-n} - a_n$ is an element of the maximal ideal, $v(x\pi^{-n} - a_n) > 0$. Let $m = v(x - a_n\pi^n)$. Then

$$m = v(\pi^m) = v(x - a_n\pi^n) > n$$

and $v(\pi^{-m}(x - a_n\pi^n)) = 0$. Let $a_m \in R$ be in the residue class of $\pi^{-m}(x - a_n\pi^n)$. Set $k = v(x - a_n\pi^n - a_m\pi^m)$. Then

$$k = v(\pi^k) = v(x - a_n\pi^n - a_m\pi^m) > m$$

and $v(\pi^{-k}(x - a_n\pi^n - a_m\pi^m)) = 0$. We proceed similarly. Either the method terminates or we get an infinite series representation $x = \sum_{i=n}^\infty a_i \pi^i$. The series converges to x because $v(x - \sum_{i=n}^k a_i \pi^i) > k \to \infty$.

To see the uniqueness of this representation, assume that

$$\sum_{i=n}^\infty a_i \pi^i = \sum_{i=n}^\infty b_i \pi^i,$$

where $a_i, b_i \in R$ for all $i \geq n$ and there exists an i such that $a_i \neq b_i$. Let i_0 be the least i such that $a_i \neq b_i$. Then $a_{i_0} - b_{i_0} \in V^\times$. Hence,

$$v((a_{i_0} - b_{i_0})\pi^{i_0}) = i_0$$

and for all $i > i_0$, $v((a_i - b_i)\pi^i) \geq i > i_0$. Therefore, for every $l > i_0$,

$$v((a_{i_0} - b_{i_0})\pi^{i_0} + \sum_{i>i_0}^{l}(a_i - b_i)\pi^i) = i_0.$$

It follows that $v(\sum_{i=n}^{\infty}(a_i - b_i)\pi^i) \neq \infty$. This is a contradiction.

The last part of the result follows from the fact that the sequence of partial sums $\{\sum_{i=n}^{k} a_i\pi^i : k \geq n\}$ of the series $\sum_{i=n}^{\infty} a_i\pi^i$ is Cauchy. $\square$

Corollary C.4.2 *Every p-adic real number $x \in \mathbb{Q}_p$ has a unique p-adic expansion $x = \sum_{i=n}^{\infty} x_i p^i$ with each $x_i \in \{0, 1, \ldots, p-1\}$. Its valuation subring $\mathbb{Z}_p$ consists of all $x = \sum_{i=n}^{\infty} x_i p^i$ with $n \geq 0$ and each $x_i \in \{0, 1, \ldots, p-1\}$. It follows that $\mathbb{Z}_p$ is the closure of $\mathbb{Z}$ in $\mathbb{Q}_p$. The maximal ideal of $\mathbb{Z}_p$ is the set of all $x = \sum_{i=1}^{\infty} x_i p^i$ with $x_i \in \{0, 1, \ldots, p-1\}$.*

Corollary C.4.3 *The completion of $(\mathbb{F}(X), v_X)$ is the field of all Laurent series $\mathbb{F}((X))$. Its valuation subring is the ring of power series $\mathbb{F}[[X]]$ with the maximal ideal the set of all power series $\sum_{n=1}^{\infty} a_i X^i$ with $a_1, a_2, \ldots \in \mathbb{F}$.*

For $f(X) = \sum_{i=0}^{n} a_i X^i$, $f'(X) = \sum_{i=1}^{n} i a_i X^{i-1}$ is called the formal derivative of f.

Theorem C.4.4 *Let v be a non-Archimedean, complete valuation on a field $\mathbb{K}$ with V the corresponding valuation subring. Suppose $f \in V[X]$ and $a_0 \in V$ satisfies $v(f(a_0)) > 2v(f'(a_0))$. Then there exists an $a \in V$ with $f(a) = 0$ and $v(a - a_0) > v(f'(a_0))$.*

Proof Let $f(X) = \sum_{i=0}^{n} p_i X^i$. Set $b_0 = f'(a_0) \in V$. By the hypothesis, $v(b_0) < \infty$. So, $b_0 \neq 0$. Set $c_0 = \frac{f(a_0)}{b_0^2}$. Then $f(a_0) = c_0 b_0^2$. Further, $v(c_0) > 0$. Choose $\epsilon > 0$ such that $v(f(a_0)) \geq 2v(b_0) + \epsilon$. Set $a_1 = a_0 - b_0 c_0$. Then

$$\begin{aligned}
f(a_0 - b_0 c_0) &= \sum_{i=0}^{n} p_i (a_0 - b_0 c_0)^i \\
&= \sum_{i=0}^{n} \sum_{j=0}^{i} (-1)^j {}_i c_j p_i a_0^{i-j} (b_0 c_0)^j \\
&= f(a_0) + \sum_{i=0}^{n} \sum_{j=1}^{i} (-1)^j {}_i c_j p_i a_0^{i-j} (b_0 c_0)^j \\
&= f(a_0) - b_0 c_0 f'(a_0) + (b_0 c_0)^2 d_0,
\end{aligned}$$

where $d_0 \in V$. Thus, $f(a_1) = (b_0 c_0)^2 d_0$. Hence,

$$\begin{aligned}
v(f(a_1)) &\geq v(b_0^2) + 2v(\tfrac{f(a_0)}{b_0^2}) \\
&= 2v(f(a_0)) - v(b_0^2) \\
&\geq 2v(b_0^2) + 2\epsilon - v(b_0^2) \\
&= v(b_0^2) + 2\epsilon.
\end{aligned}$$

By similar computation, there exists a $b \in V$ such that $f'(a_1) = f'(a_0) - b_0c_0b = b_1$, say. Since c_0 is in the maximal ideal of V, so is c_0b. Hence, $v(bc_0) > 0$. Therefore, $v(1 - bc_0) = v(1) = 0$ by Lemma C.1.9. Hence, $v(b_1) = v(b_0)$.

Set $c_1 = \frac{f(a_1)}{b_1^2}$. Then $v(c_1) \geq 2\epsilon$. Set $a_2 = a_1 - b_1c_1$ and repeat the same argument with a_1 replaced by a_2, b_0 by b_1 and c_0 by c_1. Set $b_2 = f'(a_2)$ and $c_2 = \frac{f(a_2)}{b_2^2}$. Proceeding inductively, we get a sequence (a_n) in V such that $a_{n+1} = a_n - b_nc_n$, where $b_n = f'(a_n)$ and $c_n = \frac{f(a_n)}{b_n^2}$ with $v(b_n) = v(b_0)$ and $v(c_n) \geq 2^n\epsilon$.

We now show that (a_n) is Cauchy. Let $n < m$. Then

$$v(a_m - a_n) = v\left(\sum_{i=n}^{m-1}(a_{i+1} - a_i)\right)$$
$$\geq \min\{v(a_{i+1} - a_i) : i = n, \ldots, m-1\}$$
$$\geq v(b_0) + 2^n\epsilon \to \infty.$$

Let $a_n \to a$. Since each $a_n \in V$, $a \in V$. Note that

$$v(f(a_n)) = v(c_n) + 2v(b_n) = v(c_n) + 2v(b_0) \to \infty.$$

So, $v(f(a)) = \infty$ implying that $f(a) = 0$.

Since for every n, $v(a_n - a_0) \geq v(b_0) + \epsilon$. Hence, $v(a - a_0) > v(f'(a_0))$. $\qquad\square$

Theorem C.4.5 (Hensel's Lemma) *Let v be a complete, non-Archimedean valuation on a field $\mathbb{K}$ with V the corresponding valuation subring. Let $\mathbb{K}^\sim$ denote the corresponding residue field. Suppose $f \in V[X]$ and $a_0 \in V$ are such that $f^\sim([a_0]) = 0$ and $(f')^\sim([a_0]) \neq 0$. Then there exists an $a \in V$ such that $f(a) = 0$ and $[a] = [a_0]$. In particular, if*

$$f(X) = X^n + b_{n-1}X^{n-1} + b_{n-2}X^{n-2} + b_1X + b_0 \in V[X]$$

with $b_{n-1} \in V^\times$, $b_0, \ldots, b_{n-2} \in M$, then f has a root in V.

Proof Under the first part of the hypothesis, $f(a_0) \in M$ and $f'(a_0) \in V^\times$. Thus, $v(f'(a_0)) = 0$ and $v(f(a_0)) > 0$, implying that $v(f(a_0)) > 2v(f'(a_0))$. So, by the above proposition, there exists an $a \in V$ such that $f(a) = 0$ and $v(a - a_0) > v(f'(a_0)) = 0$. This implies that $a - a_0$ belongs to the maximal ideal. Hence, $[a] = [a_0]$.

Now let f be as in the second part of the hypothesis. Then

$$f^\sim(x) = X^{n-1}(X + [b_{n-1}])$$

and

$$f'^\sim(X) = X^{n-2}(nX + (n-1)[b_{n-1}]).$$

Then $f^\sim(-[b_{n-1}]) = 0$ and $f'^\sim([-b_{n-1}]) = [-b_{n-1}]^{n-1} \neq 0$. Now we use the first part to conclude our claim. $\qquad\square$

C.5   Algebraic Extensions of Valued Fields

We now turn our attention to extensions of valued fields. Suppose $(\mathbb{K}, V_{\mathbb{K}})$ is a valued field and $\mathbb{F}$ a subfield of $\mathbb{K}$. Set $V_{\mathbb{F}} = V_{\mathbb{K}} \cap \mathbb{F}$. Then $V_{\mathbb{F}}$ is a valuation subring of $\mathbb{F}$. In this case, we call $V_{\mathbb{K}}$ an *extension* of $V_{\mathbb{F}}$.

Note that $V_{\mathbb{F}}^{\times} = V_{\mathbb{K}}^{\times} \cap \mathbb{F}$. So, $M_{\mathbb{F}} = M_{\mathbb{K}} \cap V_{\mathbb{F}}$.

We have canonical morphisms $V_{\mathbb{F}} \hookrightarrow V_{\mathbb{K}} \to \mathbb{K}^{\sim}$ with kernel of the composition $M_{\mathbb{F}}$. So, we have a monomorphism $\mathbb{F}^{\sim} \to \mathbb{K}^{\sim}$. Thus, $\mathbb{F}^{\sim}$ can be regarded as a subfield of $\mathbb{K}^{\sim}$. We call $f = [\mathbb{K}^{\sim} : \mathbb{F}^{\sim}]$ the *residue index*.

As we saw, the valuation groups can be taken to be $\Gamma_1 = \mathbb{F}^{\times}/V_{\mathbb{F}}^{\times}$ and $\Gamma_2 = \mathbb{K}^{\times}/V_{\mathbb{K}}^{\times}$. Clearly, we can assume that the ordered group Γ_1 is embedded in the ordered group Γ_2. The index $[\Gamma_2 : \Gamma_1]$ is called the *ramification index*, and is denoted by e.

Theorem C.5.1 *Let $(\mathbb{F}, V)$ be a field and $\mathbb{K}$ an extension of $\mathbb{F}$. Then there is a valuation subring R of $\mathbb{K}$ extending V.*

Proof Let M be the unique maximal ideal of V. We are going to use Zorn's lemma to prove our result. Let $\mathbb{P}$ be

$$\{R : \mathbb{K} \supset R \supset V, R \text{ a ring and the ideal of } R \text{ generated by } M \text{ is proper}\}.$$

Since $V \in \mathbb{P}$, $\mathbb{P} \neq \emptyset$. The hypothesis of Zorn's lemma is satisfied with $\mathbb{P}$ equipped with inclusion relation. Let R be a maximal element of $\mathbb{P}$ and M_R the ideal in R generated by M.

Claim 1. $V = R \cap \mathbb{F}$.

If possible, suppose there exists an $x \in (R \cap \mathbb{F}) \setminus V$. Since V is a valuation ring, $x^{-1} \in V$. Since $x \notin V$, x^{-1} is not a unit in V, and so belongs to M. But then $1 = x \cdot x^{-1} \in M_R$ implying that M_R is not proper. This is a contradiction.

Claim 2. R is a valuation subring of $\mathbb{K}$.

Let $x \in \mathbb{K}$. If possible, suppose $x, x^{-1} \notin R$. We shall show that $R[x]$ or $R[x^{-1}]$ belongs to $\mathbb{P}$. This will contradict the maximality of R and our claim will be proved.

Suppose neither $R[x]$ nor $R[x^{-1}]$ belong to $\mathbb{P}$. Then 1 belongs to ideal in $R[x]$ generated by M as well as to the ideal in $R[x^{-1}]$ generated by M. So, we have

$$1 = \sum_{i=0}^{n} a_i x^i = \sum_{j=0}^{m} b_j x^{-j},$$

where a's and b's belong to M_R. We choose m and n minimum possible. Further, without any loss of generality, we assume that $1 \leq m \leq n$.

We have

$$1 - b_0 = (1 - b_0) \sum_{i=0}^{n} a_i x^i$$

and

$$a_n x^n = a_n x^n \sum_{j=0}^{m} b_j x^{-j}.$$

Thus,

$$(1 - b_0) a_n x^n = a_n \sum_{j=1}^{m} b_j x^{n-j}.$$

Hence,

$$1 - b_0 = (1 - b_0) \sum_{i=0}^{n-1} a_i x^i + a_n \sum_{j=1}^{m} b_j x^{n-j},$$

or

$$1 = b_0 + (1 - b_0) \sum_{i=0}^{n-1} a_i x^i + a_n \sum_{j=1}^{m} b_j x^{n-j}.$$

Since all the coefficients are in M_R, we get the contradiction of the minimality of n. $\qquad\square$

Our next result is to show that if a valued field $(\mathbb{F}_2, V_2)$ is an algebraic extension of $(\mathbb{F}_1, V_1)$, then the residue field $\tilde{\mathbb{F}_2}$ is an algebraic extension of $\tilde{\mathbb{F}_1}$. This will be shown easily by showing that if $\mathbb{F}_2$ is a finite extension of $\mathbb{F}_1$, then $\tilde{\mathbb{F}_2}$ is a finite extension of $\tilde{\mathbb{F}_1}$.

Let $(\mathbb{F}_1, V_1) \subset (\mathbb{F}_2, V_2)$ be valued fields with V_2 an extension of V_1, i.e. $V_1 = V_2 \cap \mathbb{F}_1$. Let $v_i : \mathbb{F}_i \to \Gamma_i \cup \{\infty\}$ be compatible valuations, $i = 1, 2$. We assume that Γ_1 is an ordered subgroup of Γ_2 and v_2 an extension of v_1. Let M_i be the unique maximal ideal of V_i, $i = 1, 2$. Set

$$\tilde{\mathbb{F}_i} = V_i / M_i, \quad i = 1, 2,$$

the corresponding residue fields.

Lemma C.5.2 *Let $\omega_1, \ldots, \omega_f \in V_2$ be such that $[\omega_1], \ldots, [\omega_f] \in \tilde{\mathbb{F}_2}$ are independent over $\tilde{\mathbb{F}_1}$ and $\pi_1, \ldots, \pi_e \in \mathbb{F}_2^{\times}$ such that $v_2(\pi_1), \ldots, v_2(\pi_e)$ are representatives of different cosets of Γ_2 / Γ_1. Then, for every $a_{ij} \in \mathbb{F}_1, 1 \le i \le f, 1 \le j \le e$,*

$$v_2 \left(\sum_{ij} a_{ij} \omega_i \pi_j \right) = \min \{ v_2(a_{ij} \omega_i \pi_j) \}.$$

In particular, $\{\omega_i \pi_j\}$ is an independent set over $\mathbb{F}_1$.

Proof Since each $[\omega_i] \neq 0$, $\omega_i \in V_2^\times$. So, $v_2(\omega_i) = 0$ for all i. Thus, for every i, j, $v_2(a_{ij}\omega_i\pi_j) = v_2(a_{ij}\pi_j)$.

Without loss of generality, assume that not all $a_{ij} = 0$. Choose $1 \leq I \leq f$ and $1 \leq J \leq e$ such that

$$v_2(a_{IJ}\pi_J) = \min\{v_2(a_{ij}\pi_j) : 1 \leq i \leq f, 1 \leq j \leq e\}.$$

Note that for every i, $v_2(a_{iJ}\pi_J) \geq v_2(a_{IJ}\pi_J)$. So, $v_2(a_{iJ}) \geq v_2(a_{IJ})$. Thus, $a_{iJ}a_{IJ}^{-1} \in V_1$. Also, note the following:

$$v_2(a_{IJ}\omega_I\pi_J) = \min\{v_2(a_{ij}\omega_i\pi_j) : 1 \leq i \leq f, 1 \leq j \leq e\}.$$

We claim that $v_2(a_{IJ}\pi_J) = v_2(a_{IJ}\omega_I\pi_J) < v_2(a_{ij}\omega_i\pi_j) = v_2(a_{ij}\pi_j)$ for all $j \neq J$ and for all i: Otherwise,

$$v_2(\pi_J) - v_2(\pi_j) = v_2(a_{ij}) - v_2(a_{IJ}) \in \Gamma_1.$$

This implies that

$$\forall i \forall j \neq J(a_{ij}\omega_i\pi_j(a_{IJ}\omega_I\pi_J)^{-1} \in M_2).$$

Hence,

$$\sum_i \sum_{j \neq J} a_{ij}\omega_i\pi_j(a_{IJ}\omega_I\pi_J)^{-1} \in M_2.$$

Set $z = \sum_{ij} a_{ij}\omega_i\pi_j$. If possible, suppose $v_2(z) > v_2(a_{IJ}\omega_I\pi_J)$. Then, $z(a_{IJ}\omega_I\pi_J)^{-1} \in M_2$. By subtracting, we get

$$\sum_i a_{iJ}(a_{IJ})^{-1}\omega_i\omega_I^{-1} \in M_2.$$

Since $\omega_I \in V_2$, we have

$$\sum_i a_{iJ}(a_{IJ})^{-1}\omega_i \in M_2.$$

We also have $a_{ij}\pi_j(a_{IJ}\pi_J)^{-1} \in M_2$ whenever $j \neq J$. We have observed that each $a_{iJ}(a_{IJ})^{-1} \in V_1$. Further, $a_{IJ}a_{IJ}^{-1} = 1$. So, $[\omega_1], \ldots, [\omega_f]$ are not independent over $\mathbb{F}_1^\sim$. This contradiction proves the first part of the result.

To show the independence of $\{\omega_i\pi_j\}$, let $a_{ij} \in \mathbb{F}_1$ be so chosen that $z = 0$. Then $v_2(a_{ij}\omega_i\pi_j) = v_2(a_{ij}) + v_2(\omega_i) + v_2(\pi_j) = \infty$ for all i, j. Since $v_2(\omega_i) = 0$, it follows that $v_2(a_{ij}) + v_2(\pi_j) = \infty$. But $v_2(\pi_j) \in \Gamma_2$. So, $v_2(a_{ij}) = \infty$ implying that $a_{ij} = 0$. $\square$

Corollary C.5.3 (Chevalley's Fundamental Inequality) *Let* $f = [\mathbb{F}_2^\sim : \mathbb{F}_1^\sim]$, $e = [\Gamma_2 : \Gamma_1]$ *and* $n = [\mathbb{F}_2 : \mathbb{F}_1]$. *If* $n < \infty$, *then*

$$e \cdot f \le n.$$

In particular, if $\mathbb{F}_2$ is a finite extension of $\mathbb{F}_1$, then $\mathbb{F}_2^{\sim}$ is a finite extension of $\mathbb{F}_1^{\sim}$ and Γ_1 is a subgroup of Γ_2 of finite index.

Theorem C.5.4 *Suppose* $(\mathbb{F}_1, V_1)$ *and* $(\mathbb{F}_2, V_2)$ *are valued fields with* $\mathbb{F}_1$ *a subfield of* $\mathbb{F}_2$ *and* V_2 *an extension of* V_1. *Let* Γ_1 *denote the value group of* $\mathbb{F}_1$ *and* Γ_2 *that of* $\mathbb{F}_2$. *If* $\mathbb{F}_2$ *is algebraic over* $\mathbb{F}_1$, $\mathbb{F}_2^{\sim}$ *is also algebraic over* $\mathbb{F}_1^{\sim}$ *and* Γ_2/Γ_1 *is a torsion group.*

Proof Let $x \in (V_2)^{\times}$. To show that $\mathbb{F}_1^{\sim}[[x]]$ is a finite extension of $\mathbb{F}_1^{\sim}$. Since x is algebraic over $\mathbb{F}_1$, $\mathbb{F}_1[x] = \mathbb{F}_1(x)$ is a field, say $\mathbb{L}$. Further, $\mathbb{L} \subset \mathbb{F}_2$ is a finite extension of $\mathbb{F}_1$. Set $V_{\mathbb{L}} = V_2 \cap \mathbb{L}$. Then $V_{\mathbb{L}}$ is a valuation subring of $\mathbb{L}$ and an extension of V_1. The unique maximal ideal of $V_{\mathbb{L}}$ is given by

$$M_{\mathbb{L}} = M_2 \cap \mathbb{L},$$

where M_2 the unique maximal ideal of V_2.

Since $\mathbb{L}$ is a finite extension of $\mathbb{F}_1$, by Chevalley's fundamental inequality (Corollary C.5.3), $\mathbb{L}^{\sim} = V_{\mathbb{L}}/M_{\mathbb{L}}$ is a finite extension of $\mathbb{F}_1^{\sim}$. In particular, $[x]_{\mathbb{L}}$ is algebraic over $\mathbb{F}_1^{\sim}$.

To show our result, it is sufficient to get $a_0, \dots, a_n \in V_1$, not all in M_1—the unique maximal ideal of V_1, such that $\sum_{i=0}^{n} a_i x^i \in M_{\mathbb{L}}$, or equivalently in M_2. Since $[x]^{\mathbb{L}}$ is algebraic over $\mathbb{F}_1^{\sim}$, there exist $a_0, \dots, a_n \in V_1$, not all in M_1, such that

$$\sum_{i=0}^{n} [a_i]^{\mathbb{L}}([x]^{\mathbb{L}})^i = 0,$$

i.e.

$$\sum_{i=0}^{n} a_i x^i \in M_{\mathbb{L}}.$$

Now take a $x \in \mathbb{F}_2 \setminus \mathbb{F}_1$. Since $\mathbb{F}_2$ is an algebraic extension of $\mathbb{F}_1$, $[\mathbb{F}_1[x] : \mathbb{F}_1] < \infty$. Let $\Gamma = v_2(\mathbb{F}_1[x])$. Then, by Chevalley's fundamental inequality (Corollary C.5.3), $[\Gamma : \Gamma_1] < \infty$. This, in particular, implies that there exists a $m > 1$ such that $m v_2(x) \in \Gamma_1$. Thus, we have proved that Γ_2/Γ_1 is a torsion group. $\square$

Theorem C.5.5 *Let* $(\mathbb{N}, V')$ *be an algebraic extension of* $(\mathbb{F}, V)$, v' *a compatible valuation on* $\mathbb{N}$ *and* $\sigma \in G(\mathbb{N}, \mathbb{F})$. *Then*

1. $v' \circ \sigma$ *is a valuation on* $\mathbb{N}$ *with value ring* $\sigma^{-1}(V')$.
2. *If* $\sigma(V') = V'$, *then* $v' \circ \sigma = v'$.

Proof The first part of the result is trivially verified. We proceed to prove the second part.

Let Γ' be the value group of $(\mathbb{N}, V')$ and $\Gamma \subset \Gamma'$ that of $(\mathbb{F}, V)$. Since $v' \circ \sigma$ is a valuation on $\mathbb{N}$ compatible with V', there exists an isomorphism $\rho : \Gamma' \to \Gamma'$ such that $\rho \circ v' = v' \circ \sigma$. To complete the proof, we show $\rho = id_{\Gamma'}$.

Step 1. $\rho|\Gamma = id_\Gamma$.

First, take $\gamma \in \Gamma$. Get $x \in \mathbb{F}^\times$ such that $\gamma = v'(x)$. But $\sigma|\mathbb{F} = id_\mathbb{F}$. So,

$$\rho(\gamma) = \rho(v'(x)) = v'(\sigma(x)) = v'(x) = \gamma.$$

Thus, $\rho|\Gamma = id_\Gamma$.

Now take any $x \in \mathbb{N} \setminus \mathbb{F}$. We need to show that $\rho(v'(x)) = v'(x)$. Since $\mathbb{N}$ is an algebraic extension of $\mathbb{F}$, $\mathbb{F}[x]$ is a finite extension of $\mathbb{F}$. Set $\Gamma'' = v'(\mathbb{F}[x])$. Then $[\Gamma'' : \Gamma] < \infty$ by Chevalley's fundamental inequality (Corollary C.5.3). This implies that there exists a positive integer n such that $nv'(x) \in \Gamma$. Hence, $n\rho(v'(x)) = \rho(nv'(x)) = nv'(x)$. Since Γ' is torsion-free, it follows that $\rho(v'(x)) = v'(x)$.   □

Lemma C.5.6 *Let $(\mathbb{F}_1, V_1)$ be a valued field, $\mathbb{F}_2$ an algebraic extension of $\mathbb{F}_1$, and V_2, V_2' valuation subrings of $\mathbb{F}_2$ extending V_1. Then, $V_2 \subset V_2'$ implies that $V_2 = V_2'$.*

Proof Assume that $V_2 \subset V_2'$. We have $M_2' \subset M_2 \subset V_2$. To see this take a $x \in M_2'$. If possible, suppose $x \notin V_2$. Then $x^{-1} \in V_2 \subset V_2'$. Since $x \in M_2'$, we get $1 \in M_2'$ contradicting that M_2' is a proper ideal. Thus $x \in V_2$. If $x \notin M_2$, $x^{-1} \in V_2 \subset V_2'$. This implies that $1 \in M_2'$ as before.

We see that V_2/M_2' is a subring of V_2'/M_2'.

Observation 1. V_2/M_2' is a valuation subring of $(\mathbb{F}_2')^\sim$.

Let $[x] \neq 0$ be an element of V_2'/M_2'. If $x \in V_2$, $[x] \in V_2/M_2'$. Otherwise, $x^{-1} \in V_2$. But then $[x]^{-1} \in V_2/M_2'$.

Observation 2. V_2/M_2' is a field.

For $x \in V_1$, let $[x] \in V_1/M_1$ and $[x]' \in V_2/M_2'$ be corresponding cosets. This defines a natural embedding of $\mathbb{F}_1^\sim = V_1/M_1$ into V_2/M_2'.

Let $0 \neq [x]' \in V_2/M_2' \subset V_2'/M_2'$. If $x \in V_1$, $[x]^{-1}$ exists in V_1/M_1. In particular, $[x]'$ has an inverse in V_2/M_2'. Now consider the case when $x \notin V_1$. Then $\mathbb{F}_1^\sim[[x]'] \subset V_2/M_2'$ is also a field because V_2/M_2' is an algebraic extension of $\mathbb{F}_1^\sim$. Thus, $[x]'$ has an inverse in V_2/M_2'. Since V_2/M_2' is already a ring, our claim is proved now.

Since V_2/M_2' is a field and a valuation subring of V_2'/M_2', $V_2/M_2' = V_2'/M_2'$. Now let $x \in V_2'$. Then, there exists a $y \in V_2$ such that $x - y \in M_2' \subset M_2 \subset V_2$. So, $x \in V_2$.   □

We shall use the following version of Chinese remainder theorem for valued fields repeatedly.

Theorem C.5.7 (Chinese Remainder Theorem) *Let $V_1, \ldots, V_k$ be valuation subrings of a field $\mathbb{F}$ with $M_1, \ldots, M_k$ their respective unique maximal ideals. Suppose $V_i \not\subset V_j$ whenever $1 \le i \ne j \le k$. Then for every $(a_1, \ldots, a_k) \in \times_{i=1}^{k} V_i$, there exists an $a \in \cap_{i=1}^{k} V_i$ such that $a = a_i (\bmod\ M_i)$, $1 \le i \le k$.*

Remark C.5.8 Let $V_1 \subset V_2$ be valuation subrings of a field $\mathbb{F}$ with $V_1 \ne V_2$ and M_1, M_2 the maximal ideals of V_1, V_2 respectively. In the proof of Lemma C.5.6, we saw that $M_2 \subset M_1$. Choose $a_1 \in V_1^{\times}$ and $a_2 \in M_2$. If possible, suppose there exists an $a \in V_1 \cap V_2$ such that $a - a_1 \in M_1$ and $a - a_2 \in M_2 \subset M_1$. Then $a_1 - a_2 \in M_1$. So, as $a_2 \in M_2 \subset M_1$, $a_1 \in M_1$. This is a contradiction. Hence, the result is not true without the condition $\forall 1 \le i \ne j \le n (V_i \not\subset V_j)$.

Proof Set $R = \cap_i V_i$. We need to show that the canonical morphism

$$R \to V_1/M_1 \times \cdots \times V_n/M_n$$

is an epimorphism. Set

$$P_i = R \cap M_i, \ 1 \le i \le n.$$

Then, P_i is a prime ideal of R, $1 \le i \le n$: Let $a \cdot b \in P_i = R \cap M_i$ and $a, b \in V_i \setminus M_i$. Then, $a^{-1}, b^{-1} \in V_i$. So, $1 = (a \cdot b) \cdot (b^{-1} \cdot a^{-1}) \in M_i$, contradicting that M_i is proper.

Our main result will be proved by proving the following two statements:

Claim I: The canonical morphism $R/P_i \to V_i/M_i$ is an isomorphism.

Claim II: The canonical morphism $R \to R/P_1 \times \cdots \times R/P_n$ is an epimorphism.

Since P_i is a prime ideal in R, $R \setminus P_i$ is a multiplicative set. Hence, the localization of R at P_i is possible. As usual, let

$$R_{P_i} = \{a/b : a, b \in R\ \&\ b \notin P_i\}$$

denote the localization of R at P_i. Note that if $b \in R \setminus P_i$, $b \in V_i^{\times}$, $a/b = ab^{-1}/1$. So, we treat R_{P_i} as a subring of V_i.

Fix $1 \le i \le n$.

The following is the main observation to prove our result.

Main Observation. $V_i = R_{P_i}$.

We draw several corollaries first.

(1) For all $1 \leq i \neq j \leq k$, $P_i \not\subset P_j$. Otherwise,

$$V_j = R_{P_j} \subset R_{P_i} \subset V_i,$$

contradicting our hypothesis

(2) For all $1 \leq j \leq k$,

$$\forall 1 \leq j \leq k (\cap_{i \neq j} P_i \not\subset P_j).$$

For each $1 \leq i \neq j \leq k$, choose $b_i \in P_i \setminus P_j$. These exist by (1). Then

$$c_j = \Pi_{i \neq j} b_i \in \cap_{i \neq j} P_i \setminus P_j.$$

(3) Every proper ideal in R is contained in some $P_1, \ldots, P_k$.

Let I be an ideal not contained in any of $P_1, \ldots, P_k$. For each $1 \leq j \leq k$, choose an $a_j \in I \setminus P_j$ and $c_j \in \cap_{i \neq j} P_i \setminus P_j$ as in (2).

Consider $d = \sum_j a_j c_j$. We claim that for every $1 \leq i \leq k$, $d \notin P_i$: Let $1 \leq k$, $d \in P_i$. But for all $j \neq i$, $a_j c_j \in P_i$. So, $a_i c_i \in P_i$. Since $a_i, c_i \notin P_i$ and P_i is prime, we have a contradiction.

Since $d \in I \subset R = \cap_l V_i$, it follows that $d^{-1} \in V_i$ for all i, i.e. $d^{-1} \in R$. But then $1 = dd^{-1} \in I$. So, I is not a proper ideal.

From (1) and (3), we get the following.

(4) $P_1, \ldots, P_k$ are distinct and are all the maximal ideals of R.

We assume these facts and complete the proof first.

Proof of claim I. Consider $R/P_i \hookrightarrow V_i/M_i = R_{P_i}/M_i$. Since $P_i = R \cap M_i$, $\hookrightarrow$ is a monomorphism.

Claim: Let $b \in R \setminus P_i$. Then the class of b^{-1} in V_i/M_i belongs to the range of $\hookrightarrow$.

This will complete the proof as follows: Let $ab^{-1} \in R_{P_i}$. Then the class $[ab^{-1}] = [a][b^{-1}]$ of ab^{-1} in R_{P_i}/M_i is in the range of $\hookrightarrow$.

Since P_i is a maximal ideal in R by observation (4) above, R/P_i is a field. So, there exists a $c \in R$ such that $[bc] = 1$, i.e. $1 - bc \in P_i \subset M_i$. This shows that the image of $[c]$ under $\hookrightarrow$ is $[b^{-1}]$. Hence, claim I is proved.

Proof of claim II. Choose $a_1, \ldots, a_n \in R$. To get an $a \in R$ such that $\forall i (a - a_i \in P_i)$. We prove this by induction on n.

Let $n = 2$. Since P_1, P_2 are distinct maximal ideals in R and $P_1 + P_2$ is an ideal containing P_1 and P_2, $P_1 + P_2 = R$. Given $a_1, a_2 \in R$, get $p_1 \in P_1$ and $p_2 \in P_2$ such that $a_1 - a_2 = p_1 - p_2$. Take $a = a_1 - p_1 = a_2 - p_2$. Thus, $a - a_1 \in P_1$ and $a - a_2 \in P_2$.

Inductive step. Let $a_1, \ldots, a_m \in R$. By induction hypothesis, suppose there exists $b_{m-1} \in R$ such that $b_{m-1} - a_i \in P_i$ for all $1 \leq i < m$. We shall produce

a $b_m \in R$ such that $b_m - b_{m-1} \in \cap_{i<m} P_i$ and $b_m - a_m \in P_m$. Then for $i < m$,
$b_m - a_i = (b_m - b_{m-1}) + (b_{m-1} - a_i) \in P_i$.

We saw in (2) that $\cap_{i<m} P_i \not\subset P_m$ and P_m is a maximal ideal. So, $(\cap_{i<m} P_i) + P_m = R$. Hence, there exist $q \in P_m$ and $p \in \cap_{i<m} P_i$ such that $b_{m-1} - a_m = p - q$. Set $b_m = b_{m-1} - p = a_m - q \in R$. Thus, $b_m - a_m \in P_m$ and $b_m - b_{m-1} \in \cap_{i<m} P_i$.

Proof of the main observation. We have already seen that $R_{P_i} \subset V_i$. It remains to show that $V_i \subset R_{P_i}$. Take any $a \in V_i$ and set

$$I_a = \{j : a \in V_j\}.$$

For $j \in I_a$, let α_j denote the class of a in V_j/M_j. Now choose a prime number $p > 2$ such that

$$\forall j \in I_a (p > \text{char}(V_j/M_j) \,\&\, \alpha_j \neq 1 \Rightarrow \alpha_j^p \neq 1).$$

Since for any field $\mathbb{K}$, if $x \neq 1$ is in $\mathbb{K}$, $x^p = 1$ for at most one prime p, such a prime p exists.

Now set $b = \sum_{k=0}^{p-1} a^k$.

We shall observe the following.

  (i) $\forall j \in I_a (b \in V_j)$. (This follows from the definition of I_a.)
 (ii) $\forall j \in I_a (b \notin M_j)$.
(iii) $\forall j \in I_a (b^{-1} \in V_j)$. (This follows from (i) and (ii).)
 (iv) $\forall j \notin I_a (b^{-1} \in V_j)$.
  (v) $b^{-1} \in R$. (This follows from (iii) and (iv).)
 (vi) $\forall j \in I_a (b^{-1} \in R \setminus P_j)$. (This follows from (i), (iii) and (v).)
(vii) $ab^{-1} \in R$.

Assuming these, note that $a = (ab^{-1})(b^{-1})^{-1} \in R_{P_i}$.

Proof of (ii). Fix $j \in I_a$. Then

$$[b] = \sum_{k=0}^{p-1} \alpha_j^k.$$

If $\alpha_j = 1$, $[b] = p \neq 0$ because $p > \text{char}(V_j/M_j)$. This implies that $b \notin M_j$ if $\alpha_j = 1$. Now assume that $\alpha_j \neq 1$. Then

$$[b] = \sum_{k=0}^{p-1} \alpha_j^k = \frac{1 - \alpha_j^p}{1 - \alpha_j} \neq 0.$$

So, $b \notin M_j$ in this case too.

Proof of (iv). Take a $j \notin I_a$. So, $a \notin V_j$. Since V_j is a valuation ring, it follows that $a^{-1} \in V_j$. Since $a \notin V_j$, a^{-1} is not a unit in V_j and so belongs to M_j. Since M_j is proper, it follows that

$$\sum_{k=0}^{p-1} a^{-k} \in V_j \setminus M_j = V_j^{\times}.$$

As $a^{-1} \in V_j$ and

$$b = \sum_{k=0}^{p-1} a^k = a^{p-1} \sum_{k=0}^{p-1} a^{-k},$$

$b^{-1} \in V_j$.

Proof of (vii). For $j \in I_a$, $ab^{-1} \in V_j$ by (iii). If $j \notin I_a$,

$$ab^{-1} = a^{-(p-2)} \left(\sum_{k=0}^{p-1} a^{-k} \right)^{-1} \in V_j$$

because $p > 2$. Thus, $ab^{-1} \in R$.   □

Theorem C.5.9 *Let $\mathbb{K}$ be an algebraic extension of $\mathbb{F}$, V a valuation subring of $\mathbb{F}$ and $\mathbb{F}^s$ the separable closure of $\mathbb{F}$. If $[\mathbb{K} \cap \mathbb{F}^s : \mathbb{F}] = n$, then $\mathbb{K}$ has at most n valuation subrings extending V.*

Proof Let $V_1, \ldots, V_m$ be distinct valuation subrings of $\mathbb{K}$, each extending V. Then $\forall 1 \leq i \neq j \leq n (V_i \not\subset V_j)$ by Lemma C.5.6. By Chinese remainder theorem (Theorem C.5.7), for each $1 \leq i \leq m$ there exists a $c_i \in \cap_{j=1}^m V_j$ such that $c_i - 1 \in M_i$ and for all other $1 \leq j \neq i \leq m$, $c_i \in M_j$.

Since $\mathbb{K} \subset \overline{\mathbb{F}}$, each c_i is algebraic over $\mathbb{F}$. So, if $\mathbb{F}$ is of characteristic 0, each $c_1, \ldots, c_m$ is separable over $\mathbb{F}$. Otherwise, there is a natural number k such that each $c_1^{p^k}, \ldots, c_m^{p^k}$ is separable over $\mathbb{F}$.

We now show that $c_1^{p^k}, \ldots, c_m^{p^k}$ are independent over $\mathbb{F}$. It will then follow that $m \leq n$ and the proof will be complete.

Let

$$\sum_{i=1}^m a_i c_i^{p^k} = 0,$$

$a_1, \ldots, a_m \in \mathbb{F}$ not all 0.

Choose j so that $v(a_j) \leq v(a_i)$ for all i. So, $v(a_j) < \infty$ implying that $a_j \neq 0$. Further,

$$c_j^{p^k} = -\sum_{i \neq j} a_i a_j^{-1} c_i^{p^k} \in M_j.$$

Since M_j is maximal, it is prime. Hence, $c_j \in M_j$. We also have $1 - c_j \in M_j$. Thus, $1 \in M_j$ contradicting the maximality of M_j. $\qquad\square$

We now proceed to prove one of our main theorems.

Proposition C.5.10 *Let* $(\mathbb{F}, V)$ *be a valued field,* $\mathbb{K}$ *a finite, normal, separable, extension of* $\mathbb{F}$ *and* V_1, V_2 *valuation subrings of* $\mathbb{K}$, *both extensions of* V. *Then there is an* $\mathbb{F}$-*automorphism* σ *of* $\mathbb{K}$ *such that* $\sigma(V_1) = V_2$.

Proof Let $G = G(\mathbb{K}, \mathbb{F})$. Since $\mathbb{K}$ is a finite, normal, separable, extension of $\mathbb{F}$, $|G| = |[\mathbb{K} : \mathbb{F}]| < \infty$.

Set

$$H_1 = \{\sigma \in G : \sigma(V_1) = V_1\}$$

and

$$H_2 = \{\sigma \in G : \sigma(V_2) = V_2\}.$$

We partition G into the cosets of H_1 and into the set of cosets of H_2:

$$G = \cup_{i=1}^{n} H_1 \sigma_i^{-1} = \cup_{j=1}^{m} H_2 \tau_j^{-1}.$$

Claim: $\exists i, j(\sigma_i(V_1) = \tau_j(V_2))$.

Then $\sigma = \tau_j^{-1} \circ \sigma_i$ will do the job.

Suppose $\forall i, j(\sigma_i(V_1) \neq \tau_j(V_2))$. Since $\sigma_1, \dots, \sigma_n$ are pairwise inequivalent modulo H_1, $\sigma_i(V_1) \neq \sigma_{i'}(V_1)$ for all $1 \leq i \neq i' \leq n$. Similarly, $\tau_j(V_2) \neq \tau_{j'}(V_2)$ for all $1 \leq j \neq j' \leq m$.

So, $\sigma_1(V_1), \dots, \sigma_n(V_1), \tau_1(V_2), \dots, \tau_m(V_2)$ are distinct extensions of V to $\mathbb{K}$. Let M_1 be the unique maximal ideal of V_1 and M_2 that of V_2. Hence, $\sigma_i(M_1)$ is the unique maximal ideal of $\sigma_i(V_1)$ for all $1 \leq i \leq n$, and $\tau_j(M_2)$ is the unique maximal ideal of $\tau_j(V_2)$ for all $1 \leq j \leq m$.

By the Chinese remainder theorem (Theorem C.5.7), there exists $a \in \cap_i \sigma_i(V_1) \cap \cap_j \tau_j(V_2)$ such that $a - 1 \in \cap_i \sigma_i(M_1)$ and $a \in \cap_j \tau_j(M_2)$.

Claim. $\forall \sigma \in G(\sigma(a - 1) \in M_1 \ \& \ \sigma(a) \in M_2)$.

Let $\sigma = \rho \circ \sigma_i^{-1} = \delta \circ \tau_j^{-1}$ where $\rho \in H_1$ and $\delta \in H_2$. Then $\sigma(a - 1) = \rho(\sigma_i^{-1}(a - 1))$. So, $\sigma(a - 1) \in \rho(M_1) = M_1$. Thus, $\sigma(a) - 1 \in M_1$ for all $\sigma \in G$. Similarly, we prove that $\sigma(a) \in M_2$.

Now let $N(a)$ be the product of all $\sigma(a)$, $\sigma \in G$. Then $N(a) \in M_2$ and $N(a) - 1 \in M_1$.

Claim. $N(a) \in \mathbb{F}$.

(This will imply that $N(a) \in M_2 \cap \mathbb{F} = M$ and $N(a) - 1 \in M_1 \cap \mathbb{F} = M$. Thus, $1 \in M$ contradicting the maximality of M and the proof will be complete.)

Let f be the minimal polynomial of a over $\mathbb{F}$. Since $\mathbb{K}$ is a normal extension of $\mathbb{F}$ and $a \in \mathbb{K}$, all its roots belong to $\mathbb{K}$. As seen before, $\{\sigma(a) : \sigma \in G\}$ is the set of all roots of f. So, $N(a)$ is the constant term of f. Hence, $N(a) \in \mathbb{F}$. $\qquad \Box$

Corollary C.5.11 *The condition of separability can be dropped from Proposition C.5.10.*

Proof Let $\mathbb{K}$ be a finite normal extension of $\mathbb{F}$ and V_1, V_2 valuation subrings of $\mathbb{K}$, both extending V.

Set $\mathbb{L} = \mathbb{F}^s \cap \mathbb{K}$. Then $\mathbb{F} \subset \mathbb{L} \subset \mathbb{K}$. So, $\mathbb{L}$ is a finite, separable extension of $\mathbb{F}$. Since $\mathbb{F} \subset \mathbb{L} \subset \mathbb{F}^s$, $\mathbb{L}^s = \mathbb{F}^s$. So, $\mathbb{L} = \mathbb{L}^s \cap \mathbb{K}$.

We now show that $\mathbb{L}$ is a normal extension of $\mathbb{F}$: Let $f(X) \in \mathbb{F}[X]$ be monic and irreducible with a root, say a, in $\mathbb{L}$. Since a is separable over $\mathbb{F}$, $f' \neq 0$. Since f is irreducible, it is the minimal polynomial of all its roots. Since $f' \neq 0$, all its roots are separable over $\mathbb{F}$. Since $\mathbb{K}$ is a normal extension of $\mathbb{F}$, it contains all roots of f because it contains a. Thus, $\mathbb{L}$ is a finite, separable, normal extension of $\mathbb{F}$.

By our assumption, there is an $\mathbb{F}$-automorphism σ of $\mathbb{L}$ such that $\sigma(V_1 \cap \mathbb{L}) = V_2 \cap \mathbb{L}$. Get an $\mathbb{F}$-automorphism τ of $\bar{\mathbb{F}}$ extending σ.

Then $\tau(\mathbb{K}) = \mathbb{K}$: Since $\mathbb{K}$ is a finite normal extension of $\mathbb{F}$, it is the splitting field of a polynomial $g(X) \in \mathbb{F}[X]$. So, τ maps roots of g to roots of g. The assertion is seen now.

$[\mathbb{K} \cap \mathbb{L}^s : \mathbb{L}] = [\mathbb{L} : \mathbb{L}] = 1$. Hence, by Theorem C.5.9, a valuation subring of $\mathbb{L}$ has exactly one extension to $\mathbb{K}$. Now $\tau(V_1) \supset V_2 \cap \mathbb{L}$ as well as $V_2 \supset V_2 \cap \mathbb{L}$. So, $\tau(V_1) = V_2$. $\qquad \Box$

Proposition C.5.12 *Proposition C.5.10 is true for all normal extensions of $\mathbb{F}$. In particular, this is true of $\mathbb{K} = \bar{\mathbb{F}}$.*

Proof Let $\mathbb{K}$ be a normal extension of $\mathbb{F}$ and V_1, V_2 valuation subrings of $\mathbb{K}$ extending V. We are going to use Zorn's lemma. Consider

$$\mathbb{P} = \{(\mathbb{L}, \tau) : \mathbb{F} \subset \mathbb{L} \subset \mathbb{K} \;\&\; \tau \in G(\mathbb{L}, \mathbb{F}) \;\&\; \tau(V_1 \cap \mathbb{L}) = V_2 \cap \mathbb{L}\}.$$

Then $(\mathbb{F}, id) \in \mathbb{P}$. By Zorn's lemma, $\mathbb{P}$ has a maximal element, say $(\mathbb{L}, \tau)$.

Claim. $\mathbb{L} = \mathbb{K}$.

Suppose not. Take $\alpha \in \mathbb{K} \setminus \mathbb{L}$. Let f be the minimal polynomial of α over $\mathbb{F}$ and $\mathbb{L}'$ the splitting field of f over $\mathbb{L}$. We extend τ to an $\mathbb{F}$-automorphism of $\bar{\mathbb{F}}$ and denote the extension by τ itself. Since τ keeps $\mathbb{F}$ fixed and f is a polynomial over $\mathbb{F}$, τ permutes the roots of f. Thus $\tau(\mathbb{L}') = \mathbb{L}'$.

Consider $V_1' = V_1 \cap \mathbb{L}'$ and $V_1'' = \tau^{-1}(V_2 \cap \mathbb{L}')$. Note that $\mathbb{L}'$ is a finite, normal extension of $\mathbb{L}$. Hence, by the finite case, there is an $\mathbb{L}$-automorphism ρ of $\mathbb{L}'$ such that $\rho(V_1') = V_1''$. Now, set $\sigma = \tau \circ \rho$. Then σ is an $\mathbb{F}$-automorphism of $\mathbb{L}'$ extending τ and $(\mathbb{L}', \sigma) \in \mathbb{P}$. This contradicts the maximality of $(\mathbb{L}, \tau)$. $\qquad \Box$

C.6   Henselian Valued Fields

A valued field $(\mathbb{F}, V)$ is called *Henselian* if for every algebraic extension $\mathbb{L}$ of $\mathbb{F}$, V has a unique extension to $\mathbb{L}$.

Proposition C.6.1 *The following are equivalent:*

(a) $(\mathbb{F}, V)$ *is Henselian.*
(b) V *has a unique extension to* $\overline{\overline{\mathbb{F}}}$.
(c) V *has a unique extension to* $\mathbb{F}^s$.
(d) V *has a unique extension to every finite, normal, separable extension of* $\mathbb{F}$.

Proof Clearly (a) implies (b), (c) and (d).

Suppose there is a $\mathbb{K} \subset \overline{\mathbb{F}}$ to which V has extensions $V_1 \neq V_2$. By Theorem C.5.1, these can be extended to $\overline{\mathbb{F}}$. Thus (b) implies (a).

Suppose $V_1 \neq V_2$ are two extensions of V to $\overline{\mathbb{F}}$. Since $(\mathbb{F}^s)^s = \mathbb{F}^s \subset \overline{\mathbb{F}}$, we have $(\mathbb{F}^s)^s \cap \overline{\mathbb{F}} = \mathbb{F}^s$, by Theorem C.5.9, $V_1 \cap \mathbb{F}^s \neq V_2 \cap \mathbb{F}^s$ and both are extensions of V to $\mathbb{F}^s$. Thus, (c) implies (b).

Suppose $V_1 \neq V_2$ be extensions of V to $\mathbb{F}^s$. Then by Lemma C.5.6, $V_1 \not\subset V_2$. Let $\alpha \in V_1 \setminus V_2, f$ the minimal polynomial of α over $\mathbb{F}$ and $\mathbb{K}$ its splitting field over $\mathbb{F}$. Then $\mathbb{K}$ is a finite, normal, separable extension of $\mathbb{F}$ and $\alpha \in (V_1 \cap \mathbb{K}) \setminus (V_2 \cap \mathbb{K})$. Thus, (d) implies (c). $\square$

Let $(\mathbb{F}, V)$ be a valued field with $v : \mathbb{F} \to \Gamma \cup \{\infty\}$ a compatible valuation. Note that we can extend this valuation to a valuation on the field of rational functions $\mathbb{F}(X)$ as follows:

$$w(a_0 + a_1 X + \cdots + a_n X^n) = \min\{v(a_0), \ldots, v(a_n)\}.$$

and

$$w(f/g) = w(f) - w(g).$$

Call an $f \in \mathbb{F}[X]$ *primitive* if $w(f) = 0$. We have the following facts:

(1) If f is primitive, $f \in V[X]$.
(2) $f \cdot g$ is primitive whenever f and g are primitive.
(3) Every $f \in \mathbb{F}[X]$ is of the form af_1 where $a \in \mathbb{F}$ and f_1 primitive: take a to be a coefficient of f with minimum valuation.
(4) Let $f \in V[X] \subset \mathbb{F}[X]$ and $f = g_1 \ldots g_n$ with g_i's irreducible in $\mathbb{F}[X]$. Then there exists a constant multiple $h_i \in V[X]$ of g_i, $1 \leq i \leq n$, such that $f = h_1 \ldots h_n$. (Write $f = af_1$ with f_1 primitive and $g_i = b_i h_i$ with h_i primitive, $1 \leq i \leq n$. Then

$$f = af_1 = bh_1 \ldots h_n,$$

where $b = b_1 \ldots b_n$. But then $v(b) = v(a) \geq 0$. Now replace h_1 by bh_1.)
(5) Thus, very polynomial over V is a product of polynomials over V which are irreducible in $\mathbb{F}[X]$. In particular, if $\mathbb{F}$ is algebraically closed, so is $\mathbb{F}^\sim$.

(6) Assume that $f \in V[X]$ is monic and $f = h_1 \ldots h_n$ with each $h_i \in V[X]$ and irreducible in $\mathbb{F}[X]$. Let b_i be the leading coefficient of h_i, $1 \le i \le n$. Then $1 = b_1 \ldots b_n$. Hence, each b_i is a unit in V. Now take $g_i = b_i^{-1} h_i$. Then, each $g_i \in V[X]$, is monic, irreducible in $\mathbb{F}[X]$, and $f = g_1 \ldots g_n$.

Theorem C.6.2 *Let* $(\mathbb{F}, V)$ *be a valued field. The following are equivalent.*

(a) $(\mathbb{F}, V)$ *is Henselian.*

(a') *If* $f \in V[X]$ *is monic and irreducible in* $\mathbb{F}[X]$, *then there exists a* $g \in V[X]$ *monic with* $g^{\sim} \in \mathbb{F}^{\sim}[X]$ *irreducible such that* $f^{\sim} = (g^{\sim})^s$ *for some* $s \ge 1$.

(b) *Let* $f, g, h \in V[X]$ *be monic and* $f^{\sim} = g^{\sim} h^{\sim}$ *with* $g^{\sim}$, $h^{\sim}$ *relatively prime in* $\mathbb{F}^{\sim}[X]$. *Then there exist monic* $g_1, h_1 \in V[X]$ *with* $f = g_1 \cdot h_1$ *and* $g_1^{\sim} = g^{\sim}$, $h_1^{\sim} = h^{\sim}$. *In particular,* $deg(g) = deg(g^{\sim}) = deg(g_1^{\sim}) = deg(g_1)$ *and* $deg(h) = deg(h^{\sim}) = deg(h_1^{\sim}) = deg(h_1)$.

(c) *(Hensel Lemma) Suppose* $f \in V[X]$ *is monic and* $f^{\sim}$ *has a simple root* $a^{\sim} \in \mathbb{F}^{\sim}$. *Then there exists* $a_1 \in V$ *such that* $a_1^{\sim} = a^{\sim}$ *and* $f(a_1) = 0$.

(c') *(Eisenstein Criterion) Suppose* $f(x) = X^n + a_{n-1}X^{n-1} + a_{n-2}X^{n-2} + \cdots + a_1 X + a_0 \in V[X]$ *with* $a_{n-1} \notin M$ *and* $a_{n-2}, \ldots, a_0 \in M$. *Then* f *has a root in* $(V \subset) \mathbb{F}$.

(c") *(c') under the additional hypothesis that* f *has all roots distinct in* $\overline{\mathbb{F}}$.

Proof (a) implies (a'): Let $(\mathbb{F}, V)$ be Henselian and $f \in V[X]$ be monic and irreducible in $\mathbb{F}[X]$. Let $\overline{V}$ be the extension of V to the algebraic closure $\overline{\mathbb{F}}$ and v a compatible valuation on $\overline{\mathbb{F}}$. Note that there is a canonical embedding of $\mathbb{F}^{\sim}$ into $\overline{V}/\overline{M}$.

Write
$$f(x) = \Pi_{i=1}^n (x - x_i),$$

where $x_i \in \overline{\mathbb{F}}$, $1 \le i \le n$. The product $x_1 \ldots x_n$, being the constant term of f, is in V. Therefore,
$$v(x_1 \ldots x_n) = v(x_1) + \cdots + v(x_n) \ge 0.$$

Hence, at least one $x_i \in \overline{V}$. Since f is irreducible, for every $1 \le j \le n$, there is an $\mathbb{F}$-automorphism σ of $\overline{\mathbb{F}}$ such that $\sigma(x_i) = x_j$. But $\sigma(\overline{V})$ is an extension of V to $\overline{\mathbb{F}}$ and $\mathbb{F}$ is Henselian. Hence, $\sigma(\overline{V}) = \overline{V}$. Therefore, every $x_j \in \overline{V}$. Also, either all x_i or no x_i belong to the unique maximal ideal $\overline{M}$ of $\overline{V}$.

If all $x_i \in \overline{M}$,
$$f^{\sim}(X) = \pi_{i=1}^n(X - [x_i]) = X^n,$$

and X is irreducible.

Now consider the case when each x_i is a unit in $\overline{V}$.

Claim: $f^{\sim}$ cannot be written as the product of two relatively prime polynomials in $\mathbb{F}^{\sim}[X]$.

This will imply that $f^\sim$ is a power of an irreducible polynomial $g^\sim \in \mathbb{F}^\sim[X]$. To get g monic, note that, without any loss of generality, we can assume that the coefficients of g are units in V. We write $g = ag_1$ with $g_1 \in V[X]$ monic and a a unit in V. If $f^\sim = (g^\sim)^s$, $s \geq 1$, $f^\sim = (a^\sim)^s(g_1^\sim)^s$. But $f^\sim$ and $g_1^\sim$ are monic. So, $(a^\sim)^s = 1$ and (a') will be proved.

Proof of the claim. If possible, suppose

$$f^\sim = \Pi_{i=1}^n (X - [x_i]) = g^\sim \cdot h^\sim \in \mathbb{F}^\sim[X] \subset \overline{V/M}[X].$$

So, $[x_i]$'s give all the roots of $g^\sim$ and $h^\sim$. These being relatively prime, their sets of roots are disjoint. Say, $[x_j]$'s are roots of $g^\sim$ and $[x_k]$'s that of $h^\sim$.

Since f is irreducible, there is an $\mathbb{F}$-automorphism σ of $\overline{\mathbb{F}}$ with $\sigma(x_j) = x_k$. Since $[x_j]$ is a root of $g^\sim$, we have $g(x_j) \in \overline{M}$. Then $g(x_k) = \sigma(g(x_j)) \in \sigma(\overline{M}) = \overline{M}$ because $(\mathbb{F}, V)$ is Henselian. This is a contradiction.

<u>Proof of (a') implies (b):</u> Let $f, g, h \in V[X]$ be monic, $g^\sim, h^\sim \in \mathbb{F}^\sim[X]$ be relatively prime and $f^\sim = g^\sim \cdot h^\sim$. To get monic $g_1, h_1 \in V[X]$ such that $g^\sim = g_1^\sim$, $h^\sim = h_1^\sim$ and $f = g_1 \cdot h_1$.

By observation (6) made above, let $f = f_1 \dots f_n$, where $f_1, \dots, f_n \in V[X]$ are monic and irreducible in $\mathbb{F}[X]$. For each $1 \leq i \leq n$, get monic $g_i \in V[X]$ and $s_i \geq 1$ such that $g_i^\sim$ is irreducible in $\mathbb{F}^\sim[X]$ and $f_i^\sim = (g_i^\sim)^{s_i}$. Thus,

$$f^\sim = f_1^\sim \dots f_n^\sim = \Pi_{i=1}^n (g_i^\sim)^{s_i} = g^\sim \cdot h^\sim.$$

Since $g^\sim$ and $h^\sim$ are relatively prime, there exists $I \subset \{1, \dots, n\}$ such that

$$g^\sim = \Pi_{i \in I}(g_i^\sim)^{s_i} \ \& \ h^\sim = \Pi_{i \notin I}(g_i^\sim)^{s_i}.$$

Take $g_1 = \Pi_{i \in I} f_i$ and $h_1 = \Pi_{i \notin I} f_i$.

<u>Proof of (b) implies (c):</u> Let $f \in V[X]$ be monic and $f^\sim$ have a simple root $a^\sim \in \mathbb{F}^\sim$. So,

$$f^\sim(X) = (X - a^\sim)g^\sim = (X - a)^\sim g^\sim$$

with $(X - a)^\sim$ and $g^\sim$ relatively prime. Note that $g^\sim$ is also monic. We can write $f = g_1 \cdot h_1$, $g_1, h_1 \in V[X]$ monic and $g_1^\sim = X - a^\sim$. We must have $g_1 = X - a_1$. Thus $f(a_1) = 0$ and $a^\sim = a_1^\sim$.

<u>Proof of (c) implies (c'):</u> Let

$$f(X) = X^n + a_{n-1}X^{n-1} + a_{n-2}X^{n-2} + \cdots + a_1 X + a_0$$

with $a_{n-1} \notin M$ and $a_{n-2}, \dots, a_1, a_0 \in M$. So,

$$f^\sim = X^{n-1}(X + a_{n-1}^\sim).$$

Thus, $-a_{n-1}^{\sim}$ is a simple root of $f^{\sim}$. Hence, f has a root in $V \subset \mathbb{F}$ by (c).

Proof of (c") implies (a): Let $(\mathbb{F}, V)$ satisfy (c") and be not Henselian. Let $\mathbb{K}$ be a finite, normal, separable extension of $\mathbb{F}$ with V having more than one extensions to $\mathbb{K}$. Say, $V_1, \ldots, V_m, m > 1$, are all the extensions of V to $\mathbb{K}$.

Consider
$$H = \{\sigma \in G(\mathbb{K}, \mathbb{F}) : \sigma(V_1) = V_1\}$$

and $\mathbb{L} = F(H) \supset \mathbb{F}$. By Galois correspondence (Theorem B.1.9), $G(\mathbb{K}, \mathbb{L}) = H$.

This inclusion is proper: If not, then $F(H) = \mathbb{F} = F(G(\mathbb{K}, \mathbb{F}))$. Since Galois correspondence is one-to-one (Theorem B.1.9), $H = G(\mathbb{K}, \mathbb{F})$. But by Proposition C.5.10, there exists a $\sigma \in G(\mathbb{K}, \mathbb{F}) = H$ such that $\sigma(V_1) = V_2$. This is a contradiction because $V_1 \neq V_2$.

Set $V_i' = V_i \cap \mathbb{L}$, $i = 1, \ldots, m$.

Claim: $V_1' \neq V_i'$ for all $i \neq 1$.

If possible, suppose there exists an $i > 1$ such that $V_1' = V_i' = V_{\mathbb{L}}$, say. So, V_1 and V_i are two extensions of $V_{\mathbb{L}}$. Also, $\mathbb{K}$ is a normal extension of $\mathbb{L}$. So, by the last Corollary C.5.11, there exists a $\sigma \in G(\mathbb{K}, \mathbb{L}) = H$ such that $\sigma(V_1) = V_i$. This is a contradiction.

Without any loss of generality, assume that $V_1', \ldots, V_t'$ are all the distinct V_i''s and let $R = \cap V_i'$. By the Chinese remainder theorem (Theorem C.5.7), there exists a $\beta \in R$ such that $\beta - 1 \in M_1'$ and $\beta \in M_j'$ for all $1 < j \leq t$. Suppose $k > t$. Then there is a $1 < j \leq t$ such that $V_j' = V_k'$ implying that $M_k' = M_j'$. Thus, $\beta - 1 \in M_1'$, $\beta \in \cap_{i>1} M_i'$ and $\beta \in R \subset \mathbb{L}$.

Let f be the minimal polynomial of β over $\mathbb{F}$. If possible, suppose $\beta \in \mathbb{F}$. So, $\beta - 1 \in M_1' \cap \mathbb{F} = M$ and also $\beta \in M_2' \cap \mathbb{F} = M$, contradicting that M is proper. Thus, $\beta \notin \mathbb{F}$. Hence, since f is minimal, f has no root in $\mathbb{F}$.

Let
$$f(X) = X^n + a_{n-1}X^{n-1} + \cdots + a_1 X + a_0$$

and $\beta = \beta_1, \ldots, \beta_n$ be all the roots of f. Note that no root of f belongs to $\mathbb{F}$. Since $\mathbb{K}$ is a normal separable extension of $\mathbb{F}$, all β_i's belong to $\mathbb{K}$ and are distinct. We shall arrive at a contradiction of (c") by showing that $a_{n-1} \in V^\times$ and $a_0, \ldots, a_{n-2} \in M$.

Claim. $\beta_2, \ldots, \beta_n \in M_1$.

Let $j > 1$. Let $\tau \in G(\mathbb{K}, \mathbb{F})$ such that $\tau(\beta_1) = \beta_j$. So, $\tau \notin H = G(\mathbb{K}, \mathbb{L})$. Since $\tau^{-1} \notin H$, $\tau^{-1}(V_1) \neq V_1$. Let $\tau^{-1}(V_1) = V_i$ for some $i > 1$. Hence, $\tau(M_i) = M_1$. Since $\beta_1 \in M_i$, $\beta_j = \tau(\beta_1) \in M_1$.

By simple theory of equations, $a_0, \ldots, a_{n-2} \in M_1 \cap \mathbb{F} = M$.

We have $1 - \beta_1 \in M_1$. But
$$a_{n-1} = -(\beta_1 + \cdots + \beta_n).$$

So,

$$1 + a_{n-1} = 1 - \beta_1 - (\beta_2 + \cdots + \beta_n) \in M_1.$$

But $a_{n-1} \in V$. So, $1 + a_{n-1} \in M$. Since M is proper, this implies that a_{n-1} is a unit in V.   □

Corollary C.6.3 *The fields $\mathbb{Q}_p$ of p-adic reals and of Laurent series $\mathbb{F}_p((X))$ over $\mathbb{F}_p$, p any prime, are Henselian.*

It is a standard fact that these two valued fields satisfy (c).

Corollary C.6.4 *Let $(\mathbb{F}, V)$ be Henselian and $\mathbb{K}$ a subfield of $\mathbb{F}$. Set $\mathbb{L} = \mathbb{K}^s \cap \mathbb{F}$ and $V_{\mathbb{L}} = V \cap \mathbb{L}$. Then $(\mathbb{L}, V_{\mathbb{L}})$ is Heneselian. In particular, if $\mathbb{K}$ is a separably closed subfield of $\mathbb{F}$, i.e. $\mathbb{K}^s \cap \mathbb{F} = \mathbb{K}$, then $(\mathbb{K}, V \cap \mathbb{K})$ is Henselian.*

Proof First assume that $\mathbb{L} = \mathbb{K}$. Let

$$f(x) = x^n + a_{n-1}x^{n-1} + \cdots + a_1 x + a_0 \in V_{\mathbb{L}}[x]$$

with $a_{n-1} \notin M$ and $a_0, \ldots, a_{n-2} \in M$ and f have all roots distinct. Then, by (c"), f has a root $\alpha \in \mathbb{F}$. Suppose $\alpha \notin \mathbb{L}$. Let g be the minimal polynomial of α over $\mathbb{K}$. Then $g | f$. So, g has all roots distinct. This shows that α is separable over $\mathbb{K}$. Thus, $\alpha \in \mathbb{K}^s \cap \mathbb{F} = \mathbb{L}$. Hence, $(\mathbb{L}, V \cap \mathbb{L})$ is Henselian by (c").

In the general case, note that $\mathbb{K} \subset \mathbb{L} \subset \mathbb{K}^s$. So, $\mathbb{L}^s = \mathbb{K}^s$. Hence, $\mathbb{L}^s \cap \mathbb{F} = \mathbb{K}^s \cap \mathbb{F} = \mathbb{L}$. So, $(\mathbb{L}, V_{\mathbb{L}})$ is Henselian by the above case.   □

C.7 Henselization of a Valued Field

An extension $(\mathbb{F}', V')$ of $(\mathbb{F}, V)$ is called a *Henselization* of $(\mathbb{F}, V)$ if the following two conditions are satisfied:

1. $(\mathbb{F}', V')$ is Henselian.
2. If $(\mathbb{F}'', V'')$ is a Henselian extension of $(\mathbb{F}, V)$, then there exists a unique $\mathbb{F}$-embedding α of $\mathbb{F}'$ into $\mathbb{F}''$ such that $\alpha(V') \subset V''$.

We now proceed to prove the existence of a Henselization of a valued field $(\mathbb{F}, V)$ which is clearly unique modulo $\mathbb{F}$-isomorphism.

Let V^s be an extension of V to $\mathbb{F}^s$. Set

$$G^h = G^h(V^s) = \{\sigma \in G(\mathbb{F}^s, \mathbb{F}) : \sigma(V^s) = V^s\}$$

and

$$\mathbb{F}^h = \mathbb{F}^h(V^s) = F(G^h),$$

the fixed field of G^h. We also set

$$V^h = \mathbb{F}^h \cap V^s.$$

We shall prove that $(\mathbb{F}^h, V^h)$ is a Henselization of $(\mathbb{F}, V)$.

(A) G^h is closed in $G(\mathbb{F}^s, \mathbb{F})$.

Proof Let $\sigma \in G(\mathbb{F}^s, \mathbb{F}) \setminus G^h$. We have to show that there exists a finite, normal, separable extension $\mathbb{L}$ of $\mathbb{F}$ such that

$$\sigma \cdot G(\mathbb{F}^s, \mathbb{L}) \cap G^h = \emptyset.$$

Since $\sigma \notin G^h$, $\sigma(V^s) \neq V^s$. If $\sigma(V^s) \subset V^s$, since both are extensions of V and $\mathbb{F}^s$ is an algebraic extension of $\mathbb{F}$, they will be equal (Lemma C.5.6). Hence, there exists an $\alpha \in V^s$ such that $\sigma(\alpha) \notin V^s$.

Let f be the minimal polynomial of α over $\mathbb{F}$ and $\mathbb{L}$ the splitting field of f. Since $\alpha \in \mathbb{F}^s$, $\mathbb{L}$ is a finite, normal, separable extension of $\mathbb{F}$.

Now suppose $\tau \in G(\mathbb{F}^s, \mathbb{F})$ such that $\tau|\mathbb{L} = \sigma|\mathbb{L}$. But then $\tau(\alpha) = \sigma(\alpha) \notin V^s$. Hence, $\tau(\alpha) \in \tau(V^s) \setminus V^s$ implying that $V^s \neq \tau(V^s)$. Thus, $\sigma \cdot G(\mathbb{F}^s, \mathbb{L}) \cap G^h = \emptyset$.

(B) $(\mathbb{F}^h, V^h)$ is Henselian.

Proof Since G^h is closed, by Galois correspondence (Theorem B.1.9),

$$G^h(V^s) = G(\mathbb{F}^s, \mathbb{F}^h).$$

If possible, suppose $(\mathbb{F}^h, V^h)$ is not Henselian. Then there exists an extension V'^s of V^h to $(\mathbb{F}^h)^s = \mathbb{F}^s$ such that $V'^s \neq V^s$. By Corollary C.5.11, there is a $\rho \in G(\mathbb{F}^s, \mathbb{F}^h) = G^h(V^s)$ such that $\rho(V^s) = V'^s$. This is a contradiction.

(C) $(\mathbb{F}, V)$ is Henselian iff $\mathbb{F}^h = \mathbb{F}$.

Proof The if part is the assertion (B). Conversely, if $(\mathbb{F}, V)$ is Henselian, V^s is the only extension of V to $\mathbb{F}^s$. So,

$$G^h(V^s) = \{\sigma \in G(\mathbb{F}^s, \mathbb{F}) : \sigma(V^s) = V^s\} = G(\mathbb{F}^s, \mathbb{F}).$$

Hence, the fixed field of $G^h(V^s)$ is $\mathbb{F}$, i.e. $\mathbb{F}^h = \mathbb{F}$.

(D) Let V'^s be another extension of V to $\mathbb{F}^s$ and $\rho \in G(\mathbb{F}^s, \mathbb{F})$ such that $\rho(V^s) = V'^s$. Then

$$\rho G^h(V^s)\rho^{-1} = G^h(V'^s)$$

and

$$\rho(\mathbb{F}^h(V^s)) = \mathbb{F}^h(V'^s).$$

Proof Let $\sigma \in G^h(V^s)$. Then

$$\rho(\sigma(\rho^{-1}(V'^s))) = \rho(\sigma(V^s)) = \rho(V^s) = V'^s.$$

Thus,

$$\rho G^h(V^s)\rho^{-1} \subset G^h(V'^s).$$

By the same argument,

$$\rho^{-1} G^h(V'^s)\rho \subset G^h(V^s).$$

Hence,

$$\rho G^h(V^s)\rho^{-1} = G^h(V'^s).$$

Now take $x \in \mathbb{F}^h(V^s)$ and $\sigma \in G^h(V'^s)$. There is a $\tau \in G^h(V^s)$ such that $\sigma = \rho\tau\rho^{-1}$. So,

$$\sigma(\rho(x)) = \rho(\tau(\rho^{-1}(\rho(x)))) = \rho(\tau(x)) = \rho(x).$$

Thus, $\rho(\mathbb{F}^h(V^s)) \subset \mathbb{F}^h(V'^s)$. Similarly, we prove that $\rho^{-1}(\mathbb{F}^h(V'^s)) \subset \mathbb{F}^h(V^s)$. Hence, $\rho(\mathbb{F}^h(V^s)) = \mathbb{F}^h(V'^s)$.

(E) Let $(\mathbb{F}_1, V_1)$ be a Henselian extension of $(\mathbb{F}, V)$. Then there is an $\mathbb{F}$-embedding from $(\mathbb{F}^h, V^h)$ into $(\mathbb{F}_1, V_1)$.

Proof By Corollary C.6.4, $(\mathbb{F}^s \cap \mathbb{F}1, V_1 \cap \mathbb{F}^s)$ is Henselian. Hence, replacing $\mathbb{F}_1$ by $\mathbb{F}^s \cap \mathbb{F}_1$ and restricting V_1 to it, without any loss of generality, we assume that $\mathbb{F}_1 \subset \mathbb{F}^s$. Let V_1^s be an extension of V_1 to $\mathbb{F}^s$. Since $(\mathbb{F}_1, V_1)$ is Henselian, V_1^s is unique.

Note that V^s is the unique extension of V^h to $\mathbb{F}^s$. So, both V^s and V_1^s are extensions of V to $\mathbb{F}^s$. Hence, by Corollary C.5.11, there is a $\rho \in G(\mathbb{F}^s, \mathbb{F})$ such that $\rho(V^s) = V_1^s$. Then

$$\rho(\mathbb{F}^h) = \rho(\mathbb{F}^h(V^s)) = \mathbb{F}^h(V_1^s) = \mathbb{F}_1^h = \mathbb{F}_1. \qquad (*)$$

The second equality holds by (D) and the last one by (C).

Now

$$\rho(V^h) = \rho(V^s \cap \mathbb{F}^h) = \rho(V^s) \cap \rho(\mathbb{F}^h) = V_1^s \cap \mathbb{F}_1 = V_1.$$

(F) The embedding ρ obtained in (E) is unique.

Proof Let $\tau : (\mathbb{F}^h, V^h) \to (\mathbb{F}_1, V_1)$ be another $\mathbb{F}$-embedding. We need to show that $\tau = \rho|\mathbb{F}^h$. We extend τ to an $\mathbb{F}$-automorphism of $\mathbb{F}^s$. Denote the extension by τ itself. (Extend τ to $\overline{\mathbb{F}}$ and then observe that $\tau(\mathbb{F}^s) = \mathbb{F}^s$.)

We are required to show that $\rho^{-1} \circ \tau|\mathbb{F}^h = id|\mathbb{F}^h$. But $\mathbb{F}^h$ is the fixed field of $G^h(V^s)$. Hence, suffices to show that $\rho^{-1} \circ \tau \in G^h(V^s)$, i.e. $\rho^{-1}(\tau(V^s)) = V^s$.

Now $\tau(V^h) \subset V_1 \subset V_1^s$ and since $V^h \subset V^s$, $\tau(V^h) \subset \tau(V^s)$. Since $(\mathbb{F}^h, V^h)$ Henselian, so is $(\tau(\mathbb{F}^h), \tau(V^h))$. Hence, $\tau(V^s) = V_1^s = \rho(V^s)$. Thus, $\rho^{-1}(\tau(V^s)) = V^s$.

(G) Let $(\mathbb{F}', V')$ be a Henselization of $(\mathbb{F}, V)$. Then there exists an extension V^s of V to $\mathbb{F}^s$ such that $(\mathbb{F}', V') = (\mathbb{F}^h, V^h)$.

Proof Since there is a $\mathbb{F}$-isomorphism from $\mathbb{F}^h$ onto $\mathbb{F}'$ and $\mathbb{F}^h \subset \mathbb{F}^s$, $\mathbb{F}' \subset \mathbb{F}^s$. Let V^s be the unique extension of V' to $\mathbb{F}^s$. We have

$$G^h(V^s) = \{\sigma \in G(\mathbb{F}^s, \mathbb{F}) : \sigma(V^s) = V^s\} \supset G(\mathbb{F}^s, \mathbb{F}').$$

(If $\tau \in G(\mathbb{F}^s, \mathbb{F}')$, $\tau(V') = V' \subset V^s$. As $V' \subset V^s$, $V' = \tau(V') \subset \tau(V^s)$. So, both V^s and $\tau(V^s)$ are extensions of V'. Since $(\mathbb{F}', V')$ is Henselian, $\tau(V^s) = V^s$.)

By Theorem B.1.9, the above inclusion implies that

$$\mathbb{F}^h(V^s) = F(G^h(V^s)) \subset F(G(\mathbb{F}^s, \mathbb{F}')) = \mathbb{F}'.$$

As proved earlier $(\mathbb{F}^h, V^h)$ is a Henselization of $(\mathbb{F}, V)$ and so is $(\mathbb{F}', V')$.

Let $i : (\mathbb{F}^h(V^s), V^h) \hookrightarrow (\mathbb{F}', V')$ be the inclusion map and $\rho : (\mathbb{F}', V') \to (\mathbb{F}^h, V^h)$ is an embedding. So, $i \circ \rho : (\mathbb{F}', V') \to (\mathbb{F}', V')$ is an embedding. Hence, by uniqueness, $i \circ \rho = id_{\mathbb{F}'}$. Hence, the two Henselizations are the same.

We have thus proved the following theorem.

Theorem C.7.1 *Every valued field* $(\mathbb{F}, V)$ *admits a Henselization and every Henselization of* $(\mathbb{F}, V)$ *is of the form* $(\mathbb{F}^h, V^h)$ *for some extension* V^s *of* V *to* $\mathbb{F}^s$.

An extension $(\mathbb{F}_2, V_2)$ of $(\mathbb{F}_1, V_1)$ is called an *immediate extension* of $(\mathbb{F}_1, V_1)$ if the following two conditions are satisfied:

1. $\forall x \in V_2 \exists y \in V_1 (x - y \in M_2)$.
2. $\forall x \in \mathbb{F}_2^\times \exists y \in \mathbb{F}_1^\times (x \cdot y^{-1} \in V_2^\times)$.

Clearly, the first condition can be relaxed to $\forall x \in V_2 \exists y \in \mathbb{F}_1 (x - y \in M_2)$. There is a canonical embedding $\widetilde{\mathbb{F}_1} \hookrightarrow \widetilde{\mathbb{F}_2}$. The first condition says that this map is onto. So, $f = [\widetilde{\mathbb{F}_2} : \widetilde{\mathbb{F}_1}] = 1$.

The second condition says that if $\Gamma_2 = \mathbb{F}_2^\times / V_2^\times$ and $\Gamma_1 = \mathbb{F}_1^\times / V_1^\times$ are corresponding value groups, then the canonical embedding $\Gamma_1 \hookrightarrow \Gamma_2$ is an isomorphism. Thus, $e = [\Gamma_2 : \Gamma_1] = 1$. Let v_2 be a compatible valuation on $\mathbb{F}_2$. The second condition is equivalent to the statement $\forall x \in \mathbb{F}_2 \exists y \in \mathbb{F}_1 (v_2(y) = v_2(x))$.

Our next result states that the Henselization $(\mathbb{F}^h, V^h)$ is an immediate extension of $(\mathbb{F}, V)$.

Proposition C.7.2 *Let* $(\mathbb{N}, V_1)$ *be a finite, normal, separable extension of* $(\mathbb{F}, V)$. *Suppose*

$$H = \{\sigma \in G(\mathbb{N}, \mathbb{F}) : \sigma(V_1) = V_1\}$$

and $\mathbb{L} = F(H)$, *the fixed field of* H. *Then* $(\mathbb{L}, V_1 \cap \mathbb{L})$ *is an immediate extension of* $\mathbb{F}$.

Proof Let $V_1, \ldots, V_m$ be all the extensions of V to $\mathbb{N}$ and $V_i' = V_i \cap \mathbb{L}$. Then

Observation 1. $\forall i > 1(V_1' \neq V_i')$.

Proof Suppose there exists an $i > 1$ such that $V_i' = V_1'$. So, V_1 and V_i are extensions to $\mathbb{N}$ of a valuation subring of $\mathbb{L}$. Hence, there exists a $\sigma \in G(\mathbb{N}, \mathbb{L})$ such that $\sigma(V_1) = V_i \neq V_1$. But $G(\mathbb{N}, \mathbb{L}) = H$. We have arrived at a contradiction.

Step 1. Let $x \in V_1'$. Set $R = \cap_i V_i'$. By Chinese remainder theorem (Theorem C.5.7), there exists a $y \in R \subset \mathbb{L}$ such that $x - y \in M_1$ and $y \in M_i$ for all $i > 1$. If $y \in \mathbb{F}$, we have verified the first condition for x. So, assume that $y \notin \mathbb{F}$. Let $y = y_1, y_2, \ldots, y_n$ be all the roots of the minimal polynomial of y over $\mathbb{F}$. Then $y_1, \ldots, y_n$ are distinct.

Observation 2. $\forall j > 1(y_j \in M_1)$.

Let $j > 1$ and $\sigma \in G(\mathbb{N}, \mathbb{F})$ be such that $\sigma(y_1) = y_j \neq y_1$. So, $\sigma \notin G(\mathbb{N}, \mathbb{L}) = H$. So, $\sigma^{-1} \notin H$. Therefore, $V_1 \neq \sigma^{-1}(V_1) = V_i$ for some i. Hence, $y_j = \sigma(y_1) \in \sigma(M_i) = M_1$.

Take $z = y_1 + \cdots + y_n \in \mathbb{F}$. Then $x - z = (x - y_1) - y_2 - \cdots - y_n \in M_1$.

Step 2. Let ω be a valuation on $\mathbb{L}$ and $x \in \mathbb{L}^\times$. Then there exists a $y \in \mathbb{F}$ such that $\omega(x) = \omega(y)$.

Proof By Chinese remainder theorem (Theorem C.5.7), there exists a $\beta \in R$ such that $\beta - 1 \in M_1$ and $\beta \in M_i$ for all $i > 1$. Since M_1 is proper, it follows that $\beta \in V_1 \setminus M_1$. Hence

$$\omega(\beta) = 0. \tag{C.1}$$

If $\tau \in H = G(\mathbb{N}, \mathbb{L})$, $\tau(\beta) = \beta$. Hence,

$$\forall \tau \in H(\omega(\tau(\beta)) = \omega(\beta) = 0). \tag{C.2}$$

Now let $\tau \in G(\mathbb{N}, \mathbb{F}) \setminus H$. Get $i > 1$ such that $\tau(V_i) = V_1$. Then $\tau(M_i) = M_1$. Hence, $\omega(\tau(\beta)) > 0$. Thus,

$$\forall \tau \in G(\mathbb{N}, \mathbb{F}) \setminus H(\omega(\tau(\beta)) > 0). \tag{C.3}$$

Fix an $x \in \mathbb{L}^\times$. We now make our final observation to complete the proof.

Observation 3. $\exists k > 0 \forall \tau \in G(\mathbb{N}, \mathbb{F}) \setminus H(\omega(\beta^k x) \neq \omega(\tau(\beta^k x)))$.

By (1), for every positive integer k, $\omega(\beta^k x) = \omega(x)$. Also, since the value group is torsion-free, by (2), for positive integers k, l

$$\omega(\tau(\beta^k x)) = \omega(\tau(\beta^l x)) \Rightarrow k\omega(\tau(\beta)) = l\omega(\tau(\beta)) \Rightarrow k = l.$$

So, $\omega(\beta^k x) = \omega(x) = \omega(\tau(\beta^k x))$ for at most one positive integer k. Hence, our observation follows because $G(\mathbb{N}, \mathbb{F})$ is finite.

Now take $z = \beta^k x \in \mathbb{L}^\times$. Then $\omega(z) = \omega(x)$. If $z \in \mathbb{F}$, we are done. Otherwise, let

$$f(t) = t^m + a_{m-1} t^{m-1} + \cdots + a_1 t + a_0$$

be the minimal polynomial of z over $\mathbb{F}$ and $z = z_1, \ldots, z_m$ be all the roots of f which are distinct and belong to $\mathbb{N}$.

Now let $j > 1$ and $\tau \in G(\mathbb{N}, \mathbb{F})$ be such that $\tau(z_1) = z_j \neq z_1$. Then, $\tau \notin G(\mathbb{N}, \mathbb{L}) = H$. Thus

$$\forall j > 1(\omega(z_1) \neq \omega(z_j)). \tag{C.4}$$

First assume that there is no $j > 1$ such that $\omega(z_j) < \omega(z_1)$. Then $\forall j > 1(\omega(z_1) < \omega(z_j))$. Since ω is a valuation,

$$\omega(x) = \omega(z_1) = \omega(z_1 + \cdots + z_m) = \omega(-a_1).$$

We then take $y = -a_1$. In the other case, let $z_{j_1}, \ldots, z_{j_r}$ be all the z_j such that $\omega(z_j) < \omega(z_1)$. Since ω is a valuation, it follows that

$$\omega(a_r) = \omega(\pm z_{j_1} \ldots z_{j_r})$$

and

$$\omega(a_{r+1}) = \omega(\mp z_1 \cdot z_{j_1} \ldots z_{j_r}).$$

So, $y = \frac{a_{r+1}}{a_r}$ has the desired properties. $\qquad\square$

Theorem C.7.3 $(\mathbb{F}^h, V^h)$ *is an immediate extension of* $(\mathbb{F}, V)$.

Proof Take any $x \in \mathbb{F}^h \setminus \mathbb{F}$. Then $\mathbb{F}[x]$ is a finite extension of $\mathbb{F}$ and

$$x \in \mathbb{F}[x] \subset \mathbb{F}^h \subset \mathbb{F}^s.$$

Let $\mathbb{K}$ be the splitting field of the minimal polynomial f of x over $\mathbb{F}$. Then $\mathbb{K}$ is a finite, separable, normal extension of $\mathbb{F}$. Set

$$H = \{\tau \in G(\mathbb{K}, \mathbb{F}) : \tau(V^s \cap \mathbb{K}) = V^s \cap \mathbb{K}\}.$$

Since $x \in \mathbb{F}^h \cap \mathbb{K}$, by the last proposition, our result will be proved if we show the following:

Main Step. $\mathbb{F}^h \cap \mathbb{K} = F(H)$.

Proof We have $G^h = \{\sigma \in G(\mathbb{F}^s, \mathbb{F}) : \sigma(V^s) = V^s\}$.

Claim. $H = \{\sigma|\mathbb{K} : \sigma \in G^h\}$.

Assume the claim for the time being. Then

$$F(H) = F(G^h) \cap \mathbb{K} = \mathbb{F}^h \cap \mathbb{K}.$$

Proof of the claim. Now $\mathbb{K}$ is the splitting field of f. Suppose $f(\alpha) = 0$. Then, for every $\sigma \in G(\mathbb{F}^s, \mathbb{F})$, $f(\sigma(\alpha)) = \sigma(f(\alpha)) = 0$. Further, if $\sigma(V^s) = V^s$, we get that $\sigma(V^s \cap \mathbb{K}) \subset V^s \cap \mathbb{K}$. Since both $V^s \cap \mathbb{K}$ and $\sigma(V^s \cap \mathbb{K})$ are extensions of V to $\mathbb{K}$ and $\mathbb{K}$ is an algebraic extension of $\mathbb{F}$, $\sigma(V^s \cap \mathbb{K}) = V^s \cap \mathbb{K}$ (Lemma C.5.6). Hence,

$$H \supset \{\sigma|\mathbb{K} : \sigma \in G^h\}.$$

For the opposite inclusion, let $\tau \in H$ and σ an $\mathbb{F}$-automorphism of $\mathbb{F}^s$ extending τ. Then $\sigma(V^s)$ and V^s are extensions of $V^s \cap \mathbb{K}$. So, there exists a $\rho \in G(\mathbb{F}^s, \mathbb{K})$ such that $\rho(\sigma(V^s)) = V^s$. In other words, $\rho \circ \sigma \in G^h$ and

$$\rho \circ \sigma|\mathbb{K} = \rho|\mathbb{K} \circ \sigma|\mathbb{K} = id_{\mathbb{K}} \circ \tau = \tau.$$

$\square$

A valued field $(\mathbb{F}, V)$ is called *algebraically maximal* if it has no proper, algebraic, immediate extension.

Corollary C.7.4 *If $(\mathbb{F}, V)$ is algebraically maximal, it is Henselian.*

The converse of this corollary is not true. A Henselian field will be algebraically maximal under an additional condition which we describe now.

A valued field $(\mathbb{F}, V)$ with valuation $v : \mathbb{F}^{\times} \to \Gamma$ is called *finitely ramified* if one of the following two conditions are satisfied.

1. $char(\mathbb{F}^{\sim}) = 0$.
2. $char(\mathbb{F}^{\sim}) = p > 0$ and there are only finitely many elements of Γ between 0 and $v(p)$.

Remark C.7.5 If $(\mathbb{F}_1, V_1)$ is an immediate extension of $(\mathbb{F}, V)$, then $(\mathbb{F}, V)$ is finitely ramified if and only if $(\mathbb{F}_1, V_1)$ is finitely ramified. It follows that if $(\mathbb{F}, V)$ is finitely ramified, then its Henselization $(\mathbb{F}^h, V^h)$ is also finitely ramified.

Proposition C.7.6 *Let* $(\mathbb{F}, V)$ *be a finitely ramified valued field. Then*

(a) $char(\mathbb{F}) = 0$.
(b) *If* $char(\mathbb{F}^\sim) = p > 0$, *then for every positive integer* n, *there are only finitely many elements between* 0 *and* $v(n)$.

Proof If $char(\mathbb{F}^\sim) = 0$, then $char(\mathbb{F}) = 0$. Now suppose $char(\mathbb{F}^\sim) = p > 0$. If possible, suppose $char(\mathbb{F}) > 0$. But then $char(\mathbb{F}) = p$, i.e. $p = 0$ in $\mathbb{F}$. Hence. $v(p) = \infty$. So, there are infinitely many elements of Γ between 0 and $v(p)$. This contradicts that $(\mathbb{F}, V)$ is finitely ramified. Therefore, the characteristic of $\mathbb{F}$ must be 0.

Now assume that $char(\mathbb{F}^\sim) = p > 0$ and $n = p^k m$ be a positive integer with $p \nmid m$. So, $m \neq 0$ in $\mathbb{F}^\sim$ implying that $m \notin M$. So, $v(m) = 0$. Hence, $v(n) = kv(p)$. Since there are only finitely many elements of Γ between 0 and $v(p)$, we now see that there are only finitely many elements of Γ between 0 and $v(n)$. $\square$

Proposition C.7.7 *The following statements are equivalent.*

(i) *If* $(\mathbb{F}, V)$ *is finitely ramified, then* $(\mathbb{F}^h, V^h)$ *is algebraically maximal.*
(ii) *Every finitely ramified, Henselian valued field* $(\mathbb{F}, V)$ *is algebraically maximal.*

Proof (i) implies (ii): If $(\mathbb{F}, V)$ is Henselian, $(\mathbb{F}, V) = (\mathbb{F}^h, V^h)$. Thus, (ii) follows from (i).

(ii) implies (i): Assume that $(\mathbb{F}, V)$ is finitely ramified. Since $(\mathbb{F}^h, V^h)$ is an immediate extension of $(\mathbb{F}, V)$, $(\mathbb{F}^h, V^h)$ is also finitely ramified and Henselian. Hence, $(\mathbb{F}^h, V^h)$ is algebraically maximal by (ii). $\square$

Theorem C.7.8 *Every finitely ramified, Henselian valued field* $(\mathbb{F}, V)$ *is algebraically maximal.*

Proof Let $(\mathbb{F}_1, V_1)$ be a finitely ramified, Henselian valued field that has a proper immediate algebraic extension, say $(\mathbb{F}_2, V_2)$. We shall arrive at a contradiction. Without any loss of generality, we assume that $\mathbb{F}_2 = \mathbb{F}_1[y]$ for some $y \in \mathbb{F}_2 \setminus \mathbb{F}_1$. Let $\mathbb{F}_3$ be the splitting field of the minimal polynomial f of y over $\mathbb{F}_1$. Let V_3 be an extension of V_2 to $\mathbb{F}_3$ and v a compatible valuation on $\mathbb{F}_3$.

Let $y = y_1, \ldots, y_n$ be all the roots of f. So, $\sum_i y_i \in \mathbb{F}_1$. For each i, fix a $\sigma_i \in G(\mathbb{F}_3, \mathbb{F}_1)$ such that $\sigma_i(y) = y_i$.

Observation 1. $\forall b \in \mathbb{F}_1 \forall i (v(\sigma_i(y) - b) = v(y - b))$.

Let $\sigma_i \in G(\mathbb{F}_3, \mathbb{F}_1)$. Note that $\sigma_i(V_3)$ and V_3 are extensions of V_1 to $\mathbb{F}_3$ which is an algebraic extension of $\mathbb{F}_1$. Since $(\mathbb{F}_1, V_1)$ is Henselian, it follows that $\sigma_i(V_3) = V_3$. Hence, as observed earlier, $v \circ \sigma_i = v$. Therefore, for any $b \in \mathbb{F}_1$

$$(v(\sigma_i(y) - b) = v(\sigma_i(y - b)) = v(y - b)).$$

Set $a = \frac{\sum_i y_i}{n}$.

Observation 2. $\exists b \in \mathbb{F}_1(v(y - a) + v(n) < v(y - b))$.

Set $b_0 = a$. Since $y \in \mathbb{F}_2 \setminus \mathbb{F}_1$ and $b_0 \in \mathbb{F}_1$, $y - b_0 \in \mathbb{F}_2^\times$. Since $(\mathbb{F}_2, V_2)$ is an immediate extension of $(\mathbb{F}_1, V_1)$, there exists a $c \in \mathbb{F}_1^\times$ such that $(y - b_0)c^{-1} \in V_2^\times$. So, $v((y - b_0)c^{-1}) = v(y - b_0) - v(c) = 0$. Thus, $v(y - b_0) = v(c)$.

Again since $(\mathbb{F}_2, V_2)$ is an immediate extension of $(\mathbb{F}_1, V_1)$, there exists a $d \in V_1$ such that $(y - b_0)c^{-1} - d \in M_2$. Therefore, $v((y - b_0)c^{-1} - d) > 0$. Set $b_1 = b_0 + cd \in \mathbb{F}_1$. Then

$$v(y - b_0) = v(c) < v(c((y - b_0)c^{-1} - d)) = v(y - b_1).$$

Proceeding similarly we show that there exists a sequence $\{b_n\}$ in $\mathbb{F}_1$ such that

$$v(y - b_0) < v(y - b_1) < v(y - b_2) < \ldots.$$

Since there are only finitely many elements of Γ between 0 and $v(n)$, it follows that

$$v(y - a) + v(n) < v(y - b_k)$$

for some k.

Using these two observations, we now show a contradiction. Fix a $b \in \mathbb{F}_1$ as in observation 2. By Observation 1,

$$v(y - b) \leq v\left(\sum_i (\sigma_i(y) - b)\right) = v(n(a - b)) = v(a - b) + v(n).$$

So, by observation 2,

$$v(y - a) + v(n) < v(y - b) \leq v(a - b) + v(n).$$

Hence,

$$v(y - a) < v(a - b). \tag{C.5}$$

By observation 2,

$$v(y - a) < v(y - b). \tag{C.6}$$

By (1) and (2),

$$v(y - a) < v((y - b) - (a - b)) = v(y - a),$$

a contradiction. $\qquad\qquad\square$

Proposition C.7.9 *Let* $(\mathbb{K}, V, v)$ *be a Henselian valued field with value group* Γ *such that* $char(\mathbb{K}^\sim) = 0$. *Then for every countable subfield* $\mathbb{F}$ *of* $\mathbb{K}$ *there is a countable extension* $\mathbb{F}'$ *of* $\mathbb{F}$ *in* $\mathbb{K}$ *such that* $v(\mathbb{F}')$ *is pure in* Γ, *i.e.* $\Gamma/v(\mathbb{F}')$ *is torsion-free.*

Proof Set

$$\Gamma' = \{\gamma \in \Gamma : \exists m \geq 1(m\gamma \in v(\mathbb{F}))\}.$$

Since Γ is ordered, it is torsion-free. Therefore, for every $\delta \in v(\mathbb{F})$ and every $m \geq 1$, there exists at most one $\gamma \in \Gamma$ such that $m\gamma = \delta$. Since $v(\mathbb{F})$ is countable, it follows that Γ' is countable. Also, note that Γ' is pure in Γ. Enumerate $\Gamma' = \{\gamma_n : n \in \mathbb{K}\}$.

Inductively we shall define a sequence $\{\mathbb{F}_n\}$ of countable subfields of $\mathbb{K}$ such that $\mathbb{F}_0 = \mathbb{F}$ and for every n, $\gamma_n \in v(\mathbb{F}_{n+1}) \subset \Gamma'$ and $\mathbb{F}_n \subset \mathbb{F}_{n+1}$,. Then, $\mathbb{F}' = \cup_n \mathbb{F}_n$ will have all the desired properties.

Assume that we have defined $\mathbb{F}_n$. Get an $x \in \mathbb{K}^\times$ such that $v(x) = \gamma_n$. By the definition of Γ', there exists an $m \geq 1$ such that $m\gamma_n \in v(\mathbb{F})$. Let $a \in \mathbb{F}^\times$ be such that

$$v(a) = m\gamma_n = v(x^m).$$

Set $d = \frac{x^m}{a} \in V^\times$, V the valuation subring of $\mathbb{K}$. We first show that

$$\mathbb{F}_n(d)^\sim = \mathbb{F}_n^\sim([d]) \text{ and } v(\mathbb{F}_n(d)) = v(\mathbb{F}_n).$$

Case 1. $[d]$ is transcendental over $\mathbb{F}_n^\sim$.

If d were algebraic over $\mathbb{F}_n$, clearly it would be algebraic over $V \cap \mathbb{F}_n$. But then $[d]$ would be algebraic over $\mathbb{F}_n^\sim$. Thus, d is transcendental over $\mathbb{F}_n$. We also have $v(d) = 0$. By Theorem C.2.2, in this case, $\mathbb{F}_n(d)^\sim = \mathbb{F}_n^\sim([d])$ and $v(\mathbb{F}_n(d)) = v(\mathbb{F}_n)$.

Case 2. $[d]$ is algebraic over $\mathbb{F}_n^\sim$.

Let

$$f(X) = X^m + a_{m-1}X^{m-1} + \cdots + a_1 X + a_0 \in (V \cap \mathbb{F}_n)[X]$$

be an irreducible polynomial such that

$$f^\sim(X) = X^m + [a_{m-1}]X^{m-1} + \cdots + [a_1]X + [a_0]$$

is the minimal polynomial of $[d]$ over $\mathbb{F}_n^\sim$. We then have $f(d) \in M$. So, $v(f(d)) > 0$. Also, $f'(d) \in V \setminus M$. As $\mathbb{K}$ is Henselian, by Hensel's lemma (Theorem C.4.5), f has a root in V with residue class $[d]$. We replace d by one such root and denote it by d itself.

We clearly have

$$[\mathbb{F}_n^\sim[[d]] : \mathbb{F}_n^\sim] = degree(f^\sim) = degree(f) = [\mathbb{F}_n(d) : \mathbb{F}_n].$$

Now note that $\mathbb{F}_n^{\sim}[[d]] \subset \mathbb{F}_n(d)^{\sim}$. Hence, by Chevalley's fundamental inequality (Corollary C.5.3),

$$
\begin{aligned}
degree(f^{\sim}) &\leq [\mathbb{F}_n^{\sim}[[d]] : \mathbb{F}_n^{\sim}][v(\mathbb{F}_n(d)) : v(\mathbb{F}_n)] \\
&\leq [\mathbb{F}_n(d)^{\sim} : \mathbb{F}_n^{\sim}][v(\mathbb{F}_n(d)) : v(\mathbb{F}_n)] \\
&\leq [\mathbb{F}_n(d) : \mathbb{F}_n] \\
&= degree(f^{\sim}).
\end{aligned}
$$

This implies that $\mathbb{F}_n(d)^{\sim} = \mathbb{F}_n^{\sim}([d])$ and $v(\mathbb{F}_n(d)) = v(\mathbb{F}_n)$ in this case also. Now consider the polynomial

$$
g(Z) = Z^m - \frac{x^m}{ad} \in V[Z].
$$

Since $char(\mathbb{K}^{\sim}) = 0$, $v(g'(1)) = v(m) = 0$. Since $\frac{x^m}{a}$ and d have the same residue, $v(\frac{x^m}{a} - d) > 0$. Therefore,

$$
v(g(1)) = v\left(1 - \frac{x^m}{ad}\right) = v\left(\frac{x^m}{a} - d\right) - v(d) = v\left(\frac{x^m}{a} - d\right) > 0.
$$

Thus, by Hensel's lemma (Theorem C.4.5), g has a root $z \in V^{\times}$ with residue class 1. In particular, $v(z) = 0$.

We define $\mathbb{F}_{n+1} = \mathbb{F}_n(d, \frac{x}{z})$.

So, $\mathbb{F}_{n+1}$ is algebraic over $\mathbb{F}_n(d)$. Hence, by Theorem C.5.4, $v(\mathbb{F}_{n+1})$ is torsion over $v(\mathbb{F}_n(d)) = v(\mathbb{F}_n) \subset \Gamma'$. Therefore, $v(\mathbb{F}_{n+1}) \subset \Gamma'$. Further,

$$
\gamma_n = v(x) = v\left(\frac{x}{z}\right) \in v(\mathbb{F}_{n+1}).
$$

$\square$

Bibliography

1. J. Ax, The elementary theory of finite fields. Ann. Math. **88**, 103–115 (1968)
2. J. Ax, S. Kochen, Diophantine problems over local fields I. Am. J. Math. **87**, 605–630 (1965a)
3. J. Ax, S. Kochen, Diophantine problems over local fields II: A complete set of axioms for p-adic number theory. Am. J. Math. **87**, 631–648 (1965b)
4. J. Ax, S. Kochen, Diophantine problems over local fields III: Decidable fields. Ann. Math. **83**, 437–456 (1966)
5. J.T. Baldwin, A.H. Lachlan, On strongly minimal sets. J. Symb. Logic **36**, 79–96 (1971)
6. J. Bochnak, M. Coste, M.-F. Roy, *Real Algebraic Geometry*, A Series of Modern Surveys in Mathematics, vol. 36 (Springer, 1998)
7. E. Casanovas, *Simple Theories and Hyperimaginaries*, Lecture Notes in Logic, vol. 39 (Cambridge University Press, 2011)
8. E. Casanovas, D. Lascar, A. Pillay, M. Ziegler, Galois groups of first order theories. J. Math. Logic **1**, 305–319 (2001)
9. C. C. Chang, H. J. Keisler, *Model Theory* (North-Holland, 1990)
10. C. N. Delzell, A. Prestel, *Mathematical Logic and Model Theory*, UTX (Springer, 2010)
11. A. Ehrenfeucht, A. Mostowski, Models of axiomatic theories admitting automorphisms. Fund. Math. **44**, 241–248 (1956)
12. E. Engeler, A characterization of theories with isomorphic denumerable models. Abstract, Notices Am. Math. Soc. **6**, 161 (1959)
13. D. Flath, S. Wagon, How to pick out integers in the rationals: An application of number theory to logic. Am. Math. Monthly **98**, 812–823 (1991)
14. J. Gismatullin, K. Krupinski, On model-theoretic connected components in some group extensions. J. Math. Logic **15**, 1550009–1550059 (2015)
15. K. Gödel, Die vollständigkeit der Axiome des logischen Funktionenkalküls. Monatshefte für Math. **37**, 349–360 (1930)
16. N.S. Gopala Krishnan, *Commutative Algebra*, 2nd edn. (University Press (India), 2016)
17. B. Hart, B. Kim, A. Pillay, Coordinatization and canonical bases in simple theories. J. Symb. Logic **65**, 293–309 (2000)
18. L.A. Henkin, The completeness of the first-order functional calculus. J. Symb. Logic **14**, 159–166 (1949)
19. L.A. Henkin, A generalization of the concept of ω-consistency. J. Symb. Logic **19**, 183–196 (1954)
20. W. Hodges, *Model Theory* (Cambridge University Press, 1993)
21. K. Ireland, M. Rosen, *A Classical Introduction to Modern Number Theory*, 2nd edn. (Springer International Edition, 1990)

© Springer Nature Singapore Pte Ltd. 2017
H. Sarbadhikari and S.M. Srivastava, *A Course on Basic Model Theory*,
DOI 10.1007/978-981-10-5098-5

22. T. Jech, *Set Theory*, 3rd Millenium edn. (Springer, 2002)
23. I. Kaplan, B. Miller, An embedding theorem of E_0 with model-theoretic applications. J. Math. Logic **14**, 1450010 (2014)
24. I. Kaplan, B. Miller, P. Simon, The Borel cardinality of Lascar strong types. J. Lond. Math. Soc. **90**, 609–630 (2014)
25. B. Kim, A. Pillay, Simple theories. Ann. Pure Appl. Logic **88**, 149–164 (1997)
26. K. Krupiński, A. Pillay, Generalized Bohr compactification and model-theoretic connected components. *Mathematical Proceedings of the Cambridge Philosophical Society* to appear
27. K. Krupiński, A. Pillay, Amenability, definable groups and automorphism groups, Submitted
28. K. Krupiński, A. Pillay, T. Rzepecki, Topological dynamics and the complexity of strong types, submitted
29. K. Krupiński, T. Rzepecki, Smoothness of bounded invariant equivalence relations. J. Math. Logic **81**, 326–356 (2016)
30. K. Krupiński, A. Pillay, S. Solecki, Borel equivalence relations and Lascar strong types. J. Math. Logic **13**, 1350008–1350044 (2013)
31. S. Lang, *Algebra*, Revised 3rd edn. (Springer, 2002)
32. S. Lang, *Introduction to Algebraic Geometry* (Interscience, 1958)
33. D. Lascar, On the category of models of a complete theory. J. Symb. Logic **47**, 249–266 (1982)
34. D. Lascar, A. Pillay, Hyperimaginaries and automorphism groups. J. Symb. Logic **66**, 127–143 (2001)
35. D. Lascar, B. Poizat, An introduction to forking. J. Symb. Logic **44**, 330–350 (1979)
36. J. Łoś, Quelques remarques, théorèmes et problèmes sur les classes définissables d'algèbres, *Mathematical Interpretations of Formal Systems* (North-Holland 1955), pp. 98–113
37. J. Łoś, On the categoricity in power of elementary deductive systems. Coll. Math. **3**, 58–62 (1954)
38. L. Löwenheim, Über Möglichkeiten in Relativkalkül. Math. Ann. **76**, 52–60 (1915)
39. A.I. Mal'tsev, Untersuchungen aus dem Gebiete der mathematischen Logik. Mat. Sbornik **1**(43), 323–336 (1936)
40. A.I. Mal'tsev, A general method for obtaining local theorems in group theory. Uchenye Zapiski Ivanon Ped. Inst. **1**(1), 3–9 (1941)
41. D. Marker, *Model Theory, An Introduction*, GTM 217 (Springer, 2002)
42. W.E. Marsh, On ω_1-categorical and not ω-categorical theories, Dissertation (Dartmouth College, 1966)
43. M. Morley, Categoricity in power. Trans. Am. Math. Soc. **114**, 514–538 (1965)
44. Ludomir Newelski, The diameter of a Lascar strong type. Fund. Math. **176**, 157–170 (2003)
45. S. Orey, On ω-consistency and related properties. J. Symb. Logic **21**, 246–252 (1956)
46. A. Pillay, *An Introduction to Stability Theory* (Oxford Science Publications, 1983)
47. A. Pillay, *Geometric Stability Theory* (Oxford Science Publications, 1996)
48. A. Pilay, C.I. Steinhorn, Definable sets in ordered structures. Bull. Am. Math. Soc. **11**, 159–162 (1984)
49. A. Pilay, C.I. Steinhorn, Definable sets in ordered structures I. Trans. Am. Math. Soc. **295**, 565–592 (1986)
50. B. Poizat, Une théorie de Galois imaginaire. J. Symb. Logic **48**, 1151–1170 (1983)
51. A. Robinson, *Complete Theories* (North-Holland, 1956)
52. J.B. Robinson, Definability and decision problems in arithmetic. J. Symb. Logic **14**, 98–114 (1949)
53. C. Ryll-Nardzewski, On the categoricity in power $\leq \aleph_0$. Bull. Acad. Polon. Sci. Sér. Sci. Math. Astronom. Phys. **7**, 545–548 (1959)
54. S. Shelah, *Classification Theory and the Number of Non-isomorphic Models* (North-Holland, 1978)
55. T. Skolem, Logische-Kombinatorische untersuchungen über die Erfüllbarkeit oder Beweisbarkeit mathematischer Sätze nebst einem Theoreme über dichte Mengen, Videnskapsselskapets Skrifter, I. Matem.-naturv klasse I, vol. 4 (1920), pp. 1–36

56. T. Skolem, Einige Bemerkungen zur axiomatischen Begründung der Mengenlehre, in *Proceedings 5th Scandinavian Mathematics Congress*, Helsinki, 1922, pp. 217–232

57. T. Skolem, Über die Nicht-Charakterisierbarkeit der Zahlenreihe mittels endlich oder abzählbar unendlicht vieler Aussangen mit ausschliesslich Zahlenvariablen. Fund. Math. **23**, 150–161 (1934)

58. S.M. Srivastava, *A Course on Borel Sets*, GTM 180 (Springer, 1998)

59. S.M. Srivastava, *A Course on Mathematical Logic*, 2nd edn. Universitext (Springer, 2008)

60. L. Svenonius, $\aleph_0$-categoricity in first-order predicate calculus. Theoria **25**, 82–94 (1959)

61. R.G. Swan, Tarski's principle and elimination of quantifiers, preprint (available on author's webpage)

62. A. Tarski, *A Decision Method for Elementary Algebra and Geometry* (University of California Press, Berkeley and Los Angeles, 1951)

63. A. Tarski, R.L. Vaught, Arithmetical extensions of relational systems. Comput. Math. **13**, 81–102 (1957)

64. K. Tent, M. Ziegler, *A Course in Model Theory* (ASL and Cambridge University Press, 2012)

65. R.L. Vaught, Applications of Löwenheim-Skolem theorem to problems of completeness and decidability. Indag. Math. **16**, 572–588 (1954)

66. R.L. Vaught, Denumerable models of complete theories, in Infinitistic Methods (Pergamon, New York, 1961)

67. F.O. Wagner, *Simple Theories*. Mathematics and its Applications, vol. 503 (Kluwer Academic Publishers, Dordrecht, 2000)

68. A. Weil, *Foundations of Algebraic Geometry*, vol. 29 (American Mathematical Society, Colloquium Publications, 1946)

Index

© Springer Nature Singapore Pte Ltd. 2017

H. Sarbadhikari and S.M. Srivastava, *A Course on Basic Model Theory*,

DOI 10.1007/978-981-10-5098-5

Printed in the United States
By Bookmasters

Printed in the United States
By Bookmasters